VOLUME ONE HUNDRED AND ONE

VITAMINS AND HORMONES

Klotho

VOLUME ONE HUNDRED AND ONE

VITAMINS AND HORMONES

Klotho

Series Editor

GERALD LITWACK, PhD
Toluca Lake, North Hollywood, California

AMSTERDAM • BOSTON • HEIDELBERG • LONDON
NEW YORK • OXFORD • PARIS • SAN DIEGO
SAN FRANCISCO • SINGAPORE • SYDNEY • TOKYO
Academic Press is an imprint of Elsevier

Cover photo credit:
Hayashi, Y., Ito, M.
Klotho-Related Protein KLrP: Structure and Functions 1
Vitamins and Hormones (2016) **101**, pp. 1–16

Academic Press is an imprint of Elsevier
50 Hampshire Street, 5th Floor, Cambridge, MA 02139, USA
525 B Street, Suite 1800, San Diego, CA 92101-4495, USA
The Boulevard, Langford Lane, Kidlington, Oxford OX5 1GB, UK
125 London Wall, London, EC2Y 5AS, UK

First edition 2016

ISBN: 978-0-12-804819-1
ISSN: 0083-6729

For information on all Academic Press publications
visit our website at https://www.elsevier.com

Publisher: Zoe Kruze
Acquisition Editor: Mary Ann. Zimmerman
Editorial Project Manager: Helene Kabes
Production Project Manager: Vignesh Tamil
Designer: Mark Rogers

Typeset by SPi Global, India

CONTENTS

CONTRIBUTORS

C.R. Abraham
Boston University School of Medicine, Boston, MA, United States

K. Akasaka-Manya
Molecular Glycobiology, Research Team for Mechanism of Aging, Tokyo Metropolitan Geriatric Hospital and Institute of Gerontology, Tokyo, Japan

P. Aljama
Instituto Maimónides de Investigación Biomédica de Córdoba (IMIBIC), Universidad de Córdoba/Hospital Universitario Reina Sofía, Córdoba, Spain

P. Buendía
Instituto Maimónides de Investigación Biomédica de Córdoba (IMIBIC), Universidad de Córdoba/Hospital Universitario Reina Sofía, Córdoba, Spain

J. Carracedo
Instituto Maimónides de Investigación Biomédica de Córdoba (IMIBIC), Universidad de Córdoba/Hospital Universitario Reina Sofía, Córdoba, Spain

C.D. Chen
Boston University School of Medicine, Boston, MA, United States

M. Dërmaku-Sopjani
University of Prishtina, Prishtinë, Republic of Kosova

T. Endo
Molecular Glycobiology, Research Team for Mechanism of Aging, Tokyo Metropolitan Geriatric Hospital and Institute of Gerontology, Tokyo, Japan

T. Fu
University of Illinois at Urbana-Champaign, Urbana, IL, United States

Y. Hayashi
Faculty of Pharma-Sciences, Teikyo University, Tokyo, Japan

M.C. Hu
Charles and Jane Pak Center for Mineral Metabolism and Clinical Research, University of Texas Southwestern Medical Center, Dallas, TX, United States

C.-L. Huang
University of Texas Southwestern Medical Center, Dallas, TX, United States

M. Ito
Faculty of Agriculture, Graduate School of Bioresource and Bioenvironmental Sciences, Kyushu University, Fukuoka, Japan

M. Kawai
Osaka Medical Center and Research Institute for Maternal and Child Health, Izumi, Japan

J.K. Kemper
University of Illinois at Urbana-Champaign, Urbana, IL, United States

D.M. Kilkenny
Institute of Biomaterials and Biomedical Engineering; Banting and Best Diabetes Centre, University of Toronto, Toronto, ON, Canada

S. Kinoshita
Osaka Medical Center and Research Institute for Maternal and Child Health, Izumi, Japan

H. Manya
Molecular Glycobiology, Research Team for Mechanism of Aging, Tokyo Metropolitan Geriatric Hospital and Institute of Gerontology, Tokyo, Japan

D. Modan-Moses
The Edmond and Lily Safra Children's Hospital, Chaim Sheba Medical Center, Tel-Hashomer, Ramat-Gan; Tel Aviv University, Tel Aviv, Israel

P.C. Mullen
Boston University School of Medicine, Boston, MA, United States

J.A. Neyra
Charles and Jane Pak Center for Mineral Metabolism and Clinical Research, University of Texas Southwestern Medical Center, Dallas, TX, United States

R. Ramírez
Alcalá de Henares University, Madrid, Spain

J.V. Rocheleau
Institute of Biomaterials and Biomedical Engineering; Banting and Best Diabetes Centre, University of Toronto; Toronto General Research Institute, University Health Network, Toronto, ON, Canada

T. Rubinek
Institute of Oncology, Tel Aviv Sourasky Medical Center, Tel Aviv, Israel

M. Sopjani
University of Prishtina, Prishtinë, Republic of Kosova

T. Tucker-Zhou
Boston University School of Medicine, Boston, MA, United States

I. Wolf
Institute of Oncology, Tel Aviv Sourasky Medical Center; Sackler Faculty of Medicine, Tel Aviv University, Tel Aviv, Israel

Y.-L. Wu
University of Texas Southwestern Medical Center, Dallas, TX, United States

J. Xie
University of Texas Southwestern Medical Center, Dallas, TX, United States

E. Zeldich
Boston University School of Medicine, Boston, MA, United States

PREFACE

Klotho (feminine) means "spinner" in Greek and refers to one of the three fates (Moirai) that spin the thread of life. Klotho exists as an insoluble, membrane form or as soluble forms generated by enzymatic cleavage of the membrane form. The membrane full-length form (mass of 130 kDa) has two extracellular glycosyl hydrolase domains (KL1 and KL2) plus a 20-amino acid transmembrane domain and a 9-amino acid intracellular domain. The soluble forms that can circulate in the bloodstream have either KL1 or Kl1 plus KL2 domains and, once in the circulation, they function as hormones. There is also a secreted Klotho that is generated by alternative splicing of mRNA; it contains 549 amino acids and has a mass of 65 kDa. Both the soluble and the secreted forms of Klotho regulate the TRPV5 channel and the kidney medullary potassium channel 1 (ROMK1). Likely, there exist still other forms of Klotho.

In the mouse, *Klotho* gene mutation is the single gene mutation known to generate premature aging. Klotho is synthesized in several tissues, primarily in the kidney and brain choroid plexus. The actions of Klotho are related to other factors: vitamin D induces kidney Klotho and with the activity of vitamin D, the systems of growth, development, antioxidation, and homeostasis are maintained and promoted. Klotho interacts with other important hormones and growth factors. Presumably, Klotho lengthens the life span by delaying the chronic diseases of aging. The prospect of the use of Klotho in the treatment of human diseases, especially kidney disease and others, is enthralling.

In this volume, the X-ray structures of Klotho are reported as well as the topics described above together with its many roles in rescuing the disease processes. In the first chapter, Hayashi and Ito report on "Klotho-related protein KLrP: structure and functions." This is followed by the work of Kilkenny and Rocheleau: "The FGF21 receptor signaling complex: Klothoβ, FGFR1c, and other regulatory interactions." Furthering the basic aspects of Klotho actions, Sopjani and Dërmaku-Sopjani describe "Klotho-dependent cellular transport regulation." On the interactions of Klotho and other factors, Rubinek and Modan-Moses introduce "Klotho and the growth hormone/insulin-like growth factor 1 axis: novel insights into complex interactions." Then follows a report on "Klotho prevents translocation of NFκB" by Buendía, Ramírez, Aljama, and Carracedo. Kinoshita and

Kawai describe "The FGF23/Klotho regulatory network and its roles in human disorders." "MicroRNA-34a and impaired FGF19/21 signaling in obesity" by Fu and Kemper increases the span of Klotho involvement. In Chapter 8, Rubinek and Wolf introduce "The role of alpha-Klotho as a universal tumor suppressor." Positive actions of Klotho are emphasized in "Klotho is a neuroprotective and cognition-enhancing protein" by Abraham, Mullen, Zhou, Chen, and Zeldich. The last three chapters involve kidney disease and fallout to the heart. In the first of these, Akasaka-Manya, Manya, and Endo write on "Function and change with aging of α-Klotho in the kidney." "αKlotho and chronic kidney disease" is described by Neyra and Hu. Finally, Xie, Wu, and Huang report on "Deficiency of soluble αKlotho as an independent cause of uremic cardiomyopathy."

The illustration on the cover is the previously published version (*Journal of Biological Chemistry*) of the X-ray structure of Klotho-related protein, an alternative to Fig. 3A reported in Chapter 1 by Hayashi and Ito: "Klotho-related protein KLrP: structure and functions."

Helene Kabes of Elsevier (Oxford, UK) was, as usual, a central person in the development of the publication process. The collaboration of Reed-Elsevier, Chennai, India, in the development of galley proofs and final corrections of these proofs leading directly to publication, was invaluable.

Gerald Litwack
Toluca Lake, North Hollywood, CA
March 9, 2016

CHAPTER ONE

Klotho-Related Protein KLrP: Structure and Functions

Y. Hayashi*, M. Ito†,1

*Faculty of Pharma-Sciences, Teikyo University, Tokyo, Japan
†Faculty of Agriculture, Graduate School of Bioresource and Bioenvironmental Sciences, Kyushu University, Fukuoka, Japan
[1]Corresponding author: e-mail address: makotoi@agr.kyushu-u.ac.jp

Contents

Abstract

Klotho (KL) family proteins share one or two glycoside hydrolase (GH) motifs homologous to GH family 1. However, the biological significance of GH motifs in KL family proteins remains elusive. We describe here that KL-related protein (KLrP), which is composed of a single GH motif, is a cytosolic β-glucocerebrosidase (GCase, EC 3.2.1.145). We detected a neutral conduritol B epoxide (CBE)-insensitive glucosylceramide (GlcCer)-degrading activity in the cytosol fractions of human fibroblasts, rat brains, and zebrafish embryos. KL family proteins emerged as a potent candidate for the neutral GCase using a bioinformatics approach. Recombinant human KLrP, but not α-KL, β-KL, or KLPH, exhibited GCase activity with a neutral pH optimum in the presence of CBE. We solved the crystal structures of KLrP and a KLrP mutant (E165Q) in complex with glucose, which indicate that KLrP forms a $(\beta/\alpha)_8$TIM barrel structure with the double-displacement mechanism of the retaining β-glycosidase. Furthermore, knockdown of endogenous KLrP in CHOP cells using small interfering RNA (siRNA) decreased the CBE-insensitive neutral GCase activity and increased the cellular levels of GlcCer, which suggests that KLrP is involved in a novel GlcCer catabolism pathway. A KLrP D106N mutant was discovered in patients with severe Gaucher disease; however, this mutation did not affect the GCase activity of KLrP.

Vitamins and Hormones, Volume 101
ISSN 0083-6729
http://dx.doi.org/10.1016/bs.vh.2016.02.011

ABBREVIATIONS

CBE conduritol B epoxide
GalCer galactosylceramide
GBA glucosidase, beta, acid
GCase β-glucocerebrosidase (β-glucosylceramidase)
GD Gaucher disease
GH glycoside hydrolase
GlcCer glucosylceramide
GlcT-1 UDP-glucose: ceramide glucosyltransferase-1
GSL glycosphingolipid
KL klotho
KLrP Klotho-related protein
LacCer lactosylceramide
LPH lactase-phlorizin hydrolase
4MU 4-methylumbelliferyl
NBD 4-nitrobenzo-2-oxa-1,3-diazole
WT wild type

1. INTRODUCTION

Phenotypes resembling human aging were observed in transgenic mice that overexpressed the type I sodium–proton exchanger (Kuro-o et al., 1997). Kurosu et al. (2005) refer to this mutant as *klotho* (α-KL), who is the Greek goddess that spins the thread of life. Disrupting α-KL resulted in a shorter life span for the mice; conversely, overexpression of α-KL produces an extended life span. Therefore, α-KL is likely involved in controlling aging.

The α-KL protein is composed of two glycoside hydrolase (GH) motifs that are similar to enzymes in GH family 1. Recombinant human α-KL protein (human IgG1 Fc chimera protein) exhibits β-glucuronidase activity; however, two potentially catalytic glutamic acids in the GH motifs were mutated to asparagine and serine, respectively (Tohyama et al., 2004). Natural steroid β-glucuronides, such as β-estradiol 3-β-D-glucuronide, estrone 3-β-D-glucuronide, and estriol 3-β-D-glucuronide, competitively inhibit the glucuronidase activity of the α-KL protein. The α-KL protein might convert inactive glucuronylated steroids into their active forms after removing the terminal β-glucuronic acid. This process may be involved in maintaining calcium homeostasis.

In addition, the α-KL protein seems to feature sialidase, which catalyzes the removal of terminal sialic acids from N-linked glycans on the TRPV5 calcium channel (Cha et al., 2008). Removing the terminal sialic acids exposed the disaccharide galactose-*N*-acetyl-glucosamine, which is a galectin-1 ligand. Binding between galectin-1 and the *N*-glycans inhibits endocytosis of the TRPV calcium channel, which is retained at the plasma membrane surface. These results suggest that the α-KL protein regulates the turnover of TRPV at the plasma membrane; however, the biochemical evidence supporting the notion that the α-KL protein is a sialidase remains elusive. Leunissen et al. (2013) reported that α-KL and sialidase regulated TRPV5 membrane stabilization in a different manner, suggesting that α-KL does not possess sialidase activity.

Proteins that are structurally related to the α-KL protein have been discovered and designated the β-KL protein (Ito et al., 2000), KLPH (Ito, Fujimori, Hayashizaki, & Nabeshima, 2002), and Klotho-related protein (KLrP) (Yahata et al., 2000). The β-KL protein features two GH domains, similar to the α-KL protein, whereas KLPH and KLrP only feature one GH domain. The β-KL protein, KLPH, and KLrP show 41%, 36%, and 41% identity with α-KL at the amino acid level, respectively. KLrP (formerly referred to as GBA3) was previously identified as a human cytosolic β-glucosidase that can hydrolyze nonphysiological glycosides, such as 4-methylumbelliferyl (4MU)-glycosides, pNP-glycosides, and flavonoid glycosides (de Graaf et al., 2001). However, endogenous substrates for KLrP had not been identified until we reidentified KLrP as a neutral β-glucocerebrosidase (GCase) that can hydrolyze glucosylceramide (GlcCer) (Hayashi et al., 2007). GlcCer is a precursor for various glycosphingolipids (GSLs) and is synthesized on the cytosolic face of the Golgi apparatus with UDP-glucose: ceramide glucosyltransferase-1 (GlcT-1) (Ichikawa & Hirabayashi, 1998).

In this review article, we describe the structure and functions of KLrP as well as discuss the relationship between KLrP and Gaucher disease (GD).

2. THE FUNCTIONS AND CRYSTAL STRUCTURE OF KLrP

2.1 Metabolic Pathway for GSLs Involving acid GCase GBA1

GSLs with different glycan structures are present in vertebrate plasma membranes and are synthesized from the precursors GlcCer and galactosylceramide (GalCer) by GlcT-1 and GalCer synthase, respectively. GlcCer is ubiquitously distributed in mammalian tissues, while the GalCer distribution is restricted, eg, in the myelin oligodendrocytes of the brain. GlcT-1 is mainly located on the Golgi apparatus where the active site faces the cytoplasmic side (Ichikawa & Hirabayashi, 1998). On the other hand,

GalT-1 is mainly located on the ER where the active site faces the luminal side (Sprong et al., 1998). Thus, GlcCer is synthesized on the cytosolic face, whereas GalCer is on the luminal face. The GlcCer generated is then translocated to the luminal side of the Golgi membrane, where it is converted into lactosylceramide (LacCer) by LacCer synthase. LacCer synthesis is followed by the generation of complex GSLs through a step-by-step extension of sugar chains by corresponding glycosyltransferases. In contrast, GalCer is converted into GM4 and sulfatide by sialylation and sulfation, respectively; further sugar chain extension does not occur in mammals. Finally, GSLs are transported through the trans-Golgi network to the plasma membrane where the sugar moiety faces extracellular space, and the ceramide moiety is embedded in the upper layer of the membrane.

GSLs at the plasma membrane are then internalized in endocytic vesicles and transported to the lysosome, where the corresponding acid GHs hydrolyze GSLs through a step-by-step removal of sugar chains facilitated by specific activator proteins, so-called saposins. GBA1, which is also known as acid GCase, hydrolyzes GlcCer to the ceramide and glucose facilitated by saposin C in the lysosomes. Conduritol B epoxide (CBE) specifically and irreversibly inhibits GBA1 activity. An inherited GBA1 deficiency causes GD, which is the most common lysosomal storage disease and is characterized by GlcCer accumulation in the lysosomes of laden tissue macrophages. However, GlcCer accumulation in other cell types is not clear in patients with GD despite a significant decrease in GBA1 activity, which suggests an alternative catabolic pathway for GlcCer (Barranger & Ginns, 1989; Beutler & Grabowski, 2001).

2.2 Identification of KLrP as a Novel Cytosolic Neutral GCase

During the LacCer synthase activity assay in human fibroblasts using C6-4-nitrobenzo-2-oxa-1,3-diazole (NBD)-GlcCer as an acceptor substrate and UDP-Gal as a donor substrate at pH 6.0, C6-NBD-Cer was detected through HPLC in addition to the expected product C6-NBD-LacCer. The generation of C6-NBD-Cer was significantly lower but was not completely eliminated by adding CBE, which is a potent GBA1 inhibitor. The activity of the GlcCer-hydrolyzing enzyme reached a maximum at pH 6–7 in the presence of CBE and was mainly recovered in the cytosolic fraction of the fibroblasts. Cytosolic proteins that can hydrolyze GlcCer have not been reported. Interestingly, C6-NBD-GlcCer hydrolysis to C6-NBD-Cer in the presence of CBE was observed in lysates not only from human fibroblasts but also from rat brains and zebrafish embryos. However, activity was not detected in lysate from slime mold. Similar enzyme activity was detected in the bacterium *Paenibacillus* sp. TS12 (Sumida, Sueyoshi, & Ito, 2002); however, a sequence that is homologous to TS12 GCase has not been reported in the human gene database.

Using the CAZy database, we identified KLrP, which is a KL family protein, as a candidate for a cytosolic neutral GCase. KL family proteins, including KLrP, share a GH domain; however, their natural substrates are not clear. KLrP orthologues are present in databases for humans, rats, and zebrafish but not slime mold, which is consistent with detecting GCase activity at a neutral pH in the presence of CBE. Notably, a KLrP homologue was not observed in the database for mice; however, CBE-insensitive neutral GCase activity was detected in the mouse brains. Among the KL family proteins, KLrP features neither a transmembrane domain nor a signal peptide (Fig. 1), which suggests that KLrP is a cytosolic protein. Furthermore, KLrP features the two predicted catalytic glutamates on a GH domain; however, these glutamates are replaced with an asparagine and a serine in α-KL (Kuro-o et al., 1997), an asparagine and alanine in β-KL (Ito et al., 2000), and an aspartic acid in KLPH (Ito et al., 2002) (Fig. 1). Thus, we attempted to examine the GCase activity of KLrP in the presence of CBE at a neutral

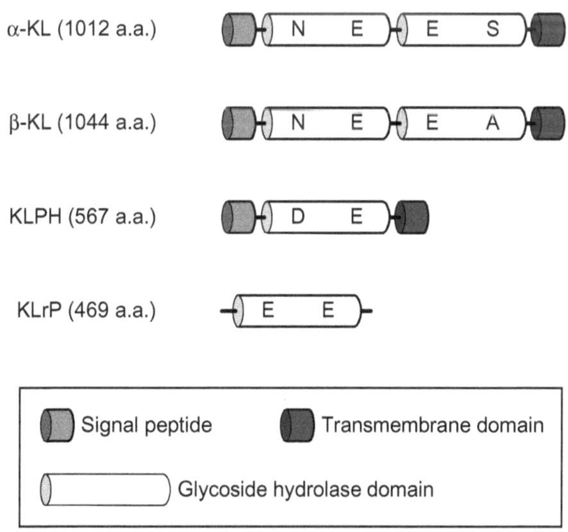

Fig. 1 The schematic structures of KL family proteins. E165 and E373 in KLrP function as an acid/base catalyst and a nucleophile, respectively. The catalytic residues in α-KL, β-KL, and KLPH that correspond to E165 and E373 in KLrP were mutated to N, D, S, or A.

pH in addition to other KL family proteins. Expectedly, neutral GCase activity increased in KLrP-overexpressing CHOP cells but not when other KL family proteins were overexpressed (Fig. 2A). This result indicates that KLrP, but not other KL proteins, is a GCase that can hydrolyze GlcCer (Fig. 2B). Myc-tagged KLrP was localized in the cytosol of CHOP cells, which confirms that KLrP is a cytosolic protein. The activity of the purified recombinant KLrP expressed in *Escherichia coli* reached a maximum at pH 6.0–7.0 when C6-NBD-GlcCer was used as a substrate, and it was not inhibited

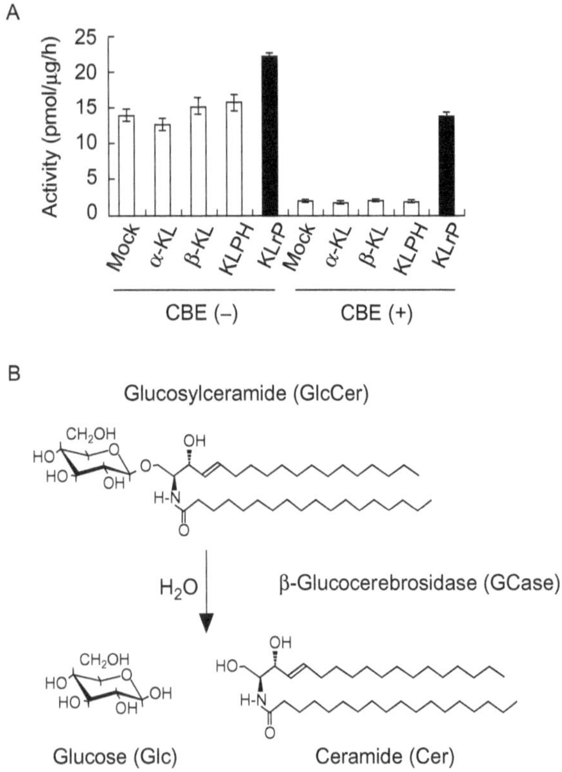

Fig. 2 Identification of KLrP as a CBE-insensitive neutral GCase. (A) GCase activity of CHOP cells transfected with cDNA encoding KL family proteins in the presence (*left panel*) or absence (*right panel*) of 0.5 mM CBE. α-KL, β-KL, KLPH, KLrP, and mock represent transfectants with cDNA encoding each KL family protein or empty vector (mock). The neutral GCase activities were measured using C6-NBD-GlcCer. (B) The reaction mode of KLrP (GCase) on GlcCer. KLrP cleaves the β-glycosidic linkage between the glucose and ceramide of GlcCer. *Panel (A) Adapted from Hayashi, Y., Okino, N., Kakuta, Y., Shikanai, T., Tani, M., Narimatsu, H., et al. (2007). Klotho-related protein is a novel cytosolic neutral beta-glycosylceramidase.* The Journal of Biological Chemistry, 282, *30889–30900.*

by CBE. GlcCer was the best substrate for KLrP followed by galactosylsphingosine (GalSph), glucosylsphingosine (GlcSph), and GalCer; however, hydrolysis of αGalCer, sulfatides, LacCer, or GM1a was not observed. The kinetic parameters of the purified KLrP using various substrates were calculated using Hanes–Woolf plots. KLrP hydrolyzed C6-NBD-GlcCer and authentic GlcCer (d18:1, C18:0) with k_{cat}/K_m values of 1.57 and 0.03, respectively, with 0.25% of sodium cholate in the reaction mixture.

KLrP knockdown using siRNA of CHOP cells decreased GCase activity in the presence of CBE and, simultaneously, increased the ^{14}C-GlcCer and ^{14}C-GSLs in HEK293 cells that were metabolically labeled with ^{14}C-Gal. These data suggest that KLrP regulates intracellular GSLs level via hydrolysis of GlcCer, which is synthesized on the cytosolic face on the Golgi apparatus. On the other hand, Dekker et al. (2011) reported that KLrP did not significantly contribute to cellular degradation of GlcCer in HuH-7 cells. They used alpha-1-C-nonyl-DIX (anDIX) as an inhibitor for both GBA1 ($IC_{50}=0.001$ μM) and KLrP ($IC_{50}=0.01$ μM) in addition to CBE ($IC_{50}=9$ μM for GBA1, but KLrP was not inhibited). They did not observe an additional increase in GlcCer in HuH-7 cells when anDIX was used with CBE compared to using CBE alone. We doubt that anDIX inhibits the activity of GlcT-1 as well as GCases because the total GlcCer in the HuH-7 cells treated with a mixture of CBE and anDIX was lower than when treated with CBE alone. Collectively, the in vivo roles of KLrP remain controversial, and further experiments are necessary to clarify whether KLrP contributes to GlcCer metabolism in vivo. KLrP knockout cells generated by the CRISPR-Cas9 system will help to solve this issue.

We stress here that KLrP hydrolyzes not only GlcCer but also GalCer, glucosylsphingosine (GlcSph), and galactosylsphingosine (GalSph); however, the hydrolysis rates for these GSLs are relatively low compared with GlcCer (Hayashi et al., 2007). This result indicates that the substrate specificity of KLrP is not strict for the C4 anomeric configuration and lipid portion, which suggests the presence of unknown substrates for KLrP in vivo. To solve this issue, KLrP-deficient cells are necessary.

2.3 KLrP Crystal Structure

To examine KLrP's catalytic mechanism at the atomic level, we determined the X-ray crystal structure using recombinant KLrP (Fig. 3A). The KLrP crystal structure in complex with glucose (KLrP/Glc) was constructed using collective synchrotron data at a 1.6 Å resolution. As expected based on the

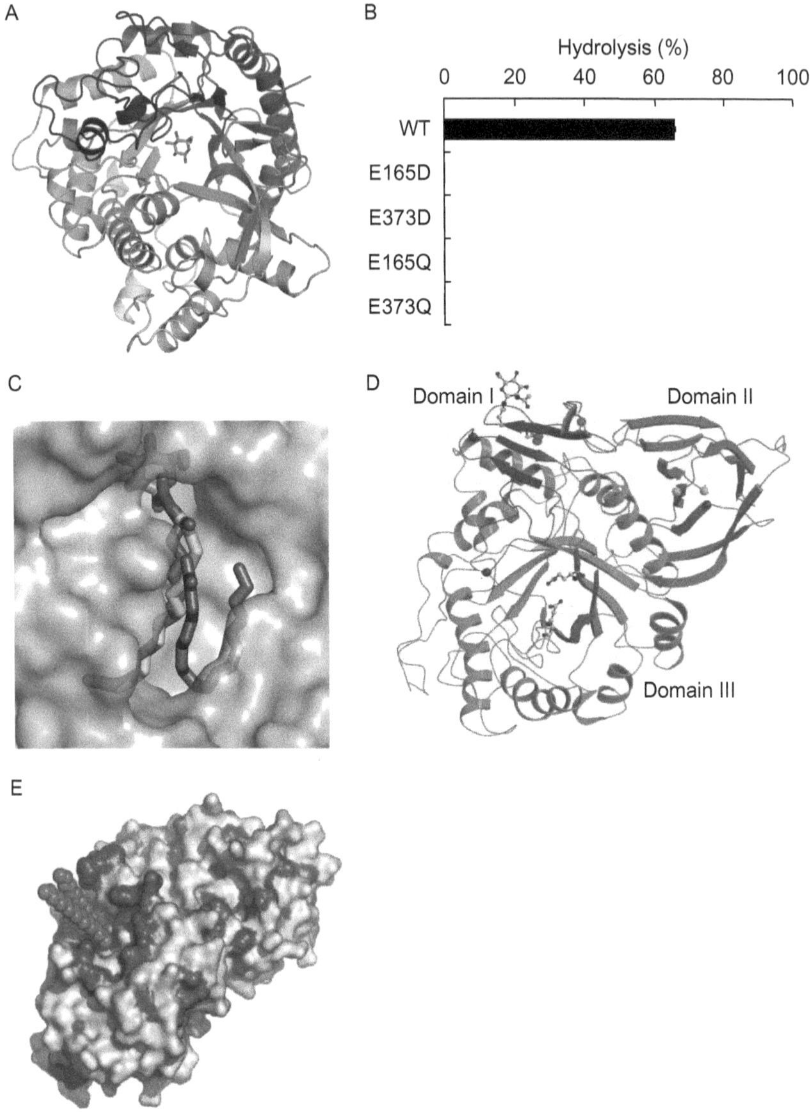

Fig. 3 KLrP and GBA1 X-ray crystal structures. (A) Structure of KLrP in complex with glucose (KLrP/Glc) illustrated using a ribbon diagram. The glucose is shown with carbon atoms in green and oxygen atoms in red. (B) Point mutations of two catalytic residues. The purified wild-type KLrP (WT) and mutants (E165D, E373D, E165Q, and E373Q) were subjected to the neutral GCase assay. For the GCase assay, 100 pmol of C6-NBD-GlcCer was incubated with 50 ng of enzyme in 50 mM MES buffer, pH 6.0, containing 0.25% sodium cholate at 37°C for 30 min. (C) A close-up view of KLrP's substrate-binding cleft. Based on the electron densities, palmitic acid is located at two different positions, form

deduced primary structure for a family of the GH-A clan, KLrP/Glc exhibits a $(\beta/\alpha)_8$TIM barrel in which Glu165 and Glu373 at the carboxyl termini of β-strands 4 and 7 could function as an acid/base catalyst and a nucleophile, respectively. Actually, the mutants E165Q and E373Q lost the neutral GCase activity (Fig. 3B). The distance between the carboxyl oxygen atoms of two catalytic residues was 5.3 Å, which indicates that the reaction proceeds through a retaining mechanism, in which the anomeric carbon is retained upon cleavage. We also solved the crystal structure for the KLrP mutant E165Q in complex with glucose, in which glucose was covalently bound to the nucleophile E373, which indicates that the enzyme reaction proceeds through a double-displacement mechanism (Noguchi et al., 2008). Collectively, KLrP is composed of a $(\beta/\alpha)_8$TIM barrel structure with the double-displacement mechanism of the retaining β-glycosidase.

The KLrP/Glc substrate-binding cleft was occupied with one molecule each of glucose, palmitic acid, and oleic acid. Two fatty acid molecules were likely derived from the host *E. coli* used to produce the recombinant GCase. Using this crystal structure, a complex model for KLrP with GlcCer was generated (Fig. 3C) in which the ceramide and glucose moieties fit with the electron densities of the fatty acids and glucose that occupy the cleft, respectively. In this model, GlcCer is entirely incorporated into the substrate-binding cavity.

The GBA1 (Fig. 3D) and KLrP (Fig. 3A) crystal structures differ. GBA1 consists of three domains: domain I at the N-terminus; domain II, which is an immunoglobulin-like domain; and domain III, which is a catalytic domain in a TIM barrel (Dvir et al., 2003). KLrP does not have domains that correspond to domains I and II in GBA1. The model of GBA1 docking with GlcCer shows that the glucose moiety and adjacent glycoside bond of GlcCer fit within the substrate-binding cavity, while ceramide moiety exists outside of the cavity (Fig. 3E). On the other hand, the model of KLrP docking with GlcCer indicates that GlcCer is completely swallowed by the

I and form II. *Green*, glucose; *blue*, palmitic acid (form I); *yellow*, palmitic acid (from II); and *purple*, oleic acid. (D and E) The GBA1 structure is illustrated using a ribbon diagram and the docked model of GBA1 with glucose. *Panel (A) Adapted from Hayashi, Y., Okino, N., Kakuta, Y., Shikanai, T., Tani, M., Narimatsu, H., et al. (2007). Klotho-related protein is a novel cytosolic neutral beta-glycosylceramidase.* The Journal of Biological Chemistry, 282, *30889–30900. Panels (D and E) Adapted from Dvir, H., Harel, M., McCarthy, A. A., Toker, L., Silman, I., Futerman, A. H., et al. (2003). X-ray structure of human acid-beta-glucosidase, the defective enzyme in Gaucher disease.* EMBO Reports, 4, *704–709.* (See the color plate.)

substrate-binding cleft (Fig. 3C). The differences between the two docking models may partially explain the in vivo roles of these enzymes in hydrolyzing GlcCer. GBA1 hydrolyzes GlcCer embedded in the lysosome membranes facilitated by activator protein saposin C (Alattia, Shaw, Yip, & Prive, 2007), while KLrP hydrolyzes nonmembrane bound GlcCer in the cytosol potentially without a specific activator protein. Saposin C likely lifts up GlcCer from the lysosome membranes to allow the enzyme access to the substrate.

2.4 Mammalian GCases Other than GBA1 and KLrP (GBA3)

Currently, at least four proteins, including GBA1 and KLrP, have been identified as mammalian GCases (Fig. 4). Knockout mice of β-glucosidase 2 (GBA2), which was thought to be involved in bile acid metabolism, have been generated. Unexpectedly, however, bile acid metabolism was normal in the GBA2 knockout mice; alternatively, GlcCer accumulation was observed in the testis, brain, and liver in the knockout mice, which indicates that GBA2 functions as a GCase in vivo (Yildiz et al., 2006). GBA2 is a

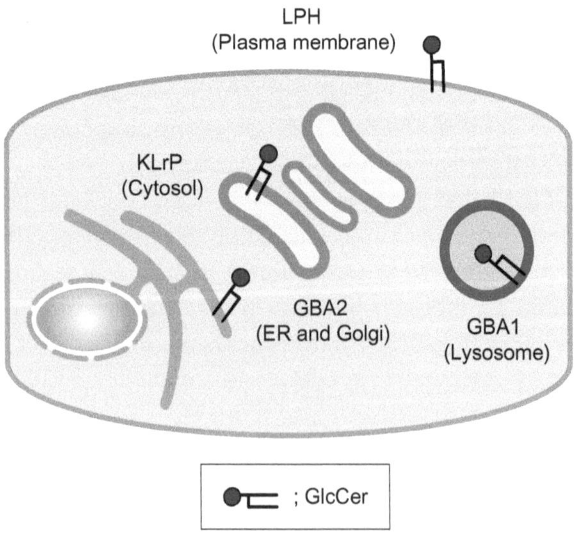

Fig. 4 Cellular localizations for four mammalian GCases. GBA1 is localized to the lysosomes, where it hydrolyzes GlcCer in an acidic environment. GBA2 and KLrP (GBA3) are a nonintegral cytosolic proteins with a neutral pH optimum; however, GBA2, but not KLrP, is strongly associated with the ER/Golgi membranes (Hayashi et al., 2007; Körschen et al., 2013). LPH is exclusively localized to the intestinal epithelial cell microvilli, where it functions as a digestive enzyme. (See the color plate.)

cytosolic protein that is strongly associated with membranes at the ER and Golgi apparatus and is likely involved in nonlysosomal degradation of GlcCer (Körschen et al., 2013). Mutation of human GBA2 decreased the GCase activity, which leads to hereditary spastic paraplegia and cerebellar ataxia (Citterio et al., 2014; Martin et al., 2013). In addition to GBA1, 2, and 3, lactase-phlorizin hydrolase (LPH) hydrolyzes GlcCer (Kobayashi & Suzuki, 1981). LPH, which is sensitive to CBE, is exclusively present in intestinal epithelial cell microvilli and may function as a digestive enzyme.

3. KLrP AND GD

GD is the most common lysosomal GSL storage disease and is caused by mutations in the gene encoding GBA1, which results in lysosomal GlcCer accumulation (Barranger & Ginns, 1989; Beutler & Grabowski, 2001). GD is divided into three major subtypes: type 1, nonneuropathic; type 2, acute neuropathic; and type 3, subacute neuropathic. Type 1 GD is characterized by anemia thrombocytopenia, hepatosplenomegaly, bone dysplasia, and an absence of neurological manifestations. Type 2 and type 3 GD are relatively more severe clinical manifestations characterized by dysfunction of the central nervous system. However, why or how GlcCer accumulation causes GD is unclear. Fig. 5 shows the working model for GD development proposed by Tony Futerman (Bodennec, Pelled, Riebeling, Trajkovic, & Futerman, 2002; Pelled et al., 2005). In this model, GlcCer accumulation in the lysosome promotes GlcCer leakage from the lysosome. The GlcCer leaked in the cytosol affects the ryanodine receptor on the ER membrane and facilitates the calcium ion release, which causes neuronal dysfunction. Furthermore, CTP:phosphatidylcholine cytidylyltransferase (CCT) is activated by the cytosolic GlcCer to increase phosphocholine (PC) synthesis, which may cause hepatomegaly and splenomegaly. Dysfunction of GBA1 is, of course, the primary cause for GD; however, the clinical variability of GD is not explained by residual activity of GBA1. Thus, researchers in this field have long considered that dysfunction of GCase(s) other than GBA1 may contribute to GD symptoms. Futerman's model can be extended with new players, KLrP and GBA2, that both may function in the cytosol to eliminate GlcCer escaped from lysosomes (Fig. 5).

The KLrP and GBA2 to GD connection have been examined by several research groups. Beutler, Beutler, and West (2004) examined whether polymorphisms in KLrP are related to GD. They found four single-nucleotide substitutions in KLrP from GD patients; however, these mutations were not

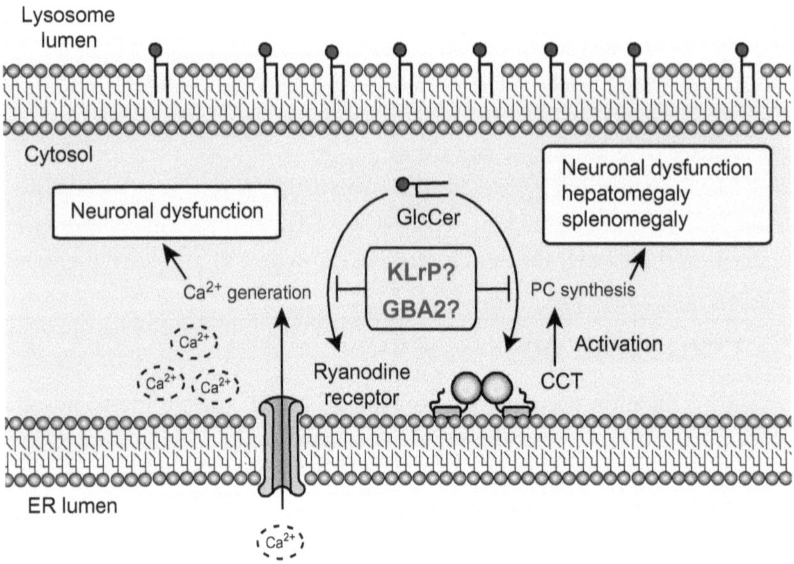

Fig. 5 Working hypothesis for how Gaucher disease (GD) is triggered by GlcCer accumulation (modified from Futerman's model). The GlcCer accumulated in the GD lysosomes escapes to the ER, where the GlcCer enhances an agonist-induced calcium release via activating the ryanodine receptor. GlcCer simultaneously increases PC synthesis through CTP: phosphatidylcholine cytidylyltransferase activation. The disturbances in cellular homeostasis induced by the GlcCer result in neuron dysfunction, hepatomegaly, and splenomegaly in GD patients. Given the cellular localizations of GBA2 and KLrP, both GCases may contribute to hydrolysis of the escaped GlcCer. (See the color plate.)

related to GD phenotypes. Aers and his associates revealed that a KLrP T1368A mutation abolished GCase activity due to the missing last α-helix of the $(\beta/\alpha)_8$ barrel. They identified individuals as wild-type (WT), heterozygous, or homozygous for this mutation in type 1 GD patients. However, no correlation was observed between this mutation and the GD phenotypes (Dekker et al., 2011).

A KLrP D106N mutant was also discovered in severe GD patients; however, the frequency of this mutation in KLrP is rare (Beutler et al., 2004). We examined the GCase activity of the purified D106N mutant using C6-NBD-GlcCer as a substrate. We did not detect a significant difference in GCase activity between the WT and mutant KLrP (Fig. 6A and B), which suggests that this mutation is not related to the GD polymorphism. Collectively, we have not obtained evidence on the connection between KLrP and GD.

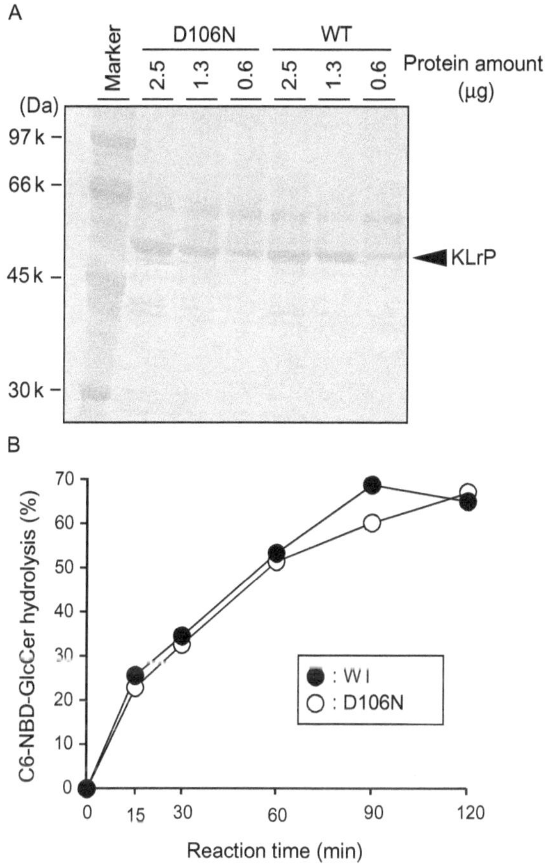

Fig. 6 The effects of a D106N mutation on the GCase activity of KLrP. (A) Purification of the WT and D106N mutation of KLrP. His-tagged KLrP and the D106N mutant were purified using HiTrap Chelating and Superdex200. (B) The GCase activities of the purified WT and D106N mutant. For the GCase assay, 100 pmol of C6-NBD-GlcCer was incubated with 25 ng of enzyme in 50 mM MES buffer, pH 6.0, containing 0.25% sodium cholate at 37°C for several minutes (0, 15, 30, 60, 90, and 120 min). The results represent the average value from two independent experiments. (See the color plate.)

In contrast, GBA2 might be involved in variable phenotypes of GD. The GBA2 activity significantly increased in GBA1-deficient mice brains (Burke et al., 2013). Similarity, the GBA2 protein and mRNA levels were elevated in GBA1-deficient murine fibroblasts (Yildiz et al., 2013). GBA2 might provide compensatory protection for GBA1 mutations; however, the SNP analysis at the GBA2 locus did not support a significant association with the severity of GD. Interestingly, Mistry et al. (2014) showed that GBA2 deletion in type I GD model mice rescued the visceral, hematologic, and skeletal phenotype, despite the increase in GlcCer and GlcSph levels. They insist that GlcCer hydrolysis by GBA2 increased the level of sphingosine, which is a toxic mediator that seriously affects type I GD pathophysiology; thus, GBA2 inhibitors might be a good target for improving type I GD clinical manifestations.

4. CONCLUSIONS AND PERSPECTIVES

KLrP is a glycosidase composed of a $(\beta/\alpha)_8$ barrel structure with double-displacement retaining mechanism. Biochemical and crystal structure analyses revealed that the KLrP substrate specificity is broad, and it hydrolyzes various β-glucosides, including GlcCer and GlcSph. KLrP can hydrolyze natural GlcCer; however, the hydrolysis rate is much lower than for C6-NBD-GlcCer or 4MU-Glc. The k_{cat} for natural GlcCer is 0.03 min^{-1} μM^{-1}, but for C6-NBD-GlcCer and 4MU-Glc it is 1.57 and 1.51 min^{-1} μM^{-1}, respectively. Notably, KLrP also hydrolyzes β-galactosides; ie, the k_{cat} for C6-NBD-GalCer is 0.75 min^{-1} μM^{-1}, and it is 0.66 min^{-1} μM^{-1} for 4MU-Gal, which suggests the presence of unknown β-galactosides susceptible to hydrolysis by KLrP in the cytosol.

Currently, four proteins, GBA1, GBA2, LPH, and KLrP, have been identified as a mammalian GCases (Fig. 4). Recently, Harzer and Yildiz (2015) compared the overall GCase activities of wild-type, GBA1-deficient, and GBA2-deficient fibroblasts when the cells were fed radioactive GlcCer. They estimated that the overall GCase activities were divided roughly into the ratio 1:1:1–2 = GBA1:GBA2:non-GBA1, 2. This result indicates that non-GBA1, 2 enzymes, including KLrP, contribute to a high proportion of overall GCase activities, which suggests a potential unknown functions for KLrP in GSL metabolism and function.

ACKNOWLEDGMENTS

We are grateful to Dr. H. Narimatsu and T. Shikanai (National Institute of Advanced Industrial Science and Technology, Japan) as well as Dr. N. Okino and Dr. Y. Kakuta (Kyushu University, Japan) for their valuable suggestions and technical support. This work is partially supported by the Grant-in-Aid for Scientific Research B (19380061 and 15H04488) (to M.I.) and Young Scientists (B) (15K18868) (to Y.H.) from the Ministry of Education, Culture, Sports, Science, and Technology of the Japanese Government.

REFERENCES

Alattia, J. R., Shaw, J. E., Yip, C. M., & Prive, G. G. (2007). Molecular imaging of membrane interfaces reveals mode of beta-glucosidase activation by saposin C. *Proceedings of the National Academy of Sciences of the United States of America*, *104*, 17394–17399.

Barranger, J. A., & Ginns, E. I. (1989). In C. R. Scriver, A. L. Beaudet, W. S. Sly, & D. Valle (Eds.), *The metabolic basis of inherited disease II* (pp. 1677–1698). New York: McGraw-Hill Inc.

Beutler, E., Beutler, L., & West, C. (2004). Mutations in the gene encoding cytosolic beta-glucosidase in Gaucher disease. *The Journal of Laboratory and Clinical Medicine*, *144*, 65–68.

Beutler, E., & Grabowski, G. A. (2001). In C. R. Scriver, A. L. Beaudet, D. Valle, & W. S. Sly (Eds.), *The metabolic and molecular bases of inherited disease III* (pp. 3635–3668). New York: McGraw-Hill Inc.

Bodennec, J., Pelled, D., Riebeling, C., Trajkovic, S., & Futerman, A. H. (2002). Phosphatidylcholine synthesis is elevated in neuronal models of Gaucher disease due to direct activation of CTP:phosphocholine cytidylyltransferase by glucosylceramide. *The FASEB Journal*, *16*, 1814–1816.

Burke, D. G., Rahim, A. A., Waddington, S. N., Karlsson, S., Enquist, I., Bhatia, K., et al. (2013). Increased glucocerebrosidase (GBA) 2 activity in GBA1 deficient mice brains and in Gaucher leucocytes. *Journal of Inherited Metabolic Disease*, *36*, 869–872.

Cha, S. K., Ortega, B., Kurosu, H., Rosenblatt, K. P., Kuro-O, M., & Huang, C. L. (2008). Removal of sialic acid involving Klotho causes cell-surface retention of TRPV5 channel via binding to galectin-1. *Proceedings of the National Academy of Sciences of the United States of America*, *105*, 9805–9810.

Citterio, A., Arnoldi, A., Panzeri, E., D'Angelo, M. G., Filosto, M., Dilena, R., et al. (2014). Mutations in CYP2U1, DDHD2 and GBA2 genes are rare causes of complicated forms of hereditary spastic paraparesis. *Journal of Neurology*, *261*, 373–381.

de Graaf, M., van Veen, I. C., van der Meulen-Muileman, I. H., Gerritsen, W. R., Pinedo, H. M., & Haisma, H. J. (2001). Cloning and characterization of human liver cytosolic beta-glycosylhydrolase. *The Biochemical Journal*, *356*, 907–910.

Dekker, N., Voorn-Brouwer, T., Verhoek, M., Wennekes, T., Narayan, R. S., Speijer, D., et al. (2011). The cytosolic beta-glucosidase GBA3 does not influence type 1 Gaucher disease manifestation. *Blood Cells, Molecules & Diseases*, *46*, 19–26.

Dvir, H., Harel, M., McCarthy, A. A., Toker, L., Silman, I., Futerman, A. H., et al. (2003). X-ray structure of human acid-beta-glucosidase, the defective enzyme in Gaucher disease. *EMBO Reports*, *4*, 704–709.

Harzer, K., & Yildiz, Y. (2015). High β-glucosidase (GBA) activity not attributable to GBA1 and GBA2 in live normal and enzyme-deficient fibroblasts may emphasise the role of additional GBAs. *Biological Chemistry*, *396*(11), 1241–1246.

Hayashi, Y., Okino, N., Kakuta, Y., Shikanai, T., Tani, M., Narimatsu, H., et al. (2007). Klotho-related protein is a novel cytosolic neutral beta-glycosylceramidase. *The Journal of Biological Chemistry*, *282*, 30889–30900.

Ichikawa, S., & Hirabayashi, Y. (1998). Glucosylceramide synthase and glycosphingolipid synthesis. *Trends in Cell Biology*, *8*, 198–202.

Ito, S., Fujimori, T., Hayashizaki, Y., & Nabeshima, Y. (2002). Identification of a novel mouse membrane-bound family 1 glycohydrolase-like protein, which carries an atypical active site structure. *Biochimica et Biophysica Acta*, *1576*, 341–345.

Ito, S., Kinoshita, S., Shiraishi, N., Nakagawa, S., Sekine, S., Fujimori, T., et al. (2000). Molecular cloning and expression analyses of mouse betaklotho, which encodes a novel Klotho family protein. *Mechanisms of Development*, *98*, 115–119.

Kobayashi, T., & Suzuki, K. (1981). The glycosylceramidase in the murine intestine. *Purification and substrate specificity The Journal of Biological Chemistry*, *256*, 7768–7773.
Körschen, H. G., Yildiz, Y., Raju, D. N., Schonauer, S., Bonigk, W., Jansen, V., et al. (2013). The non-lysosomal beta-glucosidase GBA2 is a non-integral membrane-associated protein at the endoplasmic reticulum (ER) and Golgi. *The Journal of Biological Chemistry*, *288*, 3381–3393.
Kuro-o, M., Matsumura, Y., Aizawa, H., Kawaguchi, H., Suga, T., Utsugi, T., et al. (1997). Mutation of the mouse klotho gene leads to a syndrome resembling ageing. *Nature*, *390*, 45–51.
Kurosu, H., Yamamoto, M., Clark, J. D., Pastor, J. V., Nandi, A., Gurnani, P., et al. (2005). Suppression of aging in mice by the hormone Klotho. *Science*, *309*, 1829–1833.
Leunissen, E. H., Nair, A. V., Bull, C., Lefeber, D. J., van Delft, F. L., Bindels, R. J., et al. (2013). The epithelial calcium channel TRPV5 is regulated differentially by klotho and sialidase. *The Journal of Biological Chemistry*, *288*, 29238–29246.
Martin, E., Schule, R., Smets, K., Rastetter, A., Boukhris, A., Loureiro, J. L., et al. (2013). Loss of function of glucocerebrosidase GBA2 is responsible for motor neuron defects in hereditary spastic paraplegia. *American Journal of Human Genetics*, *92*, 238–244.
Mistry, P. K., Liu, J., Sun, L., Chuang, W. L., Yuen, T., Yang, R., et al. (2014). Glucocerebrosidase 2 gene deletion rescues type 1 Gaucher disease. *Proceedings of the National Academy of Sciences of the United States of America*, *111*, 4934–4939.
Noguchi, J., Hayashi, Y., Baba, Y., Okino, N., Kimura, M., Ito, M., et al. (2008). Crystal structure of the covalent intermediate of human cytosolic beta-glucosidase. *Biochemical and Biophysical Research Communications*, *374*, 549–552.
Pelled, D., Trajkovic-Bodennec, S., Lloyd-Evans, E., Sidransky, E., Schiffmann, R., & Futerman, A. H. (2005). Enhanced calcium release in the acute neuronopathic form of Gaucher disease. *Neurobiology of Disease*, *18*, 83–88.
Sprong, H., Kruithof, B., Leijendekker, R., Slot, J. W., van Meer, G., & van der Sluijs, P. (1998). UDP-galactose:ceramide galactosyltransferase is a class I integral membrane protein of the endoplasmic reticulum. *The Journal of Biological Chemistry*, *273*, 25880–25888.
Sumida, T., Sueyoshi, N., & Ito, M. (2002). Molecular cloning and characterization of a novel glucocerebrosidase of *Paenibacillus* sp. TS12. *Journal of Biochemistry*, *132*, 237–243.
Tohyama, O., Imura, A., Iwano, A., Freund, J. N., Henrissat, B., Fujimori, T., et al. (2004). Klotho is a novel beta-glucuronidase capable of hydrolyzing steroid beta-glucuronides. *The Journal of Biological Chemistry*, *279*, 9777–9784.
Yahata, K., Mori, K., Arai, H., Koide, S., Ogawa, Y., Mukoyama, M., et al. (2000). Molecular cloning and expression of a novel klotho-related protein. *Journal of Molecular Medicine*, *78*, 389–394.
Yildiz, Y., Hoffmann, P., Vom Dahl, S., Breiden, B., Sandhoff, R., Niederau, C., et al. (2013). Functional and genetic characterization of the non-lysosomal glucosylceramidase 2 as a modifier for Gaucher disease. *Orphanet Journal of Rare Diseases*, 8:151.
Yildiz, Y., Matern, H., Thompson, B., Allegood, J. C., Warren, R. L., Ramirez, D. M., et al. (2006). Mutation of beta-glucosidase 2 causes glycolipid storage disease and impaired male fertility. *The Journal of Clinical Investigation*, *116*, 2985–2994.

CHAPTER TWO

The FGF21 Receptor Signaling Complex: Klothoβ, FGFR1c, and Other Regulatory Interactions

D.M. Kilkenny[*,†,1], **J.V. Rocheleau**[*,†,‡,§,1]

*Institute of Biomaterials and Biomedical Engineering, University of Toronto, Toronto, ON, Canada
†Banting and Best Diabetes Centre, University of Toronto, Toronto, ON, Canada
‡University of Toronto, Toronto, ON, Canada
§Toronto General Research Institute, University Health Network, Toronto, ON, Canada
1Corresponding authors: e-mail address: dawn.kilkenny@utoronto.ca; jon.rocheleau@utoronto.ca

Contents

Abstract

Scientific evidence is quickly growing that establishes FGF21 as a cytokine that signals both locally and systemically to induce metabolic effects. The focus of this chapter is the receptor/co-receptor signaling complex formed by endocrine FGF21. We provide an introduction to the major components of the complex including the Klotho family of co-receptors, fibroblast growth factor receptors (FGFRs), and the fibroblast growth factor ligands, placing each in the context of its own family members while emphasizing

Vitamins and Hormones, Volume 101
ISSN 0083-6729
http://dx.doi.org/10.1016/bs.vh.2016.02.008

structural features that drive interaction. We subsequently focus specifically on FGF21 signaling through FGFR1c and KLB, describing what is known about each protein's structure and how this drives protein interaction and formation of the signaling complex at the plasma membrane. We subsequently explore the stoichiometry of FGFR1c and KLB at the plasma membrane before and after the addition of FGF21 ligand, comparing how unique features of the interaction could potentially affect signaling intensity. Finally, we discuss how formation of the signaling complex is potentially regulated by other regulatory interactions, including galectins, the extracellular matrix, and co-expression of FGFR5.

1. INTRODUCTION

The fibroblast growth factor (FGF) family of signaling ligands consists of 22 secreted polypeptides classified into 7 sub-families based on phylogeny, sequence identity, and function (Ornitz, 2005; Ornitz & Itoh, 2015). The majority of these factors regulate development and proliferation. The endocrine subfamily of FGFs, which includes FGF19 (mouse homolog FGF15), FGF21, and FGF23, have newly appreciated roles in whole-body physiology (Fukumoto, 2008; Jones, 2008; Kharitonenkov, 2009; Kharitonenkov & Shanafelt, 2009) including potent effects on obesity, clearance of systemic glucose and lipids, insulin sensitivity, and energy expenditure (Kharitonenkov et al., 2005, 2007; Xu, Lloyd, et al., 2009). This class of endocrine factors exhibits negligible affinity for the classical high-capacity, low-affinity heparin sulfate proteoglycan (HSPG) co-receptors, but actively directs affinity for specific FGFRs via Klotho co-receptors (Wu et al., 2011).

The majority of paracrine FGFs bind HSPGs as *co-receptors* to induce dimerization of tyrosine kinase receptors (FGFR1–4) in a signaling complex (HSPG:FGF:FGFR) (Gospodarowicz & Cheng, 1986; Plotnikov, Hubbard, Schlessinger, & Mohammadi, 2000; Schlessinger et al., 2000). In contrast, the endocrine FGFs exhibit negligible affinity for HSPGs (Asada et al., 2009; Goetz et al., 2007). Two major outcomes of this low affinity are: (i) the endocrine ligands can escape the extracellular matrix (ECM) and signal at long distances from their source of secretion (Goetz et al., 2007; Moore, 2007); and (ii) an *alternative* transmembrane co-receptor is required to initiate FGFR activation (Moore, 2007). FGF21 requires *Klotho-β* (KLB) to activate FGFR1c, FGFR3c, and FGFR4; similarly, FGF19 activates FGFR4 when co-expressed with KLB. Although the requirement of KLB for FGF21 activity in vivo has been examined using $KLB^{-/-}$ mouse models, the purity of

FGF21 must be scrutinized (Tomiyama et al., 2010) and the vast majority of other studies lead us to conclude that FGF21 signaling depends upon the co-receptor KLB (Micanovic et al., 2009; Ogawa et al., 2007).

This chapter focuses on the receptor signaling complex formed at the cell membrane in response to FGF21 stimulation. We define *KLB as the transmembrane co-receptor for FGF21* (Goetz, Ohnishi, Kir, et al., 2012). The majority of evidence supports FGF21 signaling through activation of FGFR1c and FGFR3c receptor isoforms. Due to our focus on beta-cells, which we showed predominantly express FGFR1c, this chapter focuses on FGF21 signaling through KLB and FGFR1c. This chapter therefore discusses the basic structures of KLB, FGFR1c, and FGF21 to identify molecular interactions that form the final signaling complex. We subsequently explore what is known about the receptor stoichiometry before and after the addition of FGF21 ligand. We also identify how these receptors potentially interact with other membrane receptors to modulate the FGF21-induced response. In general, FGFR signaling is very potent and must be efficiently regulated, given the extensive number of relevant but redundant proteins in the family. As a consequence of the multiple proteins required for KLB/FGFR1c complex formation, it is quite likely that FGF21 signaling is also highly regulated.

2. THE COMPONENTS OF THE SIGNALING COMPLEX

2.1 Klotho-Beta

2.1.1 The Klotho Family of Receptors

Klothoβ (Klotho-beta; beta-Klotho; βKlotho; KLB) belongs to a family of receptor proteins that includes Klotho (α-Klotho; Klotho-α) and γ-Klotho (Lctl; Lactase-phlorizin hydrolase-related protein) (Table 1). Klotho proteins were originally identified in the late 1990s as a consequence of gene mutation in a mouse model of human aging (Kuro-o et al., 1997). Klotho (α-Klotho), the first family member to be identified, was shown to form a complex with several isoforms of the fibroblast growth factor receptors (FGFRs) to significantly enhance their affinity for FGF23 (Kurosu et al., 2006) and determined to be a co-receptor for FGF signaling. This concept was reinforced in mouse models deficient for FGF23, which demonstrated a similar aging phenotype, as well as metabolic abnormalities including low blood glucose, high blood phosphate, and high active

Table 1 Comparative Sequence Information for Klotho Family of Proteins

	α-Klotho	β-Klotho	γ-Klotho	GBA3
Variant name	Klotho	KLB; Klothoβ	Lctl	cBGL1
UniProt identifier (human)	Q9UEF7	Q86Z14	Q6UWM7	Q9H227
Protein length (aa)	Variant I:1012	1044	Variant I: 567	Variant I: 469
	Variant II: 549		Variant II: 394	Variant II: 162
Molecular mass (kDa)	Variant I: 116.2	119.8	Variant I: 65.1	Variant I: 54
	Variant II: 62.1		Variant II: 45.7	Variant II: 18.3
Variant I extracellular domain (aa)	948	996	519	—
Transmembrane domain (aa)	21 (helical)	21 (helical)	21 (helical)	—
Variant I intracellular domain (aa)	10	27	5	Variant I: 469 Variant II: 162
Putative N-glycosylation sites	7 (aa 106, 159, 283, 344, 607, 612, 694)	11 (aa 71, 120, 125, 211, 3078, 391, 554, 611, 702, 7069, 971)	3 (aa 80, 171, 245)	0
Sequence identity to KLB	46%	—	41%	38%
Glycosyl hydrolase domains	DI: aa 57–506	DI: aa 77–508		—
	DII: aa 519–953	DII: aa 517–967	Inactive	
Other domains of note			aa 22–567: lactase-like protein	aa 1–469: cytosolic beta-glucosidase; aa 424–425: substrate binding

Sources: Uniprot, ProteinPlusSite, NCBI BLAST.

vitamin D levels (Shimada et al., 2004; Yoshida, Fujimori, & Nabeshima, 2002). KLB was the second family member identified based on its sequence similarity to Klotho (~46%) (Ito et al., 2000). Mice deficient for whole-body KLB expression showed increased bile acid synthesis, analogous to FGF15- and FGFR4-deficient mouse models indicating potential signal integration/overlap (Ito, Fujimori, Furuya, Satoh, & Nabeshima, 2005). In 2000, it was also reported that another novel human protein called beta-glucosidase-like protein-1 (cBGL1/GBA3) had been identified that was related to Klotho. However, GBA3 does not exhibit a transmembrane domain and is predicted to function as a cytosolic protein rather than a transmembrane protein (Yahata et al., 2000). To date, in contrast to the other protein family members, GBA3 has not been shown to interact with FGF/FGFR metabolic signaling, leaving it as the least-known protein family member. The final Klotho family member, γ-Klotho, also identified in 2000 is encoded by the Lactase-like (*Lctl*) gene (Ito, Fujimori, Hayashizaki, & Nabeshima, 2002). In contrast to Klotho and KLB, γ-Klotho constitutes a smaller protein with unknown function and unknown association with FGF signaling. The remainder of this chapter focuses on the KLB protein and its role in regulating metabolic FGF21/FGFR1 signaling.

2.1.2 KLB Structure

Like human α-Klotho and γ-Klotho, KLB is expressed as a type I single-pass transmembrane protein (Fig. 1). In contrast to the multiple variants of α-Klotho and γ-Klotho, KLB only exhibits one defined variant that is larger than all other Klotho proteins (Table 1). KLB has a significantly larger extracellular than cytoplasmic sequence (996 vs 27 aa, respectively), with the extracellular sequence containing two tandem repeats of a β-glycosidase domain (D1 = aa 77–508 and D2 = aa 517–967). β-Glycosidase enzymes play a role in the digestion of sugar moieties of substrate glucosidases (Henrissat & Davies, 1997). However, the β-glycosidase domains of KLB are predicted to be *inactive* due to the absence of essential glutamate residues at positions 241 and 889 (UniProt: Q86Z14 (KLOTB_HUMAN)). KLB also notably exhibits 11 putative extracellular N-glycosylation sites, compared to significantly fewer putative sites on Klotho (7) and γ-Klotho (3) (Table 1) (Ito et al., 2000; Kuro-o et al., 1997). As will be discussed further in this chapter, the inactive β-glycosidase domains and pattern of extensive glycosylation provide multiple mechanisms for KLB to associate with sugars in the ECM and galectin lattice to potentially modulate activity at

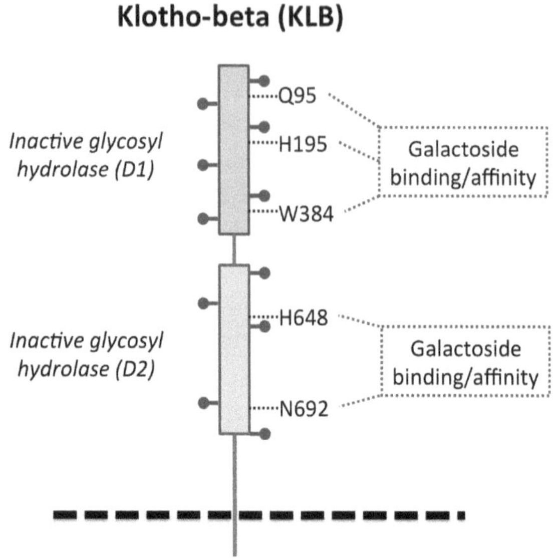

Fig. 1 KLB structure. KLB transmembrane receptor has a much larger extracellular domain (996 aa) compared to the cytoplasmic (27 aa) region. The extracellular domain contains two inactive glycosyl-hydrolase domains (D1, D2) with residues maintained for binding galactoside sugars. The extracellular region contains 11 putative N-glycosylation sites (indicated by *stick and ball markings*) spread across the D1 and D2 domains. (See the color plate.)

the cell surface. In contrast, the small intercellular sequence of KLB suggests that this receptor plays a limited role in directly activating intracellular signaling but rather simply facilitates interaction of FGF21 to FGFRs.

2.1.3 KLB Tissue Expression

KLB expression is limited to a small number of rodent target tissues, and while primarily expressed in tissue of enterohepatic origin (such as the liver and pancreas), expression levels are also high in both brown and white adipose tissue (Adams, Cheng, Coskun, & Kharitonenkov, 2012; Fon Tacer et al., 2010). KLB is also detected to a lesser degree in the hypothalamus and brainstem (Bookout et al., 2013). Mice lacking whole-body KLB are viable and fertile, but are smaller than their wild-type littermates and exhibit increased hepatic bile acid synthesis (Ito et al., 2005). An alternative study further demonstrated that the effects of FGF21 on the growth and metabolism of KLB knockout mice were negligible, suggesting that KLB was required for in vivo action of FGF21 (Adams et al., 2012). Similarly, tissue-specific knockout of KLB in adipose tissue blocked the acute insulin-sensitizing effects of FGF21 (Ding et al., 2012). The tissue-specific expression of KLB, in combination with appropriate FGFR expression, seems to define activity of FGF21 (Kurosu et al., 2007). Furthermore, expression of KLB has been shown to change during the disease state at both the transcript and protein levels. Several studies have demonstrated that during obesity, while serum levels of FGF21 and FGF19 are increased, KLB mRNA and protein decrease in both rodents and humans (Fisher et al., 2010; Gallego-Escuredo et al., 2015; Nygaard, Moller, Kievit, Grove, & Andersen, 2014). KLB expression is also reduced in cell models exposed to the proinflammatory cytokine TNFα (Diaz-Delfin et al., 2012). Combined, these data reiterate that KLB is required as part of the FGF21 signaling complex and that KLB inactivation results in metabolic disturbances.

2.2 Fibroblast Growth Factor Receptors

2.2.1 The Fibroblast Growth Factor Receptor Family

FGF signaling is mediated by a family of transmembrane tyrosine kinase fibroblast growth factor receptors (FGFR1–5). This family of low-capacity, high-affinity receptors are structurally similar and exhibit a high degree of sequence homology at the amino acid level (Haugsten, Sorensen, Brech, Olsnes, & Wesche, 2005; Mohammadi, Olsen, & Ibrahimi, 2005). These receptors are classified into five families according to the genes that encode them: FGFR1 (*flg*), FGFR2 (*bek*), FGFR3, FGFR4 (*flg2*), and FGFR5 (Dionne, Jaye, & Schlessinger, 1991; Kim, Moon, Yu, Kim, & Koh, 2001; Lee, Johnson, Cousens, Fried, & Williams, 1989; Olwin & Hauschka, 1986; Sleeman et al., 2001; Wiedemann & Trueb, 2000). The first FGFR mRNA was isolated and partially sequenced from a human placental endothelial cell library using the v-fms oncogene as a probe (von Heijne, 1986). It was determined that this receptor was encoded by *flg* (fms-like gene) (Ruta et al., 1989). A receptor closely related to human *flg* was purified from chicken embryos (Lee et al., 1989) and determined to be encoded by the chicken *cek* (chicken embryo kinase) gene (Pasquale & Singer, 1989). The first partial clone of FGFR2, called *bek*, was identified after a mouse liver expression library was screened with antibodies to phosphotyrosine (Kornbluth, Paulson, & Hanafusa, 1988). Two additional receptor genes encoding FGFR3 and FGFR4 were subsequently isolated from human leukemia cell lines (Keegan, Johnson,

Williams, & Hayman, 1991; Partanen, Makela, Alitalo, Lehvaslaiho, & Alitalo, 1990; Partanen et al., 1991). A subtractive cDNA subcloning approach in human cartilage was used to identify the newest member of this receptor protein family, FGFR5 (FGFRL1, fibroblast growth factor receptor-like 1) (Wiedemann & Trueb, 2000).

Collectively, the FGFR genes encode cell surface glycoproteins ranging in size from 55 to 150 kDa (Dionne et al., 1991; Hanneken, Ying, Ling, & Baird, 1994; Silva, Altamentova, Kilkenny, & Rocheleau, 2013). In general, it has been determined that FGFRs bind FGF ligands with high affinity (K_d 20–600 pM), leading them to be called "high-affinity" receptors (Partanen, Vainikka, Korhonen, Armstrong, & Alitalo, 1992). The FGFR1 and FGFR2 proteins are the most closely related exhibiting overall conservation of 72% at the amino acid level, while FGFR1 and FGFR4 are the least related exhibiting only 55% homology (Johnson & Williams, 1993). The FGFR1–3 genes are subject to differential exon splicing, thereby increasing the total number of receptor isoforms potentially available to direct specificity of ligand binding and enhance regulation of the signaling pathways. FGFR1 and FGFR5 share upward of 50% similarity; however, it must be considered that while FGFR5 contains a similar extracellular portion, it lacks the intracellular tyrosine kinase domains common to all of the other receptor isoforms (Steinberg et al., 2010). For the purposes of this chapter, we focus on FGFR1 as the signaling receptor for FGF21. Due to co-expression of FGFR5 in beta-cells, we later describe the potential interaction between FGFR1 and FGFR5 in modulating the response to FGF21.

2.2.2 Fibroblast Growth Factor Receptor Structure

The prototypical full-length FGFR exhibits an N-terminal hydrophobic signal sequence, three extracellular immunoglobulin-like (Ig-like) domains (D1, D2, D3), a single transmembrane domain with a long juxtamembrane sequence, and a split intracellular tyrosine kinase domain (Fig. 2) (Dionne et al., 1991; Johnson & Williams, 1993). The first and second domains (D1, D2) are separated by a short stretch of acidic residues (Eisemann, Ahn, Graziani, Tronick, & Ron, 1991). Binding of specific ligands and cofactors (ie, heparin/heparan sulfate proteoglycans) to the D2/D3 subdomains stabilizes the dimeric receptor complex and enhances catalytic activity of the tyrosine kinase domains. Specificity of receptor binding to the vast array of known FGF ligands is dictated by multiple elements in the latter portion of D3 (Bocharov et al., 2013; Lemmon & Schlessinger,

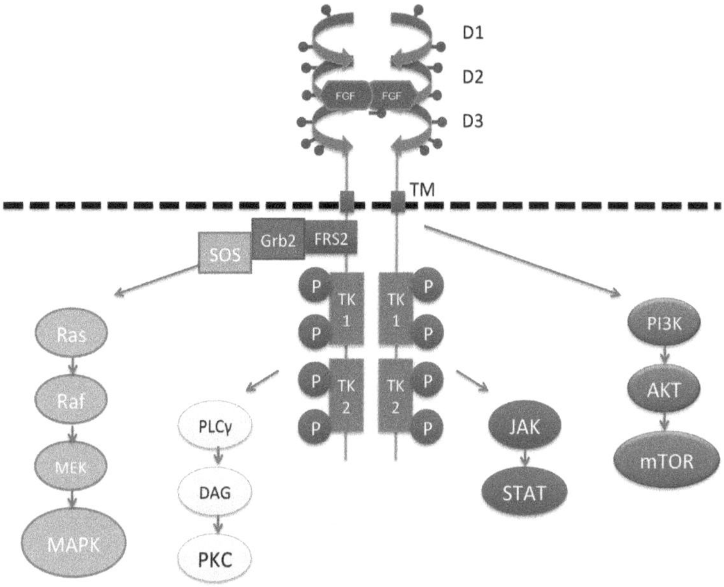

Fig. 2 FGFR signaling cascades. FGF ligand binding to the extracellular immunoglobulin-like D2–D3 domains of canonical FGFR1, which exhibits seven N-glycosylation sites (indicated by *ball and stick markings*), results in *trans* activation of the split intracellular tyrosine kinase (TK) domains of a receptor dimer. Receptor activation can initiative a multitude of intracellular signaling cascades, including direct interaction with the FRS2 docking protein, as well as downstream activation of the Ras/MAPK pathway, the PKC pathway, the Jak/STAT pathway, and the PI3K/mTOR pathway. (See the color plate.)

2010; Mohammadi et al., 2005; Zimmer, Givol, & Yayon, 1993) and also depends on ligand availability. The ligand-bound receptor initiates *trans*-autophosphorylation of the intracellular tyrosine residues and intrinsic activity of the intracellular split kinase domain (Basilico & Moscatelli, 1992; Jaye, Schlessinger, & Dionne, 1992; Powers, McLeskey, & Wellstein, 2000). The kinase domain consists of an N-terminal lobe that functions as an enzyme and a C-terminal lobe that functions as a substrate (Bae & Schlessinger, 2010; Mohammadi et al., 2005). Interestingly, the D1 sub-domain and acidic box region are not required for ligand binding (Chellaiah, Yuan, Chellaiah, & Ornitz, 1999; Crumley et al., 1991) and D1 has been shown to weakly interact with the D2/D3 domains causing receptor inactivation by autoinhibition (Kalinina et al., 2012; Mohammadi et al., 2005). An abundance of substrate and adaptor proteins (ie, FRS2, PLC, Akt, ERK1/2, and Grb2) have been identified as phosphorylation targets in FGF/FGFR-mediated signaling pathways (Zhen, Sorensen, Jin,

Suo, & Wiedlocha, 2007). The C-terminal regions of FGFRs exhibit reduced sequence identity when compared to the similarity found in the extracellular sequences, and this is thought to dictate the interaction with cellular substrates specific to each receptor kinase (Dionne et al., 1991). A heparin-binding (HB) domain is found proximal to extracellular D3, as point mutation at this site abolishes binding of both heparin and ligand to the receptor (Kan et al., 1993). While ligand binding enhances FGFR activation due to stabilization or altered structure of the dimeric receptor complex, receptors are also capable of dimerizing in the absence of ligand and can exhibit basal levels of phosphorylation activity (Belov & Mohammadi, 2012).

Ligand-binding specificity of FGFR1–3 is regulated by alternative splicing of the D3 domain (Avivi, Yayon, & Givol, 1993; Goetz & Mohammadi, 2013; Johnson, Lu, Chen, Werner, & Williams, 1991; Olsen et al., 2006; Orr-Urtreger et al., 1993; Yeh et al., 2003). In contrast, the D3 domain of FGFR4 does not undergo differential exon splicing (Partanen et al., 1991). Exon 7 (exon IIIa) of FGFR1–3 encodes the N-terminal region of D3, whereas exons 8 (exon IIIb) and 9 (exon IIIc) are responsible for variability in the C-terminal sequence. Differential exon splicing in the ligand-binding domain of these genes generates additional receptor isoforms with varying ligand affinities that are referred to as the IIIa, IIIb, and IIIc isoforms (Avivi et al., 1993; Chellaiah, McEwen, Werner, Xu, & Ornitz, 1994; Dell & Williams, 1992; Johnson et al., 1991; Johnson & Williams, 1993; Ornitz et al., 1996; Werner et al., 1992; Yeh et al., 2003). Variation in coding is tissue and development specific, with the exon IIIb isoform generally restricted to cells of epithelial lineage and the IIIc isoform generally restricted to cells of mesenchymal lineage (Miki et al., 1992; Orr-Urtreger et al., 1993). This restriction in receptor splice variant expression allows for reciprocal signaling loops between and within tissues and plays a large role in the coordinated proliferation during development and in adult tissue homeostasis (De Moerlooze et al., 2000; Zhang et al., 2006). Alternative splicing also contributes to receptors that exhibit two or three IgG-like domains: these isoforms are referred to as the "β" and "α" isoforms, respectively (Hou et al., 1991; Johnson, Lee, Lu, & Williams, 1990; Johnson et al., 1991). Differential expression of these isoforms further contributes to signaling regulation; for example, the FGFR1β isoform lacks the D1 domain and therefore presumably does not show similar autoinhibition as the FGFR1α isoform. The FGFR1α and FGFR1β isoforms also exhibit varying affinities for both ligand and heparin (Kan et al., 1993; Zimmer et al., 1993). Finally, a cDNA encoding only IgG D2–D3 of FGFR1 was shown to produce a secreted receptor protein that retained the capacity to bind ligand (Duan, Werner, & Williams, 1992).

2.2.3 FGFR1c Tissue Expression

In contrast to KLB, FGFRs are ubiquitously expressed within the body and most tissues express one or more of receptor isoforms. Differential expression of various receptors is regulated in a tissue-specific and/or developmentally specific manner, although it is not uncommon to observe co-expression (Jayson et al., 1999; Logan, Frautschy, Gonzalez, & Baird, 1992; Patstone, Pasquale, & Maher, 1993; Yazaki, Fujita, Ohta, Kawasaki, & Itoh, 1993). In this chapter, we are focused on the IIIc variants of FGFR1 (FGFR1c) which are generally expressed in tissue of mesenchymal origin (MacArthur et al., 1995; Orr-Urtreger et al., 1993; Sun, Mariani, & Martin, 2002). In contrast, it has been determined that epithelial tissues tend to express IIIb receptor variants (Bellusci, Grindley, Emoto, Itoh, & Hogan, 1997; Beyer, Werner, Dickson, & Grose, 2003; Fon Tacer et al., 2010).

A shift in expression from a 3 IgG-loop "α" to a 2 IgG-loop "β" of FGFR1 has been identified as a marker of cellular transformation and cancer (Jin, Bi, Huang, & Cote, 1999). Conversely, malignant astrocytomas have been shown to express both FGFR1α and FGFR1β isoforms, while normal fetal and adult brain express only FGFR1α (Yamaguchi, Saya, Bruner, & Morrison, 1994). Similarly, while the "long" (α) isoform of FGFR1 was expressed in embryonic mouse heart, only the "short" (β) FGFR1 isoform was detected in adult heart (Jin et al., 1994). Similarly, FGFR2α and FGFR2β isoforms have also been shown to be differentially expressed during amphibian development (Shi, Launay, Fromentoux, Feige, & Boucaut, 1994). These studies suggest that a switch in receptor isoform expression occurs during tissue development.

Exposure to various ligands (ie, FGF1 or FGF2) has also been shown to promote differential alternative splicing of exon IIIb or IIIc of FGFR2 and FGFR3 (Scotet & Houssaint, 1998), suggesting that the cell/tissue microenvironment can drive receptor isoform expression. This concept is further confirmed in studies that examined changes in mRNA splicing during cardiac transplant rejection, demonstrating that the increases in growth factors and cytokines secreted during these episodes promoted expression of various FGFR isoforms (Zhao, Frist, Yeoh, & Miller, 1994). This type of response therefore depends not only on ligand concentration but also on the degree of availability to interact with the receptors. It has been established that low-affinity, heparan proteoglycans (HSPGs) at the cell surface function as "co-receptors" that bind and protect pericellular ligand from proteolytic cleavage, thereby creating a reservoir of biologically active FGF that is retained in close proximity to the high-affinity FGFRs (Moscatelli, 1987; Quarto & Amalric, 1994). It is evident that tissue-specific expression of FGFRs is dynamic and, in combination with the appropriate co-receptor expression, defines FGF activity (ie, KLB defines activity of FGF21–FGFR1) (Kurosu et al., 2007).

2.3 Endocrine FGF21

2.3.1 The Fibroblast Growth Factor Family

FGFs compose a superfamily of 22 polypeptides, so-named due to potent stimulatory effects on fibroblast proliferation that were originally observed (Gospodarowicz, 1975). This family of peptides exhibit diverse yet redundant roles in biological processes such as cellular proliferation, differentiation, and metabolism (Basilico & Moscatelli, 1992; Powers et al., 2000). Extensive sequence alignment has been performed in an effort to discern the phylogenetic relationship of the various FGF ligands (Coulier et al., 1997; Oulion, Bertrand, & Escriva, 2012). The nucleotide coding sequence of the first eight members of the FGF (FGF1–8) family exhibits 14% conserved identity; similarly a conserved internal stretch of approximately 112 amino acids exhibits 17% homology (Baird, 1994; Hughes & Hall, 1993; Zhan, Bates, Hu, & Goldfarb, 1988). Conservation of two cysteine residues may be important for disulfide bond formation and subsequent stability of the protein tertiary structure (Zhan et al., 1988). Most FGFs exhibit a signal sequence and undergo posttranslational glycosylation characteristic of secreted proteins (McFarlane, Georgopoulou, Coughlan, Gillian, & Breen, 1999; Sareneva, Pirhonen, Cantell, Kalkkinen, & Julkunen, 1994). Signal sequences are absent however from the coding regions of FGF1, FGF2, FGF16, and FGF20, while FGF9 expresses a three-residue cleaved sequence (Bikfalvi, Klein, Pintucci, & Rifkin, 1997; Katoh & Katoh, 2005; Miyamoto et al., 1993). It has been determined that these proteins are still exported from cells by a novel energy-dependent mechanism (Florkiewicz, Majack, Buechler, & Florkiewicz, 1995). Nevertheless, all FGFs are capable of functioning extracellularly and mediate biological responses by signaling through a co-receptor system that includes high-affinity cell surface FGFRs possessing intrinsic tyrosine kinase activity, as well as low-affinity cell surface or matrix-associated HSPGs (Klagsbrun &

Baird, 1991; Roghani & Moscatelli, 1992). Temporal and tissue-specific expression of the different FGFs, as well as the varying affinities with which they bind multiple receptors, leads to an exceptionally complex pattern of ligand/receptor interactions and cellular biological responses. Despite the significant apparent redundancy in FGF and FGFR isoforms, each pairing exhibits variable affinity and therefore offers a degree of signaling regulation when/if expressed in the same cells and/or tissues.

The FGFs are grouped together according to sequence similarity and biochemical function (Table 2) (Ornitz & Itoh, 2015). Paracrine FGFs, so-named for local signaling action in neighboring cells, include members of the FGF1, FGF4, FGF7, FGF8, and FGF9 subfamilies. A hallmark feature of canonical paracrine FGFs is the ability to bind and activate FGFRs using HSPGs as a cofactor (Ornitz, 2000). Intracellular FGFs, so-named for their biological actions within the cytosol, include the FGF11 subfamily. The intracellular family of FGF "homologous factors" are interacting proteins that serve as cofactors for voltage-gated sodium channels and other molecules and, despite structural homology with paracrine FGFs, are not known to interact with FGFRs (Hennessey, Wei, & Pitt, 2013; Olsen et al., 2003; Wang, Hoch, & Pitt, 2011). Endocrine FGFs, which exhibit signaling effects at distant target cells/tissues, include the FGF15(mouse)/19(human) subfamily. FGF19, FGF21, and FGF23 have all recently emerged as potent regulators of whole animal physiology that vary in function from classical paracrine/autocrine FGFs by serving as long-distance signaling molecules. The phenotypes of multiple knockout mouse models have repeatedly demonstrated that this group of growth factors exhibits important roles in metabolic processes (Itoh, Ohta, & Konishi, 2015; Kubicky, Wu, Kharitonenkov, & De Luca, 2012; Liang et al., 2014; So, Cheng, Xu, Lam, & Leung, 2015; Yuan et al., 2014). The endocrine FGFs also exhibit structural homology to the paracrine FGFs and retain capacity to interact with FGFRs, but exhibit low affinity for HSPGs and instead require Klotho family proteins as cofactors for receptor binding and activation to occur (Mohammadi et al., 2005; Suzuki et al., 2008). For the purposes of this chapter, we will focus on FGF21.

2.3.2 FGF Structure

The FGFs are single-chain polypeptides with a conserved domain flanked by nonconserved extensions (Fig. 3). The majority of these ligands have an N-terminal signal peptide that facilitates secretion through classical

Table 2 FGF Subfamily of Ligands

Mechanism of Action	Subfamily	Ligands	Signal Sequence	Secreted Peptide	Preferred Receptor	Co-receptor
Endocrine	FGF15/19	FGF19, FGF21, FGF23	√	√	R1–3 IIIc isoforms; R4	Klothos
Paracrine	FGF1	FGF1, FGF2	X	√ (nonconventional)	FGF1 → all; FGF2 → R1c, R3c	HSPGs
	FGF4	FGF4, FGF5, FGF6	√	√	R1c	HSPGs
	FGF7	FGF3, FGF7, FGF10, FGF22	√	√	R2b	HSPGs
	FGF8	FGF8, FGF17, FGF18	√	√	R3c	HSPGs
	FGF9	FGF9, FGF16, FGF20	X	√ (nonconventional)	R3c	HSPGs
Intracellular	FGF11	FGF11, FGF12, FGF13, FGF14	X	X	—	—

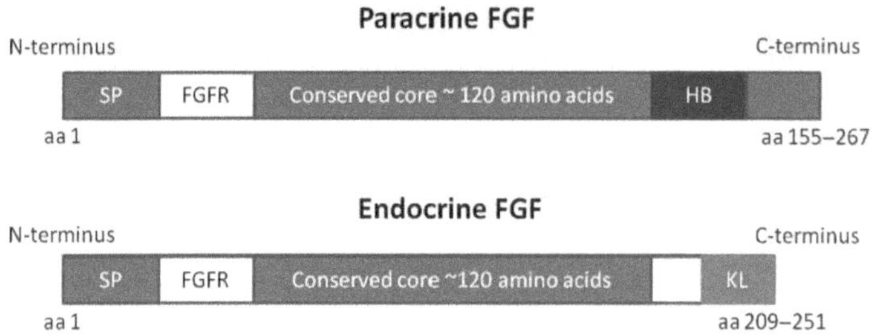

Fig. 3 Critical domains of paracrine and endocrine FGFs. Despite variability in biological function, paracrine and endocrine FGF peptides share a conserved core sequence of ~120 aa and present as a class of proteins of ~18–24 kDa molecular mass (Table 3). An N-terminal signal peptide (SP) and FGFR binding site (FGFR; directs interaction with high-affinity FGFRs) are also common sequence features. The C-terminal sequence displays some variability and is required for association with low-affinity HSPGs (HB; paracrine ligands) or klotho co-receptors (KL; endocrine ligands). (See the color plate.)

mechanisms; however, several FGFs (FGF1, FGF2, FGF9, FGF16) do not have a cleavable signal peptide and are secreted by nonconventional mechanisms. FGF proteins exhibit a homologous core region that consists of 120–130 amino acids ordered into 12 antiparallel β-strands (β1–β12) flanked by divergent N- and C-terminal tails (Beenken & Mohammadi, 2009). In general, primary sequence variation of these terminal tails accounts for the varied receptor affinity and biology of the different ligands (Mohammadi et al., 2005). A heparin binding (HB) site is formed within the core of the ligand by the β1–β2 region in combination with a portion of the region that spans β10 and β12 (Beenken & Mohammadi, 2009) (Fig. 3). This HB domain provides a contiguous, positively charged surface in the paracrine FGF proteins that promotes their binding to negatively charged heparin, thereby retaining ligand within pericellular regions to promote local receptor activation and signaling.

2.3.3 Endocrine FGF21 Structure

Structurally, FGF21 is a 181 amino acid peptide (~22.3 kDa molecular mass) derived from a 209 amino acid mature protein encoded by the *fgf21* gene located in chromosome 19 (Nishimura, Nakatake, Konishi, & Itoh, 2000) (Table 3; UniProt ID: Q9NSA1). The mature protein exhibits a 28 amino acid hydrophobic N-terminus signal peptide that directs constitutive secretion (Fig. 3). In contrast to paracrine FGFs, FGF21 lacks the capacity to bind extracellular heparin due to a modified β-trefoil structural motif at the C-terminus (Asada et al., 2009; Goetz et al., 2007). The domain

Table 3 Comparative Summary of Endocrine vs Paracrine FGFs

	Endocrine FGF21	**Canonical Paracrine FGF2**
Amino acids (mature)	181	155–210
Molecular weight	20 kDa	18–24.5 kDa
FGFR affinity[a]	FGFR1c, 3c	FGFR1c, 3c > 2c, 1b, 4
Co-receptor	KLB	Heparin sulfate proteoglycans
Major tissue of origin	Liver	Ubiquitous

[a]Data from Ornitz et al. (1996).

formed by β1–β2 and the portion of the region that spans β10 and β12 in FGF19, FGF21, and FGF23 (ie, the HB domain of paracrine FGFs) contains ridges that sterically reduce HSPG binding (Goetz et al., 2007). The reduced or negligible affinity for heparin has been specifically demonstrated by elution studies using HS affinity columns (Asada et al., 2009). Whereas FGF21 exhibited negligible affinity and passed directly through the columns, FGF19 and 23 were shown to demonstrate minimal affinity and eluted from the columns on early passes of 10 mM HCl (pH 7.4). Furthermore, addition of heparin to KLB-expressing HEK293 cells or 3T3L1–KLB cells has been shown to have no effect on FGF21-induced responses (Kharitonenkov et al., 2008). The low (or negligible) binding affinity of FGF21 allows the molecules to escape the heparin-rich ECM of secreting cells and diffuse through the matrix or into the circulation to affect distal tissues in an endocrine-like manner (Mohammadi et al., 2005; Ornitz & Itoh, 2015). Specifically, FGF21 has been shown to exhibit pleiotropic function in regulating obesity and whole-body glucose and lipid metabolism (Kharitonenkov, 2009).

2.3.4 Endocrine FGF21 Tissue Expression

Paracrine FGFs are ubiquitously expressed throughout all stages of growth and development, with particularly high expression levels evident during embryogenesis (De Moerlooze et al., 2000; Kimelman & Kirschner, 1987). In adult tissues, FGFs contribute significantly to cellular proliferation, survival, wound healing, tissue repair, migration, and differentiation and therefore may be differentially expressed depending upon the biological requirements (Turner & Grose, 2010). The liver has been identified as the main source of circulating FGF21 (Badman et al., 2007; Inagaki et al., 2007); however, it is also synthesized in other tissues associated with carbohydrate metabolism including white adipose tissue, pancreas, skeletal muscle, digestive tract, and skin (Crooks et al., 2014; Donate-Correa et al., 2015; Oishi, Konishi, Murata, & Itoh, 2011; Wente et al., 2006). FGF21 is also found in detectable levels in the plasma and has been estimated to be within the range of 0.05–2 ng/mL in healthy adult humans, depending upon the method used (generally immunologically based) to assay the samples (Adams & Kharitonenkov, 2012; Angelin, Larsson, & Rudling, 2012; Galman et al., 2008). Furthermore, FGF21 is also secreted into the circulation in diurnal rhythm (Yu et al., 2011); therefore, one must be cognizant of this secretion pattern when measuring ligand concentrations. During prolonged fasting, starvation, and other stresses, expression of FGF21 in both the liver and serum is dramatically elevated (Adams & Kharitonenkov, 2012; Fisher et al., 2011; Potthoff et al., 2009). Similarly, serum FGF21 levels are significantly increased in patients identified with Type II diabetes, suggestive of ligand association with disease progression (Cheng, Zhu, Jiang, & Fan, 2011). Several recent studies have also shown that FGF21 expression is induced in muscle (ie, skeletal, heart) under conditions that cause both local and systemic metabolic stress (Benetti et al., 2013; Crooks et al., 2014; Hojman et al., 2009). Collectively, these scenarios suggest that elevated FGF21 levels are an indicator of metabolic stress.

With regard to systemic FGF21 expression, one must consider that the peptide has a short half-life in blood (~1 h in rodents and monkeys) as a consequence of enzymatic degradation and susceptibility to renal clearance (Hager, Spahr, Xu, Salimi-Moosavi, & Hall, 2013; Kharitonenkov et al., 2007; Xu, Stanislaus, et al., 2009). A recombinant Fc-FGF21(RG) fusion protein, where the Fc portion of human IgG1 was fused to the N-terminus of human FGF21 and two amino acid substitutions were introduced to reduce peptide aggregation and proteolytic cleavage, has been shown to result in a bioactive molecule that was expressed at detectable levels within primate blood for up to 30 h (Hecht et al., 2012). This molecule becomes clinically relevant given that beyond simply subsisting for longer periods of time in the circulation, it retains the ability to bind to receptor and co-receptors molecules for longer, thereby initiating and sustaining biological signaling for extended periods of time.

In terms of whole-body physiology, FGF21 effects are mediated by KLB expression on target tissues and co-expression of FGFR1c, FGFR3c, or FGFR4, with the main target organ proposed to be adipose tissue given the effects of FGF21 on fatty acid oxidation. Critically, we (and others) have noted significant effects within the endocrine pancreatic islets, a tissue critical to glucose homeostasis, due to co-expression of FGFR1c and KLB (Kharitonenkov et al., 2007; So et al., 2013; Sun et al., 2012; Wente et al., 2006).

3. SPECIFIC INTERACTIONS DRIVING COMPLEX FORMATION

3.1 Interaction of FGF21 and KLB

The C-terminal domain of FGF21 is required for ligand binding to KLB (Goetz, Ohnishi, Kir, et al., 2012; Yie et al., 2009). Serial truncation of the C-terminal residues reduces the interaction between FGF21 and KLB (Yie et al., 2009), and removal of five C-terminal amino acids of FGF21 decreased ligand-binding affinity to KLB by up to 10-fold compared to wild type as determined by surface plasmon resonance (SPR) (Micanovic et al., 2009). Consistently, replacing the C-terminal domain of a paracrine FGF with the C-terminal domain of FGF21 established paracrine factor binding to KLB (Goetz, Ohnishi, Kir, et al., 2012).

Goetz and coworkers have also elegantly examined KLB ligand binding using SPR-based competition assays (Goetz, Ohnishi, Ding, et al., 2012). These studies determined overlapping binding sites on KLB that competitively bind FGF21 and FGF19. These studies specifically used the C-terminal tails of each growth factor as interfering species; interestingly, $FGF21^{C\text{-tail}}$ was found to be less effective than $FGF19^{C\text{-tail}}$ at preventing KLB binding to either full-length FGF19 or FGF21, suggesting that KLB exhibits variable affinity for these endocrine growth factors. These data further confirm that variable binding affinity is associated with the C-terminal tail sequence of each growth factor. Physiologically, FGF19 is likely to outcompete FGF21 for KLB binding; however, these endocrine factors do not bind with equal affinity to the same FGFR, and therefore, FGFR isoform likely directs specificity in this case. Furthermore, it has been confirmed that FGF23, which shows the highest degree of homology with FGF21 and is most highly expressed in bone tissue (Liu et al., 2003; Riminucci et al., 2003), exhibits binding preference for α-Klotho and not KLB (Kurosu et al., 2006).

3.2 Interaction of FGF21 and FGFR1

The binding of FGF21 to FGFR1c has been shown to depend on a specific region in the N-terminal domain of the growth factor (Gupte et al., 2011), with deletion resulting in reduced FGFR1c activation (Yie et al., 2009). Although these studies were not able to demonstrate direct binding between ligand and receptor, they confirmed that the effect of the N-terminus is specific to the receptor as serial deletions of the C-terminus had no effect on receptor activation. Multiple subsequent studies have physically deleted portions of the N-terminus of FGF21 to observe the effects on FGFR1c activation (Micanovic et al., 2009; Yie et al., 2012). Deletion of six amino acids from the N-terminus and four from the C-terminus each reduced FGF21 activity by 10-fold, reiterating that FGF21 activity is sensitive to terminal deletions. Interestingly, treatment of 3T3L1/KLB-expressing cells with these mutant ligands stimulated phosphorylation of ERK1/2 and prevented apoptosis, further confirming that the C-terminal domain is required for binding, but not activation, of the receptor (Micanovic et al., 2009). A separate study created an inhibitory N-terminally truncated form of FGF21 (ΔN17) that retained the ability to bind to KLB, but was unable to induce FGFR activation due to the absence of the N-terminal sequence (Adams et al., 2012). Consistently, this truncated ligand acted in an antagonistic manner in a mouse model by blocking FGF21-mediated reductions in serum glucose.

It has further been confirmed that the third extracellular domain D3 of FGFR1 plays a specific role in FGF21 binding and receptor activation. Specifically, a 6-amino acid sequence within D3 of FGFR1c was identified as prerequisite for FGF21 binding (Gupte et al., 2011). This was accomplished by generation of chimeric receptors composed of interchanged D2–D3 domains of FGFR1c and FGFR4, given that this region binds both endocrine and paracrine factors (Beenken & Mohammadi, 2009; Ogawa et al., 2007). It has also been shown that FGF21 requires FGFR1 as part of its receptor signaling complex, while FGF19 exhibits preference for FGFR4 (Yang et al., 2012). FGF21 induced activation of FGFR4 only when the FGFR1c D2–D3 domain was swapped in; conversely, FGFR1c lost the ability to be activated by FGF21 when this region was replaced by the FGFR4 sequence. It was specifically determined that the "TTKDEM" sequence of FGFR1c was responsible for receptor sensitivity to FGF21. Interestingly, this sequence falls into the FGFR1 βC′–βE loop, which is important for determination of paracrine FGF specificity by crystal structure analysis, suggesting that the interface between the FGF21 and D2 domain is mainly hydrophobic (Plotnikov et al., 2000). Overall, the general consensus appears to be that FGF21 prefers to bind to FGFR1c and FGFR3c, but affinity varies given the microenvironment (Suzuki et al., 2008).

Despite the expanding body of knowledge available, there is still some controversy as to whether KLB is actually required for FGF21 signaling via FGFRs; similarly, other studies suggest that specific preference for one FGFR isoform or another does not exist for FGF21 signaling (Fisher et al., 2011; Gallego-Escuredo et al., 2015; Potthoff et al., 2009). One wonders how many of the potential interacting molecules obscure/alter formation of the signaling complex; alternatively, the purity of lab reagents must also be assessed for adequate comparison. Perhaps a more general consensus is that expression of KLB enhances the FGF21/FGFR1c signaling response whether it be from direct interaction or some other mechanism.

3.3 Interaction of KLB and FGFR1

The interaction between KLB and FGFRs has been shown by co-immunoprecipitation (co-IP) in multiple studies. Urakawa et al. used a luciferase activity assay in conjunction with co-IP to show KLB preference for FGFR1IIIc isoform interaction in a HEK293 culture system (Urakawa et al., 2006). An alternate set of studies confirmed KLB association with FGFR1c by co-IP but also demonstrated interaction with FGFR3c and FGFR4 (association with FGFR2c was detected at a significantly lower degree) (Kurosu et al., 2006). Conversely, the binding of soluble KLB showed a preference for the "c" isoforms of FGFR1 and FGFR2, as well as FGFR4, with no appreciable binding to FGFR3 when FGFR extracellular Fc-fusion proteins were used (Kharitonenkov et al., 2008). Taken together, these results suggest that FGFR "c" isoforms more efficiently coprecipitate with KLB compared to "b" isoforms and further suggest that FGFR2 may be a less relevant binding partner.

Particularly relevant to this book chapter is that Klotho was visualized as a doublet by immunoblot analysis, indicating varying degrees of glycosylation (Imura et al., 2004). In this particular study, when Klotho was co-expressed with high-affinity FGFRs (FGFR1, FGFR3, FGFR4), the upper protein band was enriched on the blots suggesting that the sugar chains of Klotho may facilitate receptor/co-receptor interactions. This potentially enhanced interaction of FGFR:Klotho is associated with endocrine FGF23, indicative of a role in facilitating complex formation and signaling. Furthermore, KLB of ~110.2 kDa was determined to be glycosylated as a consequence of migration at ~130 kDa under nonreducing conditions (Ming et al., 2012).

Notably, the FGF8 and FGFR1c binding sites on KLB overlap, meaning that they competitively bind to the same regions. Specific mutation analyses demonstrated that Klotho proteins (α and β) engage the same hydrophobic groove in the D3 domain of FGFR1c as does FGF8b (Goetz, Ohnishi, Ding, et al., 2012). More specifically, the N-terminal domain of FGF8b was determined to bind to a groove formed by several specific residues of FGFR1c (L290, L305, P306, V308, T340, L342, L349, and H351). These data suggest that FGF8b and KLB compete for FGFR1c binding, thereby directing responses as either endocrine or paracrine depending upon ligand availability.

Goetz and colleagues also used SPR-based assays to identify the KLB binding site of multiple FGFRs, including FGFR1b, FGFR1c, FGFR2b, FGFR2c, FGFR3b, FGFR3c, and FGFR4 (Goetz, Ohnishi, Ding, et al., 2012). By this analysis, KLB exhibited the greatest affinity for FGFR4 (K_d, 84 nM), FGFR1c (K_d, 124 nM), and FGFR2c (K_d, 170 nM), however bound poorly to FGFR3c and was not observed to interact with any of the FGFR "b" isoforms. This study confirms preferential binding of KLB with the "c" (IgG D3 domain) of FGFRs. In contrast, α-Klotho demonstrated the greatest binding affinity for FGFR1c (K_d, 72 nM), followed by FGFR3c (K_d, 82 nM), and similar affinity for FGFR4 like KLB (K_d, 123 nM), but poorly to FGFR2c.

3.4 KLB Domains Relevant for Binding

Interestingly, it becomes increasingly apparent as one searches the published literature that while it has been defined that KLB interacts with both FGFR1 and FGF21, the specific sequences of the co-receptor required for these interactions remain undefined. Because the KLB crystal structure is also not available, it is also challenging to surmise structure, function, and specific sites of interaction between KLB and the ligand/receptor/complex. Although sequences can be used to make predictions, the puzzle may be much like trying to discern a specific ligand-binding site of the low-affinity HSPGs to FGFRs.

4. STOICHIOMETRY OF THE SIGNALING COMPLEX

In canonical FGFR signaling, HSPGs function as a reservoir of paracrine FGF at a cell's surface, making it available to the higher-affinity FGFR as required. FGFs bound specifically to the heparan sulfate chains (HSCs) on

HSPG are presented to FGFR1 to form a heterocomplex. Critically, this membrane signaling complex has been modeled as a 2:2:2 ratio of HSC: FGF:FGFR (Schlessinger et al., 2000). Mohammadi's group has done extensive study of these interactions and defined key residues that are involved in the hydrogen bond network that confers structural rigidity at the interface between ligand and receptor. An alternative model proposes that the complex is composed of a 1:2:2 ratio of HSC:FGF:FGFR with two FGF ligands bound to a single HSC (Harmer et al., 2004; Robinson, Harmer, Goodger, Blundell, & Gallagher, 2005). Both models have been compellingly argued; however, it should also be noted that these models are not necessarily exclusive and each may have a role in signaling at different times and contexts to ultimately tune FGFR1 signaling activity. For example, under situations where the more energetically favorable 2:2:2 model is predominantly formed, we could expect a more stable and long-lived signaling complex that results in a more potent signaling cascade of longer duration. In either model, the FGF ligand interacts with the HSCs and two FGFR molecules to stabilize the complex. It is therefore interesting to speculate what these models tell us about endocrine FGF21-induced signaling, and in particular whether KLB exhibits a similar stoichiometry and plays a similar role as HSPGs in forming the FGFR1c signaling complex.

4.1 Preformed KLB Heterodimers with FGFR1c/4

We quantitatively studied the interaction between fluorescent protein-tagged KLB and FGFR1c in living cells using fluorescence recovery after photobleaching (FRAP) and Number and Brightness (N&B) Analysis (Ming et al., 2012). Using these advanced imaging methods, we measured the stoichiometry of KLB and FGFR1c during individual or co-expression, as well as the subsequent responses to FGF21 ligand (Fig. 4). Our data suggest that when KLB and FGFR1c are individually expressed, they behave as monomers and preformed dimers, *respectively* (Fig. 4A). Identifying FGFR1c as a preformed dimer was initially surprising as it goes against the prevailing dogma in a field that considers ligand-induced receptor homodimerization a key trigger for signaling. However, we are *not* the first to propose FGFRs as preformed homodimers (McKeehan, Wang, & Kan, 1998; Uematsu et al., 2000; Wang et al., 1997). Similarly, a recent study has confirmed preformed homodimer conformation of FGFR1 using FRET with SNAP- and ACP-tag labeling (Comps-Agrar, Dunshee, Eaton, & Sonoda, 2015). More

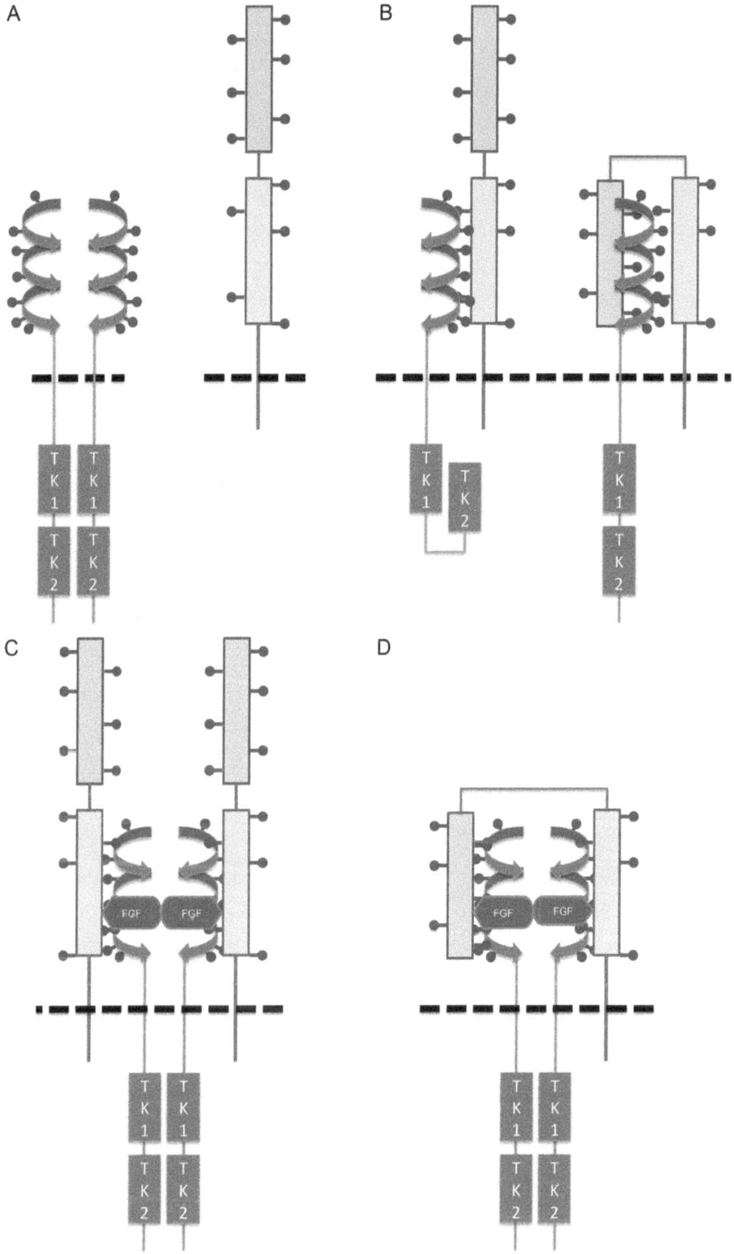

Fig. 4 Putative receptor stoichiometry. The cartoons represent putative receptor stoichiometry of FGFR1 and KLB in the absence (A, B) and presence (C, D) of FGF21. (A) Our data indicate that FGFR1c diffuses as preformed dimers, while KLB diffuses as a monomer. (B) Co-expression of FGFR1c and KLB results in formation of a 1:1 heterocomplex. This cartoon models two potential interactions highlighting both allosteric conformational changes (*left*) and multiple binding interactions (*right*). (C) A cartoon representation of a 2:2:2 receptor complex formed by FGF21:FGFR1:KLB. This putative model was anticipated after finding a preformed 1:1 heterodimer of FGFR1 and KLB, but has not been observed. (D) A putative model of the 2:2:1 receptor complex of FGF21:FGFR1:KLB. This model indicates putative interaction of one KLB with two FGFR1c receptors in an effort to explain the release of a second KLB when the complex is formed. (See the color plate.)

specifically, this study showed evidence that FGFR1c homodimerization was driven by interaction between the C-terminal tyrosine kinase domains. In general, the receptor tyrosine kinase field is moving toward models of preformed dimers as the norm rather than the exception. Ligand stimulation in these models subsequently induces allosteric changes in conformation resulting in tyrosine kinase *trans*-activation (Maruyama, 2015). We further postulate that an allosteric/rotational model fits well with the speed of ligand-induced FGFR1c signaling, where no diffusion time is required for finding a partner receptor. An allosteric/rotational model is also consistent with the tight control of FGFR1c activity. This suggests that baseline activity would be affected simply by receptor membrane diffusion, particularly for ectopic FGFR1 expression, which we have not observed in our own studies. We therefore believe that canonical FGFR1c signaling does not involve a monomer to dimer transition, but rather involves allosteric changes in the conformational state of preformed dimers (ie, the "rotation model").

A number of studies have used co-IP to show that in the absence of FGF ligand, KLB interacts with FGFR1c and FGFR4, suggesting that a preformed heterocomplex may be necessary for an endocrine FGF21-induced response (Kharitonenkov et al., 2008). We showed that co-expression of KLB and FGFR1c results specifically in a 1:1 heterocomplex (Fig. 4B) (Ming et al., 2012). This model suggests that KLB either allosterically or competitively blocks the homodimerization of FGFR1c. To visualize these two possibilities, we show two versions of the heterocomplex: (i) an allosteric model with KLB inducing a conformational change in the C-terminal of FGFR1c (*left*), and (ii) a competition model showing KLB binding that sterically blocks the FGFR1c homodimer site (*right*). Although these are relatively simplistic models, they are presented here solely for discussion purposes. In the allosteric model, interaction of KLB with FGFR1c induces a change in receptor structure. In the competition model, FGFR1c and KLB binding could involve multiple sites on KLB where one site drives heterocomplex formation, while the other site competitively blocks FGFR1c homodimerization. The competitive model further suggests that while KLB competitively blocks FGFR1c self-association, it could also slow the process of ligand-induced homodimerization. In contrast, the allosteric model leaves open the possibility that FGFR1c is still available for self-association, but predicts that an additional force (ie, ligand) would be required to drive the interaction. It should also be noted that formation of an FGFR1c:KLB heterodimer by either model increases the entropy of ligand-induced FGFR1c dimerization, thereby potentially slowing or modulating the response to FGF21. A muted response is consistent with a majority of published data that shows FGF21-induced activity (eg, ERK1/2 activation) is generally less potent than canonical ligand responses.

Based on a preformed 1:1 FGFR1:KLB heterocomplex, we initially anticipated that addition of FGF21 would induce a 2:2 FGFR1c:KLB heterocomplex (Fig. 4C). In contrast to our expectations, FGF21 induced dimerization of FGFR1c but no change in KLB aggregate size was observed, consistent with a 2:1 FGFR1c:KLB signaling complex (Fig. 4D). One caveat to this interpretation is that we used single-color N&B analysis to measure aggregation state (Digman, Dalal, Horwitz, & Gratton, 2008), which uses "moment-analysis for the measurement of the average number of molecules and brightness in each pixel in fluorescence microscopy images," rather than a "cross-correlation" method that measures direct interaction between FGFR1c dimers and KLB monomers. Therefore, our data could also be interpreted to indicate FGFR1c homodimerization causing release of the two KLB proteins. However, severing the interaction between FGFR1c and KLB is inconsistent with co-IP data, suggesting that this interaction becomes stronger with addition of FGF21 (Kharitonenkov et al., 2008). We therefore believe that our data integrate with the literature that supports an FGF21-induced 2:1 FGFR1c and KLB heterocomplex.

The transition we observed from a 1:1 to a 2:1 FGFR1:KLB complex leads to the following question: *Why does the complex release one KLB?* One intriguing way to rationalize this transition is to assume that KLB has multiple sites for FGFR1c and FGF21 binding, such that a single KLB can stabilize the signaling complex while competing away a second KLB molecule. This model does not require each of the binding sites on KLB to have high target affinity, and competition could occur due to cooperative binding. In other words, a single high-affinity binding site could subsequently trigger the occupancy of lower affinity sites by simply increasing local concentration (ie, lowering the entropy). It is intriguing that we have found no evidence in the literature for specific FGF21 and FGFR1c binding sites in the KLB sequence—we postulate that this may be a consequence of the complexity and cooperativity of multiple, and potentially redundant, binding sites. Based on this model, we therefore anticipate that multiple FGF21 and FGFR1c binding sites will be identified within KLB.

When KLB was overexpressed in our model, we observed significant co-receptor aggregation upon addition of FGF21. This is consistent with the concept that a minimum of two ligand-binding sites are present on

KLB. Curiously, KLB aggregation did not occur in the absence of the galectin lattice (see later discussion), suggesting that to some degree, multimodal binding of FGF21 involves interactions with the ECM. One potential artifact of this study is that endogenously expressed FGFR1c may have played a role in the aggregation process; however, we believe this unlikely due to (i) ectopic KLB expression in excess to endogenous FGFRs; and (ii) our previous data showed that ligand addition to cells co-expressing KLB and FGFR1c resulted only in KLB monomer formation. These data suggest that KLB can bind to FGF21 at two separate sites, making it interesting to speculate that these interactions may competitively remove KLB from the signaling complex.

5. OTHER REGULATORS OF THE SIGNALING COMPLEX FORMATION

Canonical FGFR1c signaling is very potent and tightly regulated and is influenced by mechanisms such as expression of other membrane receptors that compete for ligand or otherwise inhibit FGFR homodimerization. We therefore anticipate that FGF21-induced signaling involves significant regulation of both FGFR1c and KLB. Our lab has focused on the regulation mechanisms of FGF21 signaling in pancreatic islets with a specific focus on potential regulators expressed in or impacting beta-cells including: (i) inactivation of KLB by the galectin lattice; (ii) regulation of FGFR1c activity by the ECM; and (iii) regulation of FGFR1c:KLB interactions by FGFR5 (Fig. 5).

5.1 KLB Inactivation by the Galectin Lattice

Galectins are a family of sugar binding proteins that exhibit binding specificity for β-galactoside sugars on glycoproteins (ie, extracellular N-glycosylation sites). Recalling that KLB is heavily glycosylated with 11 putative N-glycosylation sites, this co-receptor represents an excellent candidate protein to bind to the galectin lattice. Galectin binding to sugars on the extracellular domain of KLB receptor could affect the signaling threshold, membrane residency time, and endocytosis of this protein (Brewer, Miceli, & Baum, 2002; Demetriou, Granovsky, Quaggin, & Dennis, 2001; Dennis, Lau, Demetriou, & Nabi, 2009; Lajoie, Goetz, Dennis, & Nabi, 2009). Many galectin proteins (galectin-1–4, -7–10, -12, -13) have been identified in humans; however, galectin-1 and -3 are arguably the most

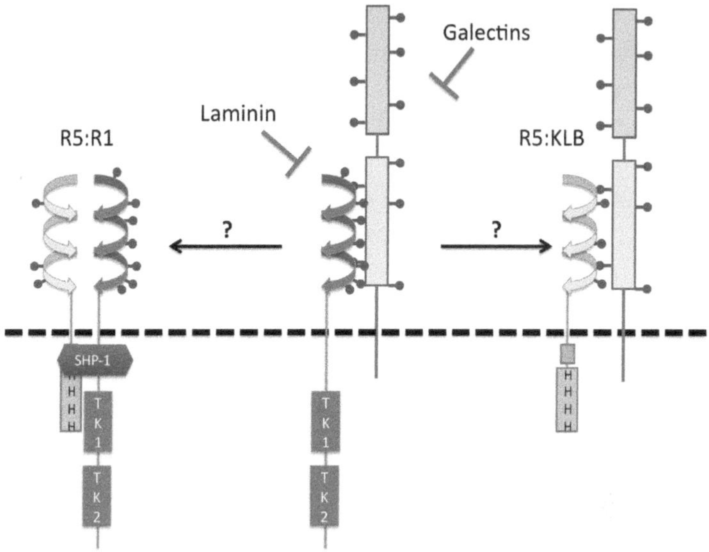

Fig. 5 Potential regulators of FGFR1:KLB signaling. The formation of FGFR1:KLB heterodimers is potentially diminished by laminin-induced downregulation of FGFR1 expression, as well as association of KLB with the galectin lattice. Given extracellular sequence similarity, the potential heterodimerization of FGFR1c with the novel FGFR5 and KLB is also proposed. We previously identified FGFR5 interaction with SHP-1 which may act as a scaffold protein to differentially activate and/or regulate FGFR1c signaling. (See the color plate.)

prominent due to potent and often opposing roles in modulating T-cell activ ity (galectin-1) as well as cancer cell inflammation, metastasis, and apoptosis (galectin-3) (Boscher, Dennis, & Nabi, 2011). At the molecular level, galectins are classified based on individual structure as well as the associated lattice generated in the ECM (dimeric, tandem dimer, or chimeric). For example, galectin-3 is a pentameric binding protein that can induce large aggregates, whereas galectin-1 will induce more weakly associated dimers. It is likely that the diverse lattice structures formed by galectins account for the varied roles in modulating receptor protein responses. More specifically, dimeric galectin-1 and chimeric galectin-3 often exhibit opposing roles in cell biology (ie, increasing and decreasing membrane receptor residency time through prevention and promotion of receptor internalization, *respectively*).

To the best of our knowledge, there is no specific literature identifying expression of galectin-1 in pancreatic islets; however, immune cell expression is well defined and associated with pancreatic islet infiltration (ie, meta-inflammation) (Perone et al., 2009). Intriguingly, the effects of galectin-3 on islet responses and the diabetic phenotype are controversial. One study examining obese LGALS3$^{(-/-)}$ mice suggests that galectin-3 deficiency accelerates high-fat diet-induced obesity and diabetes (Pejnovic et al., 2013); conversely, an alternate study using the same LGALS3$^{(-/-)}$ mouse model suggests that galectin-3 deficiency protects pancreatic islet cells from cytokine-triggered apoptosis (Saksida et al., 2013). As a corollary, the antiapoptotic effect of galectin-3 has been suggested to induce immune-mediated beta-cell damage and accelerate diabetes (Radosavljevic et al., 2012). These controversies dictate that further investigation is warranted to appreciate the impact on KLB-associated signaling.

We initially showed that the galectin lattice dampens FGF21-induced ERK1/2 responses (temporal and dynamics) in a beta-cell model (Ming et al., 2012). By using FRAP in combination with N&B analyses, we subsequently showed that the dampened response was due to KLB (and not FGFR1c) association with the galectin lattice, where KLB:FGFR1c heterodimer formation was inhibited (Fig. 5). We subsequently showed by co-IP that galectin-3 associates with KLB but anticipate that it is more likely that multiple galectins, including galectin-1, compete for these binding sites. Overall, our data suggest that association with the galectin lattice renders KLB inactive, and that changes in the lattice structure (ie, remodeling) alter KLB capacity to participate in heterodimer formation with FGFR1c to regulate downstream signaling pathways.

5.2 ECM Regulation of FGFRs

Within the pancreas, beta-cells are juxtaposed to endothelial cells (ECs), an anatomical arrangement that facilitates blood glucose sensing and the secretion of insulin directly into the blood stream. One means of communication between these cells in an effort to maintain these physiological responses is through the basement membrane. Numerous studies have shown significant impact of the basement membrane on beta-cell insulin processing, proliferation, and survival (Edamura et al., 2003; Kaido, Yebra, Cirulli, & Montgomery, 2004; Nikolova et al., 2006; Nobukini & Thomas, 2004). Upon our initial observations of FGFR1c activity in beta-cells, we postulated that EC-mediated modulation of the basement membrane provided a mechanism to control this signaling pathway. We subsequently modeled the changes in the basement membrane induced by beta-cell disease by culturing beta-cells on various matrix components (Kilkenny & Rocheleau, 2008). We observed a significant decrease in FGFR1c expression at the plasma membrane when cells were grown on a laminin substrate; conversely, FGFR1c expression increased in the presence of collagen type IV. We also observed a significant decrease in the temporal response to FGF1 signifying a decrease in FGFR1 activity. We therefore postulate that a changing beta-cell basement membrane composition has the capacity to regulate the FGF21-induced endocrine response. Whether this regulation would act in concert with changes in galectin-lattice formation to either enhance or counteract the effect could be an area of future study (Fig. 5).

5.3 Fibroblast Growth Factor Receptor 5

FGFR5 (FGFRL1) is the newest identified member of the FGFR family (Kim et al., 2001; Sleeman et al., 2001; Wiedemann & Trueb, 2000). This receptor isoform shares up to 50% similarity with FGFR1 (Steinberg et al., 2010), and when gaps in the amino acid alignments are ignored, the extracellular portion of FGFR5 was identified as being 29%, 31%, 34%, and 32% identical to the extracellular portion of FGFR1, FGFR2, FGFR3, and FGFR4, respectively (Kim et al., 2001). The extracellular domain closely resembles that of the canonical receptors; therefore, FGFR5 retains the ability to bind ligand as well as heparin (Trueb, Zhuang, Taeschler, & Wiedemann, 2003). More specifically, the D1–D3 domains were also described as being 35%, 38%, and 40% identical (41%, 48%, and 49% similar) to FGFR4 (Wiedemann & Trueb, 2000). The FGFR5 isoform is particularly distinct given its deficiency for the intracellular kinase domains common to all the other receptor isoforms. Instead, the unique intracellular domain is only ~100 amino acids in length consisting of a tandem tyrosine-based motif (PKL**Y**PKL**Y**TDI) followed by a histidine-rich region (Sleeman et al., 2001). The tandem tyrosine-based motif is predicted to play a role in either directing receptor endocytosis/transmembrane trafficking (Trueb, 2011) or functioning as a putative SH2-binding motif given sequence similarity to immunoreceptor tyrosine-based inhibition motifs (Sleeman et al., 2001). The histidine-rich region has been shown to interact with zinc and nickel ions (Zhuang, Karotki, Bruecker, & Trueb, 2009) and may therefore have capacity to directly modulate the adjacent SH2-binding motif. Due to the deficiency of intracellular tyrosine kinase domains, this receptor isoform was originally predicted to negatively regulate FGF signaling by either competitively binding pericellular ligands to minimize interactions with cell surface FGFRs, or heterodimerizing with full-length FGFRs to prevent *trans*-autophosphorylation. To date, neither of these hypotheses has been conclusively supported.

We have previously shown that overexpression of FGFR5 in a beta-cell line (βTC3) significantly increases activation of MAPK, either basally or in the presence of FGF2 (Silva et al., 2013). We further showed using immunoprecipitation that FGFR5 associates with the intracellular signaling substrate SHP-1 via the unique C-terminal domain (Silva et al., 2013). These data would suggest that despite the absence of the classical tyrosine kinase domains, FGFR5 does exhibit the ability to regulate intracellular signaling activity. It remains unknown whether SHP-1 functions as a phosphatase to negatively regulate activation of other signaling molecules, or whether it functions as a substrate scaffold for interaction with other FGFRs. Although we have observed increased Src kinase activity when AD293-R5 cells are stimulated with FGF2 (unpublished data), specific association with other defined signaling molecules has yet to be confirmed.

Heterodimerization between FGFR5 and FGFR1c is another potential mechanism by which FGFR5 may function to regulate FGF21 signaling (Fig. 5). Immunoprecipitation/immunoblot experiments have previously demonstrated that canonical FGFR1 and FGFR2 have the ability to homodimerize and heterodimerize, suggesting that these receptor proteins not only have constitutive affinity for one another in the absence of ligand but also exhibit capacity for inter-receptor signaling (Bellot et al., 1991). The concept of inter-receptor signaling is supported by the findings of a large systematic protein screen in a zebrafish model that indicated interaction of FGFR4 with FGFR5 isoforms (FGFRla and FGFR1lb) (Bushell, Sollner, Schuster-Boeckler, Bateman, & Wright, 2008). In contrast, while FGFR5 was shown to constitutively homodimerize using FRET analysis, it was not observed to heterodimerize with either FGFR3αIIIc or FGFR4 (C-terminal truncated proteins) (Rieckmann, Kotevic, & Trueb, 2008). We have also confirmed that both FGFR1c and FGFR5 exhibit constitutive homodimerization using homoFRET analysis (Ming et al., 2012; unpublished data); however, the potential for heteroaggregation remains inconclusive. Combined, these studies confirm that FGFR1c exist in dimeric form at the cell membrane and may heterodimerize selectively as a mechanism to regulate FGF21 activity.

The potential for interaction between FGFR5 and KLB has not been defined; however, it is intriguing that the extracellular D2–D3 domain of FGFR1c where KLB binds shows high sequence similarity and modeled 3D structure as FGFR5 (Plotnikov et al., 2000; Sleeman et al., 2001; Zhuang, Pandey, Villiger, & Trueb, 2015). Given that KLB engages the hydrophobic groove in the D3 domain of FGFR1c, the fact that FGFR5 is predicted to have sequence similarity in this domain warrants investigation into whether FGFR5 and KLB complex. It is interesting to note that 3D modeling predicted an extended loop in D3 of FGFR5 that was larger than the comparable loop structure of FGFR1/R2. It was determined that four specific residues of the hydrophobic pocket (L281, F303, V304, and L339), which are far apart in the primary sequence but close together to form an "active site" in the 3D structure, are shielded by the specific long loop of FGFR5 D3. This is intriguing due to the possibility that FGFR5 may play a role in regulation of the FGF21 signaling complex, given that this structural pocket presents the hypothetical binding site for KLB. There are currently no published studies demonstrating FGF21 activation of FGFR5. Although no affinity of FGFR5 for FGF21 was demonstrated by dot-blot analysis (low affinity and modest affinity for FGF19 and FGF23 were observed, respectively) (Steinberg et al., 2010), this study did not incorporate KLB; therefore, it would be interesting to observe if soluble co-receptor might alter the affinity. Finally, overlap in tissue expression of these molecules should be considered as it may provide clues as to whether FGFR5 may play a role in modulating the FGF21: FGFR1c:KLB complex. Like FGFR1c and KLB expression, high levels of FGFR5 expression have been identified in human liver, kidney, skeletal muscle, heart, and endocrine pancreas (Sleeman et al., 2001).

6. SUMMARY

Endocrine FGFs have emerged as potent regulators of whole-body physiology. In exploring the mechanisms of ligand-induced signaling, we have introduced the receptors and ligands and placed them within the context of their component families. Furthermore, we have described how each protein has been shown to interact with the other proteins in the complex. We subsequently focused on endocrine FGF21 signaling through FGFR1c and KLB as a consequence of our interest in beta-cell physiology. More specifically, we have included descriptions of our work measuring receptor stoichiometry in the complexes before and after addition of FGF21, revealing a unique transition from FGFR1c homodimers and KLB monomers to a preformed 1:1 heterocomplex. The transition to a 2:1 FGFR1c:KLB signaling complex predicts multiple binding sites are available on KLB for both FGF21 and FGFR1c proteins. Finally, we have described how these signaling complexes could potentially be modulated by further interactions with the galectin lattice, ECM, and/or FGFR5.

REFERENCES

Adams, A. C., Cheng, C. C., Coskun, T., & Kharitonenkov, A. (2012). FGF21 requires betaklotho to act in vivo. *PLoS One*, *7*(11), e49977.

Adams, A. C., & Kharitonenkov, A. (2012). FGF21: The center of a transcriptional nexus in metabolic regulation. *Current Diabetes Reviews*, *8*(4), 285–293.

Angelin, B., Larsson, T. E., & Rudling, M. (2012). Circulating fibroblast growth factors as metabolic regulators—A critical appraisal. *Cell Metabolism*, *16*(6), 693–705.

Asada, M., Shinomiya, M., Suzuki, M., Honda, E., Sugimoto, R., Ikekita, M., et al. (2009). Glycosaminoglycan affinity of the complete fibroblast growth factor family. *Biochimica et Biophysica Acta*, *1790*(1), 40–48.

Avivi, A., Yayon, A., & Givol, D. (1993). A novel form of FGF receptor-3 using an alternative exon in the immunoglobulin domain III. *FEBS Letters*, *330*(3), 249–252.

Badman, M. K., Pissios, P., Kennedy, A. R., Koukos, G., Flier, J. S., & Maratos-Flier, E. (2007). Hepatic fibroblast growth factor 21 is regulated by PPARalpha and is a key mediator of hepatic lipid metabolism in ketotic states. *Cell Metabolism*, *5*(6), 426–437.

Bae, J. H., & Schlessinger, J. (2010). Asymmetric tyrosine kinase arrangements in activation or autophosphorylation of receptor tyrosine kinases. *Molecules and Cells*, *29*(5), 443–448.

Baird, A. (1994). Fibroblast growth factors: Activities and significance of non-neurotrophin neurotrophic growth factors. *Current Opinion in Neurobiology*, *4*(1), 78–86.

Basilico, C., & Moscatelli, D. (1992). The FGF family of growth factors and oncogenes. *Advances in Cancer Research*, *59*, 115–165.

Beenken, A., & Mohammadi, M. (2009). The FGF family: Biology, pathophysiology and therapy. *Nature Reviews Drug Discovery*, *8*(3), 235–253.

Bellot, F., Crumley, G., Kaplow, J. M., Schlessinger, J., Jaye, M., & Dionne, C. A. (1991). Ligand-induced transphosphorylation between different FGF receptors. *The EMBO Journal*, *10*(10), 2849–2854.

Bellusci, S., Grindley, J., Emoto, H., Itoh, N., & Hogan, B. L. (1997). Fibroblast growth factor 10 (FGF10) and branching morphogenesis in the embryonic mouse lung. *Development*, *124*(23), 4867–4878.

Belov, A. A., & Mohammadi, M. (2012). Grb2, a double-edged sword of receptor tyrosine kinase signaling. *Science Signaling*, *5*(249), pe49.

Benetti, E., Mastrocola, R., Rogazzo, M., Chiazza, F., Aragno, M., Fantozzi, R., et al. (2013). High sugar intake and development of skeletal muscle insulin resistance and inflammation in mice: A protective role for PPAR-delta agonism. *Mediators of Inflammation*, *2013*, 509502.

Beyer, T. A., Werner, S., Dickson, C., & Grose, R. (2003). Fibroblast growth factor 22 and its potential role during skin development and repair. *Experimental Cell Research*, *287*(2), 228–236.

Bikfalvi, A., Klein, S., Pintucci, G., & Rifkin, D. B. (1997). Biological roles of fibroblast growth factor-2. *Endocrine Reviews*, *18*(1), 26–45.

Bocharov, E. V., Lesovoy, D. M., Goncharuk, S. A., Goncharuk, M. V., Hristova, K., & Arseniev, A. S. (2013). Structure of FGFR3 transmembrane domain dimer: Implications for signaling and human pathologies. *Structure*, *21*(11), 2087–2093.

Bookout, A. L., de Groot, M. H., Owen, B. M., Lee, S., Gautron, L., Lawrence, H. L., et al. (2013). FGF21 regulates metabolism and circadian behavior by acting on the nervous system. *Nature Medicine*, *19*(9), 1147–1152.

Boscher, C., Dennis, J. W., & Nabi, I. R. (2011). Glycosylation, galectins and cellular signaling. *Current Opinion in Cell Biology*, *23*(4), 383–392.

Brewer, C. F., Miceli, M. C., & Baum, L. G. (2002). Clusters, bundles, arrays and lattices: Novel mechanisms for lectin-saccharide-mediated cellular interactions. *Current Opinion in Structural Biology*, *12*(5), 616–623.

Bushell, K. M., Sollner, C., Schuster-Boeckler, B., Bateman, A., & Wright, G. J. (2008). Large-scale screening for novel low-affinity extracellular protein interactions. *Genome Research*, *18*(4), 622–630.

Chellaiah, A. T., McEwen, D. G., Werner, S., Xu, J., & Ornitz, D. M. (1994). Fibroblast growth factor receptor (FGFR) 3. Alternative splicing in immunoglobulin-like domain III creates a receptor highly specific for acidic FGF/FGF-1. *The Journal of Biological Chemistry*, *269*(15), 11620–11627.

Chellaiah, A., Yuan, W., Chellaiah, M., & Ornitz, D. M. (1999). Mapping ligand binding domains in chimeric fibroblast growth factor receptor molecules. Multiple regions determine ligand binding specificity. *The Journal of Biological Chemistry*, *274*(49), 34785–34794.

Cheng, X., Zhu, B., Jiang, F., & Fan, H. (2011). Serum FGF-21 levels in type 2 diabetic patients. *Endocrine Research*, *36*(4), 142–148.

Comps-Agrar, L., Dunshee, D. R., Eaton, D. L., & Sonoda, J. (2015). Unliganded fibroblast growth factor receptor 1 forms density-independent dimers. *The Journal of Biological Chemistry*, *290*(40), 24166–24177.

Coulier, F., Pontarotti, P., Roubin, R., Hartung, H., Goldfarb, M., & Birnbaum, D. (1997). Of worms and men: An evolutionary perspective on the fibroblast growth factor (FGF) and FGF receptor families. *Journal of Molecular Evolution*, *44*(1), 43–56.

Crooks, D. R., Natarajan, T. G., Jeong, S. Y., Chen, C., Park, S. Y., Huang, H., et al. (2014). Elevated FGF21 secretion, PGC-1alpha and ketogenic enzyme expression are hallmarks of iron-sulfur cluster depletion in human skeletal muscle. *Human Molecular Genetics*, *23*(1), 24–39.

Crumley, G., Bellot, F., Kaplow, J. M., Schlessinger, J., Jaye, M., & Dionne, C. A. (1991). High-affinity binding and activation of a truncated FGF receptor by both aFGF and bFGF. *Oncogene*, *6*(12), 2255–2262.

De Moerlooze, L., Spencer-Dene, B., Revest, J. M., Hajihosseini, M., Rosewell, I., & Dickson, C. (2000). An important role for the IIIb isoform of fibroblast growth factor receptor 2 (FGFR2) in mesenchymal-epithelial signalling during mouse organogenesis. *Development*, *127*(3), 483–492.

Dell, K. R., & Williams, L. T. (1992). A novel form of fibroblast growth factor receptor 2. Alternative splicing of the third immunoglobulin-like domain confers ligand binding specificity. *The Journal of Biological Chemistry*, *267*(29), 21225–21229.

Demetriou, M., Granovsky, M., Quaggin, S., & Dennis, J. W. (2001). Negative regulation of T-cell activation and autoimmunity by Mgat5 N-glycosylation. *Nature*, *409*(6821), 733–739.

Dennis, J. W., Lau, K. S., Demetriou, M., & Nabi, I. R. (2009). Adaptive regulation at the cell surface by N-glycosylation. *Traffic*, *10*(11), 1569–1578.

Diaz-Delfin, J., Hondares, E., Iglesias, R., Giralt, M., Caelles, C., & Villarroya, F. (2012). TNF-alpha represses beta-Klotho expression and impairs FGF21 action in adipose cells: Involvement of JNK1 in the FGF21 pathway. *Endocrinology*, *153*(9), 4238–4245.

Digman, M. A., Dalal, R., Horwitz, A. F., & Gratton, E. (2008). Mapping the number of molecules and brightness in the laser scanning microscope. *Biophysical Journal*, *94*(6), 2320–2332.

Ding, X., Boney-Montoya, J., Owen, B. M., Bookout, A. L., Coate, K. C., Mangelsdorf, D. J., et al. (2012). betaKlotho is required for fibroblast growth factor 21 effects on growth and metabolism. *Cell Metabolism*, *16*(3), 387–393.

Dionne, C. A., Jaye, M., & Schlessinger, J. (1991). Structural diversity and binding of FGF receptors. *The Annals of the New York Academy of Sciences*, *638*, 161–166.

Donate-Correa, J., Martin-Nunez, E., Delgado, N. P., de Fuentes, M. M., Arduan, A. O., Mora-Fernandez, C., et al. (2015). Implications of fibroblast growth factor/Klotho system in glucose metabolism and diabetes. *Cytokine & Growth Factor Reviews*. http://dx.doi.org/10.1016/j.cytogfr.2015.12.003.

Duan, D. S., Werner, S., & Williams, L. T. (1992). A naturally occurring secreted form of fibroblast growth factor (FGF) receptor 1 binds basic FGF in preference over acidic FGF. *The Journal of Biological Chemistry*, *267*(23), 16076–16080.

Edamura, K., Nasu, K., Iwami, Y., Ogawa, H., Sasaki, N., & Ohgawara, H. (2003). Effect of adhesion or collagen molecules on cell attachment, insulin secretion, and glucose responsiveness in the cultured adult porcine endocrine pancreas: A preliminary study. *Cell Transplantation*, *12*(4), 439–446.

Eisemann, A., Ahn, J. A., Graziani, G., Tronick, S. R., & Ron, D. (1991). Alternative splicing generates at least five different isoforms of the human basic-FGF receptor. *Oncogene*, *6*(7), 1195–1202.

Fisher, F. M., Chui, P. C., Antonellis, P. J., Bina, H. A., Kharitonenkov, A., Flier, J. S., et al. (2010). Obesity is a fibroblast growth factor 21 (FGF21)-resistant state. *Diabetes*, *59*(11), 2781–2789.

Fisher, F. M., Estall, J. L., Adams, A. C., Antonellis, P. J., Bina, H. A., Flier, J. S., et al. (2011). Integrated regulation of hepatic metabolism by fibroblast growth factor 21 (FGF21) in vivo. *Endocrinology*, *152*(8), 2996–3004.

Florkiewicz, R. Z., Majack, R. A., Buechler, R. D., & Florkiewicz, E. (1995). Quantitative export of FGF-2 occurs through an alternative, energy-dependent, non-ER/Golgi pathway. *Journal of Cellular Physiology*, *162*(3), 388–399.

Fon Tacer, K., Bookout, A. L., Ding, X., Kurosu, H., John, G. B., Wang, L., et al. (2010). Research resource: Comprehensive expression atlas of the fibroblast growth factor system in adult mouse. *Molecular Endocrinology*, *24*(10), 2050–2064.

Fukumoto, S. (2008). Actions and mode of actions of FGF19 subfamily members. *Endocrine Journal*, *55*(1), 23–31.

Gallego-Escuredo, J. M., Gomez-Ambrosi, J., Catalan, V., Domingo, P., Giralt, M., Fruhbeck, G., et al. (2015). Opposite alterations in FGF21 and FGF19 levels and disturbed expression of the receptor machinery for endocrine FGFs in obese patients. *International Journal of Obesity*, *39*(1), 121–129.

Galman, C., Lundasen, T., Kharitonenkov, A., Bina, H. A., Eriksson, M., Hafstrom, I., et al. (2008). The circulating metabolic regulator FGF21 is induced by prolonged fasting and PPARalpha activation in man. *Cell Metabolism*, *8*(2), 169–174.

Goetz, R., Beenken, A., Ibrahimi, O. A., Kalinina, J., Olsen, S. K., Eliseenkova, A. V., et al. (2007). Molecular insights into the Klotho-dependent, endocrine mode of action of fibroblast growth factor 19 subfamily members. *Molecular and Cellular Biology*, *27*(9), 3417–3428.

Goetz, R., & Mohammadi, M. (2013). Exploring mechanisms of FGF signalling through the lens of structural biology. *Nature Reviews Molecular Cell Biology*, *14*(3), 166–180.

Goetz, R., Ohnishi, M., Ding, X., Kurosu, H., Wang, L., Akiyoshi, J., et al. (2012a). Klotho coreceptors inhibit signaling by paracrine fibroblast growth factor 8 subfamily ligands. *Molecular and Cellular Biology*, *32*(10), 1944–1954.

Goetz, R., Ohnishi, M., Kir, S., Kurosu, H., Wang, L., Pastor, J., et al. (2012b). Conversion of a paracrine fibroblast growth factor into an endocrine fibroblast growth factor. *The Journal of Biological Chemistry*, *287*(34), 29134–29146.

Gospodarowicz, D. (1975). Purification of a fibroblast growth factor from bovine pituitary. *The Journal of Biological Chemistry*, *250*(7), 2515–2520.

Gospodarowicz, D., & Cheng, J. (1986). Heparin protects basic and acidic FGF from inactivation. *Journal of Cellular Physiology*, *128*(3), 475–484.

Gupte, J., Yang, L., Wu, X., Weiszmann, J., Hecht, R., Lemon, B., et al. (2011). The FGFR D3 domain determines receptor selectivity for fibroblast growth factor 21. *Journal of Molecular Biology*, *408*(3), 491–502.

Hager, T., Spahr, C., Xu, J., Salimi-Moosavi, H., & Hall, M. (2013). Differential enzyme-linked immunosorbent assay and ligand-binding mass spectrometry for analysis of biotransformation of protein therapeutics: Application to various FGF21 modalities. *Analytical Chemistry*, *85*(5), 2731–2738.

Hanneken, A., Ying, W., Ling, N., & Baird, A. (1994). Identification of soluble forms of the fibroblast growth factor receptor in blood. *Proceedings of the National Academy of Sciences of the United States of America*, *91*(19), 9170–9174.

Harmer, N. J., Ilag, L. L., Mulloy, B., Pellegrini, L., Robinson, C. V., & Blundell, T. L. (2004). Towards a resolution of the stoichiometry of the fibroblast growth factor (FGF)-FGF receptor-heparin complex. *Journal of Molecular Biology*, *339*(4), 821–834.

Haugsten, E. M., Sorensen, V., Brech, A., Olsnes, S., & Wesche, J. (2005). Different intracellular trafficking of FGF1 endocytosed by the four homologous FGF receptors. *Journal of Cell Science*, *118*(Pt 17), 3869–3881.

Hecht, R., Li, Y. S., Sun, J., Belouski, E., Hall, M., Hager, T., et al. (2012). Rationale-based engineering of a potent long-acting FGF21 analog for the treatment of type 2 diabetes. *PLoS One*, 7(11), e49345.

Hennessey, J. A., Wei, E. Q., & Pitt, G. S. (2013). Fibroblast growth factor homologous factors modulate cardiac calcium channels. *Circulation Research*, *113*(4), 381–388.

Henrissat, B., & Davies, G. (1997). Structural and sequence-based classification of glycoside hydrolases. *Current Opinion in Structural Biology*, 7(5), 637–644.

Hojman, P., Pedersen, M., Nielsen, A. R., Krogh-Madsen, R., Yfanti, C., Akerstrom, T., et al. (2009). Fibroblast growth factor-21 is induced in human skeletal muscles by hyperinsulinemia. *Diabetes*, *58*(12), 2797–2801.

Hou, J. Z., Kan, M. K., McKeehan, K., McBride, G., Adams, P., & McKeehan, W. L. (1991). Fibroblast growth factor receptors from liver vary in three structural domains. *Science*, *251*(4994), 665–668.

Hughes, S. E., & Hall, P. A. (1993). Overview of the fibroblast growth factor and receptor families: Complexity, functional diversity, and implications for future cardiovascular research. *Cardiovascular Research*, *27*(7), 1199–1203.

Imura, A., Iwano, A., Tohyama, O., Tsuji, Y., Nozaki, K., Hashimoto, N., et al. (2004). Secreted Klotho protein in sera and CSF: Implication for post-translational cleavage in release of Klotho protein from cell membrane. *FEBS Letters*, *565*(1-3), 143–147.

Inagaki, T., Dutchak, P., Zhao, G., Ding, X., Gautron, L., Parameswara, V., et al. (2007). Endocrine regulation of the fasting response by PPARalpha-mediated induction of fibroblast growth factor 21. *Cell Metabolism*, *5*(6), 415–425.

Ito, S., Fujimori, T., Furuya, A., Satoh, J., & Nabeshima, Y. (2005). Impaired negative feedback suppression of bile acid synthesis in mice lacking betaKlotho. *The Journal of Clinical Investigation*, *115*(8), 2202–2208.

Ito, S., Fujimori, T., Hayashizaki, Y., & Nabeshima, Y. (2002). Identification of a novel mouse membrane-bound family 1 glycosidase-like protein, which carries an atypical active site structure. *Biochimica et Biophysica Acta*, *1576*(3), 341–345.

Ito, S., Kinoshita, S., Shiraishi, N., Nakagawa, S., Sekine, S., Fujimori, T., et al. (2000). Molecular cloning and expression analyses of mouse betaklotho, which encodes a novel Klotho family protein. *Mechanisms of Development*, *98*(1-2), 115–119.

Itoh, N., Ohta, H., & Konishi, M. (2015). Endocrine FGFs: Evolution, physiology, pathophysiology, and pharmacotherapy. *Frontiers in Endocrinology (Lausanne)*, *6*, 154.

Jaye, M., Schlessinger, J., & Dionne, C. A. (1992). Fibroblast growth factor receptor tyrosine kinases: Molecular analysis and signal transduction. *Biochimica et Biophysica Acta*, *1135*(2), 185–199.

Jayson, G. C., Vives, C., Paraskeva, C., Schofield, K., Coutts, J., Fleetwood, A., et al. (1999). Coordinated modulation of the fibroblast growth factor dual receptor mechanism during transformation from human colon adenoma to carcinoma. *International Journal of Cancer*, *82*(2), 298–304.

Jin, W., Bi, W., Huang, E. S., & Cote, G. J. (1999). Glioblastoma cell-specific expression of fibroblast growth factor receptor-1beta requires an intronic repressor of RNA splicing. *Cancer Research, 59*(2), 316–319.

Jin, Y., Pasumarthi, K. B., Bock, M. E., Lytras, A., Kardami, E., & Cattini, P. A. (1994). Cloning and expression of fibroblast growth factor receptor-1 isoforms in the mouse heart: Evidence for isoform switching during heart development. *Journal of Molecular and Cellular Cardiology, 26*(11), 1449–1459.

Johnson, D. E., Lee, P. L., Lu, J., & Williams, L. T. (1990). Diverse forms of a receptor for acidic and basic fibroblast growth factors. *Molecular and Cellular Biology, 10*(9), 4728–4736.

Johnson, D. E., Lu, J., Chen, H., Werner, S., & Williams, L. T. (1991). The human fibroblast growth factor receptor genes: A common structural arrangement underlies the mechanisms for generating receptor forms that differ in their third immunoglobulin domain. *Molecular and Cellular Biology, 11*(9), 4627–4634.

Johnson, D. E., & Williams, L. T. (1993). Structural and functional diversity in the FGF receptor multigene family. *Advances in Cancer Research, 60*, 1–41.

Jones, S. (2008). Mini-review: Endocrine actions of fibroblast growth factor 19. *Molecular Pharmaceutics, 5*(1), 42–48.

Kaido, T., Yebra, M., Cirulli, V., & Montgomery, A. M. (2004). Regulation of human beta-cell adhesion, motility, and insulin secretion by collagen IV and its receptor alpha1beta1. *The Journal of Biological Chemistry, 279*(51), 53762–53769.

Kalinina, J., Dutta, K., Ilghari, D., Beenken, A., Goetz, R., Eliseenkova, A. V., et al. (2012). The alternatively spliced acid box region plays a key role in FGF receptor autoinhibition. *Structure, 20*(1), 77–88.

Kan, M., Wang, F., Xu, J., Crabb, J. W., Hou, J., & McKeehan, W. L. (1993). An essential heparin-binding domain in the fibroblast growth factor receptor kinase. *Science, 259*(5103), 1918–1921.

Katoh, Y., & Katoh, M. (2005). Comparative genomics on FGF16 orthologs. *International Journal of Molecular Medicine, 16*(5), 959–963.

Keegan, K., Johnson, D. E., Williams, L. T., & Hayman, M. J. (1991). Isolation of an additional member of the fibroblast growth factor receptor family, FGFR-3. *Proceedings of the National Academy of Sciences of the United States of America, 88*(4), 1095–1099.

Kharitonenkov, A. (2009). FGFs and metabolism. *Current Opinion in Pharmacology, 9*(6), 805–810.

Kharitonenkov, A., Dunbar, J. D., Bina, H. A., Bright, S., Moyers, J. S., Zhang, C., et al. (2008). FGF-21/FGF-21 receptor interaction and activation is determined by beta-Klotho. *Journal of Cellular Physiology, 215*(1), 1–7.

Kharitonenkov, A., & Shanafelt, A. B. (2009). FGF21: A novel prospect for the treatment of metabolic diseases. *Current Opinion in Investigational Drugs, 10*(4), 359–364.

Kharitonenkov, A., Shiyanova, T. L., Koester, A., Ford, A. M., Micanovic, R., Galbreath, E. J., et al. (2005). FGF-21 as a novel metabolic regulator. *The Journal of Clinical Investigation, 115*(6), 1627–1635.

Kharitonenkov, A., Wroblewski, V. J., Koester, A., Chen, Y. F., Clutinger, C. K., Tigno, X. T., et al. (2007). The metabolic state of diabetic monkeys is regulated by fibroblast growth factor-21. *Endocrinology, 148*(2), 774–781.

Kilkenny, D. M., & Rocheleau, J. V. (2008). Fibroblast growth factor receptor-1 signaling in pancreatic islet beta-cells is modulated by the extracellular matrix. *Molecular Endocrinology, 22*(1), 196–205.

Kim, I., Moon, S., Yu, K., Kim, U., & Koh, G. Y. (2001). A novel fibroblast growth factor receptor-5 preferentially expressed in the pancreas(1). *Biochimica et Biophysica Acta, 1518*(1-2), 152–156.

Kimelman, D., & Kirschner, M. (1987). Synergistic induction of mesoderm by FGF and TGF-beta and the identification of an mRNA coding for FGF in the early Xenopus embryo. *Cell*, *51*(5), 869–877.

Klagsbrun, M., & Baird, A. (1991). A dual receptor system is required for basic fibroblast growth factor activity. *Cell*, *67*(2), 229–231.

Kornbluth, S., Paulson, K. E., & Hanafusa, H. (1988). Novel tyrosine kinase identified by phosphotyrosine antibody screening of cDNA libraries. *Molecular and Cellular Biology*, *8*(12), 5541–5544.

Kubicky, R. A., Wu, S., Kharitonenkov, A., & De Luca, F. (2012). Role of fibroblast growth factor 21 (FGF21) in undernutrition-related attenuation of growth in mice. *Endocrinology*, *153*(5), 2287–2295.

Kuro-o, M., Matsumura, Y., Aizawa, H., Kawaguchi, H., Suga, T., Utsugi, T., et al. (1997). Mutation of the mouse klotho gene leads to a syndrome resembling ageing. *Nature*, *390*(6655), 45–51.

Kurosu, H., Choi, M., Ogawa, Y., Dickson, A. S., Goetz, R., Eliseenkova, A. V., et al. (2007). Tissue-specific expression of betaKlotho and fibroblast growth factor (FGF) receptor isoforms determines metabolic activity of FGF19 and FGF21. *The Journal of Biological Chemistry*, *282*(37), 26687–26695.

Kurosu, H., Ogawa, Y., Miyoshi, M., Yamamoto, M., Nandi, A., Rosenblatt, K. P., et al. (2006). Regulation of fibroblast growth factor-23 signaling by klotho. *The Journal of Biological Chemistry*, *281*(10), 6120–6123.

Lajoie, P., Goetz, J. G., Dennis, J. W., & Nabi, I. R. (2009). Lattices, rafts, and scaffolds: Domain regulation of receptor signaling at the plasma membrane. *The Journal of Cell Biology*, *185*(3), 381–385.

Lee, P. L., Johnson, D. E., Cousens, L. S., Fried, V. A., & Williams, L. T. (1989). Purification and complementary DNA cloning of a receptor for basic fibroblast growth factor. *Science*, *245*(4913), 57–60.

Lemmon, M. A., & Schlessinger, J. (2010). Cell signaling by receptor tyrosine kinases. *Cell*, *141*(7), 1117–1134.

Liang, Q., Zhong, L., Zhang, J., Wang, Y., Bornstein, S. R., Triggle, C. R., et al. (2014). FGF21 maintains glucose homeostasis by mediating the cross talk between liver and brain during prolonged fasting. *Diabetes*, *63*(12), 4064–4075.

Liu, S., Guo, R., Simpson, L. G., Xiao, Z. S., Burnham, C. E., & Quarles, L. D. (2003). Regulation of fibroblastic growth factor 23 expression but not degradation by PHEX. *The Journal of Biological Chemistry*, *278*(39), 37419–37426.

Logan, A., Frautschy, S. A., Gonzalez, A. M., & Baird, A. (1992). A time course for the focal elevation of synthesis of basic fibroblast growth factor and one of its high-affinity receptors (flg) following a localized cortical brain injury. *The Journal of Neuroscience*, *12*(10), 3828–3837.

MacArthur, C. A., Lawshe, A., Xu, J., Santos-Ocampo, S., Heikinheimo, M., Chellaiah, A. T., et al. (1995). FGF-8 isoforms activate receptor splice forms that are expressed in mesenchymal regions of mouse development. *Development*, *121*(11), 3603–3613.

Maruyama, I. N. (2015). Activation of transmembrane cell-surface receptors via a common mechanism? The "rotation model" *Bioessays*, *37*(9), 959–967.

McFarlane, I., Georgopoulou, N., Coughlan, C. M., Gillian, A. M., & Breen, K. C. (1999). The role of the protein glycosylation state in the control of cellular transport of the amyloid beta precursor protein. *Neuroscience*, *90*(1), 15–25.

McKeehan, W. L., Wang, F., & Kan, M. (1998). The heparan sulfate-fibroblast growth factor family: Diversity of structure and function. *Progress in Nucleic Acid Research and Molecular Biology*, *59*, 135–176.

Micanovic, R., Raches, D. W., Dunbar, J. D., Driver, D. A., Bina, H. A., Dickinson, C. D., et al. (2009). Different roles of N- and C-termini in the functional activity of FGF21. *Journal of Cellular Physiology, 219*(2), 227–234.

Miki, T., Bottaro, D. P., Fleming, T. P., Smith, C. L., Burgess, W. H., Chan, A. M., et al. (1992). Determination of ligand-binding specificity by alternative splicing: Two distinct growth factor receptors encoded by a single gene. *Proceedings of the National Academy of Sciences of the United States of America, 89*(1), 246–250.

Ming, A. Y., Yoo, E., Vorontsov, E. N., Altamentova, S. M., Kilkenny, D. M., & Rocheleau, J. V. (2012). Dynamics and distribution of Klothobeta (KLB) and fibroblast growth factor receptor-1 (FGFR1) in living cells reveal the fibroblast growth factor-21 (FGF21)-induced receptor complex. *The Journal of Biological Chemistry, 287*(24), 19997–20006.

Miyamoto, M., Naruo, K., Seko, C., Matsumoto, S., Kondo, T., & Kurokawa, T. (1993). Molecular cloning of a novel cytokine cDNA encoding the ninth member of the fibroblast growth factor family, which has a unique secretion property. *Molecular and Cellular Biology, 13*(7), 4251–4259.

Mohammadi, M., Olsen, S. K., & Ibrahimi, O. A. (2005). Structural basis for fibroblast growth factor receptor activation. *Cytokine & Growth Factor Reviews, 16*(2), 107–137.

Moore, D. D. (2007). Physiology. Sister act. *Science, 316*(5830), 1436–1438.

Moscatelli, D. (1987). High and low affinity binding sites for basic fibroblast growth factor on cultured cells: Absence of a role for low affinity binding in the stimulation of plasminogen activator production by bovine capillary endothelial cells. *Journal of Cellular Physiology, 131*(1), 123–130.

Nikolova, G., Jabs, N., Konstantinova, I., Domogatskaya, A., Tryggvason, K., Sorokin, L., et al. (2006). The vascular basement membrane: A niche for insulin gene expression and beta cell proliferation. *Developmental Cell, 10*(3), 397–405.

Nishimura, T., Nakatake, Y., Konishi, M., & Itoh, N. (2000). Identification of a novel FGF, FGF-21, preferentially expressed in the liver. *Biochimica et Biophysica Acta, 1492*(1), 203–206.

Nobukini, T., & Thomas, G. (2004). The mTOR/S6K signalling pathway: The role of the TSC1/2 tumour suppressor complex and the proto-oncogene Rheb. *Novartis Foundation Symposium, 262*, 148–154. discussion 154-149, 265-148.

Nygaard, E. B., Moller, C. L., Kievit, P., Grove, K. L., & Andersen, B. (2014). Increased fibroblast growth factor 21 expression in high-fat diet-sensitive non-human primates (Macaca mulatta). *International Journal of Obesity, 38*(2), 183–191.

Ogawa, Y., Kurosu, H., Yamamoto, M., Nandi, A., Rosenblatt, K. P., Goetz, R., et al. (2007). BetaKlotho is required for metabolic activity of fibroblast growth factor 21. *Proceedings of the National Academy of Sciences of the United States of America, 104*(18), 7432–7437.

Oishi, K., Konishi, M., Murata, Y., & Itoh, N. (2011). Time-imposed daily restricted feeding induces rhythmic expression of Fgf21 in white adipose tissue of mice. *Biochemical and Biophysical Research Communications, 412*(2), 396–400.

Olsen, S. K., Garbi, M., Zampieri, N., Eliseenkova, A. V., Ornitz, D. M., Goldfarb, M., et al. (2003). Fibroblast growth factor (FGF) homologous factors share structural but not functional homology with FGFs. *The Journal of Biological Chemistry, 278*(36), 34226–34236.

Olsen, S. K., Li, J. Y., Bromleigh, C., Eliseenkova, A. V., Ibrahimi, O. A., Lao, Z., et al. (2006). Structural basis by which alternative splicing modulates the organizer activity of FGF8 in the brain. *Genes & Development, 20*(2), 185–198.

Olwin, B. B., & Hauschka, S. D. (1986). Identification of the fibroblast growth factor receptor of Swiss 3 T3 cells and mouse skeletal muscle myoblasts. *Biochemistry, 25*(12), 3487–3492.

Ornitz, D. M. (2000). FGFs, heparan sulfate and FGFRs: Complex interactions essential for development. *Bioessays, 22*(2), 108–112.

Ornitz, D. M. (2005). FGF signaling in the developing endochondral skeleton. *Cytokine & Growth Factor Reviews, 16*(2), 205–213.

Ornitz, D. M., & Itoh, N. (2015). The Fibroblast Growth Factor signaling pathway. *Wiley Interdisciplinary Reviews: Developmental Biology, 4*(3), 215–266.

Ornitz, D. M., Xu, J., Colvin, J. S., McEwen, D. G., MacArthur, C. A., Coulier, F., et al. (1996). Receptor specificity of the fibroblast growth factor family. *The Journal of Biological Chemistry, 271*(25), 15292–15297.

Orr-Urtreger, A., Bedford, M. T., Burakova, T., Arman, E., Zimmer, Y., Yayon, A., et al. (1993). Developmental localization of the splicing alternatives of fibroblast growth factor receptor-2 (FGFR2). *Developmental Biology, 158*(2), 475–486.

Oulion, S., Bertrand, S., & Escriva, H. (2012). Evolution of the FGF gene family. *International Journal of Evolutionary Biology, 2012*, 298147.

Partanen, J., Makela, T. P., Alitalo, R., Lehvaslaiho, H., & Alitalo, K. (1990). Putative tyrosine kinases expressed in K-562 human leukemia cells. *Proceedings of the National Academy of Sciences of the United States of America, 87*(22), 8913–8917.

Partanen, J., Makela, T. P., Eerola, E., Korhonen, J., Hirvonen, H., Claesson-Welsh, L., et al. (1991). FGFR-4, a novel acidic fibroblast growth factor receptor with a distinct expression pattern. *The EMBO Journal, 10*(6), 1347–1354.

Partanen, J., Vainikka, S., Korhonen, J., Armstrong, E., & Alitalo, K. (1992). Diverse receptors for fibroblast growth factors. *Progress in Growth Factor Research, 4*(1), 69–83.

Pasquale, E. B., & Singer, S. J. (1989). Identification of a developmentally regulated protein-tyrosine kinase by using anti-phosphotyrosine antibodies to screen a cDNA expression library. *Proceedings of the National Academy of Sciences of the United States of America, 86*(14), 5449–5453.

Patstone, G., Pasquale, E. B., & Maher, P. A. (1993). Different members of the fibroblast growth factor receptor family are specific to distinct cell types in the developing chicken embryo. *Developmental Biology, 155*(1), 107–123.

Pejnovic, N. N., Pantic, J. M., Jovanovic, I. P., Radosavljevic, G. D., Milovanovic, M. Z., Nikolic, I. G., et al. (2013). Galectin-3 deficiency accelerates high-fat diet-induced obesity and amplifies inflammation in adipose tissue and pancreatic islets. *Diabetes, 62*(6), 1932–1944.

Perone, M. J., Bertera, S., Shufesky, W. J., Divito, S. J., Montecalvo, A., Mathers, A. R., et al. (2009). Suppression of autoimmune diabetes by soluble galectin-1. *The Journal of Immunology, 182*(5), 2641–2653.

Plotnikov, A. N., Hubbard, S. R., Schlessinger, J., & Mohammadi, M. (2000). Crystal structures of two FGF-FGFR complexes reveal the determinants of ligand-receptor specificity. *Cell, 101*(4), 413–424.

Potthoff, M. J., Inagaki, T., Satapati, S., Ding, X., He, T., Goetz, R., et al. (2009). FGF21 induces PGC-1alpha and regulates carbohydrate and fatty acid metabolism during the adaptive starvation response. *Proceedings of the National Academy of Sciences of the United States of America, 106*(26), 10853–10858.

Powers, C. J., McLeskey, S. W., & Wellstein, A. (2000). Fibroblast growth factors, their receptors and signaling. *Endocrine-Related Cancer*, 7(3), 165–197.

Quarto, N., & Amalric, F. (1994). Heparan sulfate proteoglycans as transducers of FGF-2 signalling. *Journal of Cell Science, 107*(Pt 11), 3201–3212.

Radosavljevic, G., Volarevic, V., Jovanovic, I., Milovanovic, M., Pejnovic, N., Arsenijevic, N., et al. (2012). The roles of Galectin-3 in autoimmunity and tumor progression. *Immunologic Research, 52*(1–2), 100–110.

Rieckmann, T., Kotevic, I., & Trueb, B. (2008). The cell surface receptor FGFRL1 forms constitutive dimers that promote cell adhesion. *Experimental Cell Research, 314*(5), 1071–1081.

Riminucci, M., Collins, M. T., Fedarko, N. S., Cherman, N., Corsi, A., White, K. E., et al. (2003). FGF-23 in fibrous dysplasia of bone and its relationship to renal phosphate wasting. *The Journal of Clinical Investigation, 112*(5), 683–692.

Robinson, C. J., Harmer, N. J., Goodger, S. J., Blundell, T. L., & Gallagher, J. T. (2005). Cooperative dimerization of fibroblast growth factor 1 (FGF1) upon a single heparin saccharide may drive the formation of 2:2:1 FGF1.FGFR2c.heparin ternary complexes. *The Journal of Biological Chemistry*, *280*(51), 42274–42282.
Roghani, M., & Moscatelli, D. (1992). Basic fibroblast growth factor is internalized through both receptor-mediated and heparan sulfate-mediated mechanisms. *The Journal of Biological Chemistry*, *267*(31), 22156–22162.
Ruta, M., Burgess, W., Givol, D., Epstein, J., Neiger, N., Kaplow, J., et al. (1989). Receptor for acidic fibroblast growth factor is related to the tyrosine kinase encoded by the fms-like gene (FLG). *Proceedings of the National Academy of Sciences of the United States of America*, *86*(22), 8722–8726.
Saksida, T., Nikolic, I., Vujicic, M., Nilsson, U. J., Leffler, H., Lukic, M. L., et al. (2013). Galectin-3 deficiency protects pancreatic islet cells from cytokine-triggered apoptosis in vitro. *Journal of Cellular Physiology*, *228*(7), 1568–1576.
Sareneva, T., Pirhonen, J., Cantell, K., Kalkkinen, N., & Julkunen, I. (1994). Role of N-glycosylation in the synthesis, dimerization and secretion of human interferon-gamma. *The Biochemical Journal*, *303*(Pt 3), 831–840.
Schlessinger, J., Plotnikov, A. N., Ibrahimi, O. A., Eliseenkova, A. V., Yeh, B. K., Yayon, A., et al. (2000). Crystal structure of a ternary FGF-FGFR-heparin complex reveals a dual role for heparin in FGFR binding and dimerization. *Molecular Cell*, *6*(3), 743–750.
Scotet, E., & Houssaint, E. (1998). Exon III splicing switch of fibroblast growth factor (FGF) receptor-2 and -3 can be induced by FGF-1 or FGF-2. *Oncogene*, *17*(1), 67–76.
Shi, D. L., Launay, C., Fromentoux, V., Feige, J. J., & Boucaut, J. C. (1994). Expression of fibroblast growth factor receptor-2 splice variants is developmentally and tissue-specifically regulated in the amphibian embryo. *Developmental Biology*, *164*(1), 173–182.
Shimada, T., Kakitani, M., Yamazaki, Y., Hasegawa, H., Takeuchi, Y., Fujita, T., et al. (2004). Targeted ablation of Fgf23 demonstrates an essential physiological role of FGF23 in phosphate and vitamin D metabolism. *The Journal of Clinical Investigation*, *113*(4), 561–568.
Silva, P. N., Altamentova, S. M., Kilkenny, D. M., & Rocheleau, J. V. (2013). Fibroblast growth factor receptor like-1 (FGFRL1) interacts with SHP-1 phosphatase at insulin secretory granules and induces beta-cell ERK1/2 protein activation. *The Journal of Biological Chemistry*, *288*(24), 17859–17870.
Sleeman, M., Fraser, J., McDonald, M., Yuan, S., White, D., Grandison, P., et al. (2001). Identification of a new fibroblast growth factor receptor, FGFR5. *Gene*, *271*(2), 171–182.
So, W. Y., Cheng, Q., Chen, L., Evans-Molina, C., Xu, A., Lam, K. S., et al. (2013). High glucose represses beta-klotho expression and impairs fibroblast growth factor 21 action in mouse pancreatic islets: Involvement of peroxisome proliferator-activated receptor gamma signaling. *Diabetes*, *62*(11), 3751–3759.
So, W. Y., Cheng, Q., Xu, A., Lam, K. S., & Leung, P. S. (2015). Loss of fibroblast growth factor 21 action induces insulin resistance, pancreatic islet hyperplasia and dysfunction in mice. *Cell Death & Disease*, *6*, e1707.
Steinberg, F., Zhuang, L., Beyeler, M., Kalin, R. E., Mullis, P. E., Brandli, A. W., et al. (2010). The FGFRL1 receptor is shed from cell membranes, binds fibroblast growth factors (FGFs), and antagonizes FGF signaling in Xenopus embryos. *The Journal of Biological Chemistry*, *285*(3), 2193–2202.
Sun, X., Mariani, F. V., & Martin, G. R. (2002). Functions of FGF signalling from the apical ectodermal ridge in limb development. *Nature*, *418*(6897), 501–508.
Sun, M. Y., Yoo, E., Green, B. J., Altamentova, S. M., Kilkenny, D. M., & Rocheleau, J. V. (2012). Autofluorescence imaging of living pancreatic islets reveals fibroblast growth factor-21 (FGF21)-induced metabolism. *Biophysical Journal*, *103*(11), 2379–2388.

Suzuki, M., Uehara, Y., Motomura-Matsuzaka, K., Oki, J., Koyama, Y., Kimura, M., et al. (2008). betaKlotho is required for fibroblast growth factor (FGF) 21 signaling through FGF receptor (FGFR) 1c and FGFR3c. *Molecular Endocrinology*, *22*(4), 1006–1014.

Tomiyama, K., Maeda, R., Urakawa, I., Yamazaki, Y., Tanaka, T., Ito, S., et al. (2010). Relevant use of Klotho in FGF19 subfamily signaling system in vivo. *Proceedings of the National Academy of Sciences of the United States of America*, *107*(4), 1666–1671.

Trueb, B. (2011). Biology of FGFRL1, the fifth fibroblast growth factor receptor. *Cellular and Molecular Life Sciences*, *68*(6), 951–964.

Trueb, B., Zhuang, L., Taeschler, S., & Wiedemann, M. (2003). Characterization of FGFRL1, a novel fibroblast growth factor (FGF) receptor preferentially expressed in skeletal tissues. *The Journal of Biological Chemistry*, *278*(36), 33857–33865.

Turner, N., & Grose, R. (2010). Fibroblast growth factor signalling: From development to cancer. *Nature Reviews Cancer*, *10*(2), 116–129.

Uematsu, F., Kan, M., Wang, F., Jang, J. H., Luo, Y., & McKeehan, W. L. (2000). Ligand binding properties of binary complexes of heparin and immunoglobulin-like modules of FGF receptor 2. *Biochemical and Biophysical Research Communications*, *272*(3), 830–836.

Urakawa, I., Yamazaki, Y., Shimada, T., Iijima, K., Hasegawa, H., Okawa, K., et al. (2006). Klotho converts canonical FGF receptor into a specific receptor for FGF23. *Nature*, *444*(7120), 770–774.

von Heijne, G. (1986). A new method for predicting signal sequence cleavage sites. *Nucleic Acids Research*, *14*(11), 4683–4690.

Wang, C., Hoch, E. G., & Pitt, G. S. (2011). Identification of novel interaction sites that determine specificity between fibroblast growth factor homologous factors and voltage-gated sodium channels. *The Journal of Biological Chemistry*, *286*(27), 24253–24263.

Wang, F., Kan, M., McKeehan, K., Jang, J. H., Feng, S., & McKeehan, W. L. (1997). A homeo-interaction sequence in the ectodomain of the fibroblast growth factor receptor. *The Journal of Biological Chemistry*, *272*(38), 23887–23895.

Wente, W., Efanov, A. M., Brenner, M., Kharitonenkov, A., Koster, A., Sandusky, G. E., et al. (2006). Fibroblast growth factor-21 improves pancreatic beta-cell function and survival by activation of extracellular signal-regulated kinase 1/2 and Akt signaling pathways. *Diabetes*, *55*(9), 2470–2478.

Werner, S., Duan, D. S., de Vries, C., Peters, K. G., Johnson, D. E., & Williams, L. T. (1992). Differential splicing in the extracellular region of fibroblast growth factor receptor 1 generates receptor variants with different ligand-binding specificities. *Molecular and Cellular Biology*, *12*(1), 82–88.

Wiedemann, M., & Trueb, B. (2000). Characterization of a novel protein (FGFRL1) from human cartilage related to FGF receptors. *Genomics*, *69*(2), 275–279.

Wu, A. L., Kolumam, G., Stawicki, S., Chen, Y., Li, J., Zavala-Solorio, J., et al. (2011). Amelioration of type 2 diabetes by antibody-mediated activation of fibroblast growth factor receptor 1. *Science Translational Medicine*, *3*(113), 113ra126.

Xu, J., Lloyd, D. J., Hale, C., Stanislaus, S., Chen, M., Sivits, G., et al. (2009). Fibroblast growth factor 21 reverses hepatic steatosis, increases energy expenditure, and improves insulin sensitivity in diet-induced obese mice. *Diabetes*, *58*(1), 250–259.

Xu, J., Stanislaus, S., Chinookoswong, N., Lau, Y. Y., Hager, T., Patel, J., et al. (2009). Acute glucose-lowering and insulin-sensitizing action of FGF21 in insulin-resistant mouse models—Association with liver and adipose tissue effects. *American Journal of Physiology. Endocrinology and Metabolism*, *297*(5), E1105–E1114.

Yahata, K., Mori, K., Arai, H., Koide, S., Ogawa, Y., Mukoyama, M., et al. (2000). Molecular cloning and expression of a novel klotho-related protein. *The Journal of Molecular Medicine (Berlin)*, *78*(7), 389–394.

Yamaguchi, F., Saya, H., Bruner, J. M., & Morrison, R. S. (1994). Differential expression of two fibroblast growth factor-receptor genes is associated with malignant progression in

human astrocytomas. *Proceedings of the National Academy of Sciences of the United States of America, 91*(2), 484–488.
Yang, C., Jin, C., Li, X., Wang, F., McKeehan, W. L., & Luo, Y. (2012). Differential specificity of endocrine FGF19 and FGF21 to FGFR1 and FGFR4 in complex with KLB. *PLoS One*, 7(3), e33870.
Yazaki, N., Fujita, H., Ohta, M., Kawasaki, T., & Itoh, N. (1993). The structure and expression of the FGF receptor-1 mRNA isoforms in rat tissues. *Biochimica et Biophysica Acta, 1172*(1-2), 37–42.
Yeh, B. K., Igarashi, M., Eliseenkova, A. V., Plotnikov, A. N., Sher, I., Ron, D., et al. (2003). Structural basis by which alternative splicing confers specificity in fibroblast growth factor receptors. *Proceedings of the National Academy of Sciences of the United States of America, 100*(5), 2266–2271.
Yie, J., Hecht, R., Patel, J., Stevens, J., Wang, W., Hawkins, N., et al. (2009). FGF21 N- and C-termini play different roles in receptor interaction and activation. *FEBS Letters, 583*(1), 19–24.
Yie, J., Wang, W., Deng, L., Tam, L. T., Stevens, J., Chen, M. M., et al. (2012). Understanding the physical interactions in the FGF21/FGFR/beta-Klotho complex: Structural requirements and implications in FGF21 signaling. *Chemical Biology & Drug Design, 79*(4), 398–410.
Yoshida, T., Fujimori, T., & Nabeshima, Y. (2002). Mediation of unusually high concentrations of 1,25-dihydroxyvitamin D in homozygous klotho mutant mice by increased expression of renal 1alpha-hydroxylase gene. *Endocrinology, 143*(2), 683–689.
Yu, H., Xia, F., Lam, K. S., Wang, Y., Bao, Y., Zhang, J., et al. (2011). Circadian rhythm of circulating fibroblast growth factor 21 is related to diurnal changes in fatty acids in humans. *Clinical Chemistry, 57*(5), 691–700.
Yuan, Q., Jiang, Y., Zhao, X., Sato, T., Densmore, M., Schuler, C., et al. (2014). Increased osteopontin contributes to inhibition of bone mineralization in FGF23-deficient mice. *Journal of Bone and Mineral Research, 29*(3), 693–704.
Zhan, X., Bates, B., Hu, X. G., & Goldfarb, M. (1988). The human FGF-5 oncogene encodes a novel protein related to fibroblast growth factors. *Molecular and Cellular Biology, 8*(8), 3487–3495.
Zhang, X., Stappenbeck, T. S., White, A. C., Lavine, K. J., Gordon, J. I., & Ornitz, D. M. (2006). Reciprocal epithelial-mesenchymal FGF signaling is required for cecal development. *Development, 133*(1), 173–180.
Zhao, X. M., Frist, W. H., Yeoh, T. K., & Miller, G. G. (1994). Modification of alternative messenger RNA splicing of fibroblast growth factor receptors in human cardiac allografts during rejection. *The Journal of Clinical Investigation, 94*(3), 992–1003.
Zhen, Y., Sorensen, V., Jin, Y., Suo, Z., & Wiedlocha, A. (2007). Indirubin-3'-monoxime inhibits autophosphorylation of FGFR1 and stimulates ERK1/2 activity via p38 MAPK. *Oncogene, 26*(44), 6372–6385.
Zhuang, L., Karotki, A. V., Bruecker, P., & Trueb, B. (2009). Comparison of the receptor FGFRL1 from sea urchins and humans illustrates evolution of a zinc binding motif in the intracellular domain. *BMC Biochemistry, 10*, 33.
Zhuang, L., Pandey, A. V., Villiger, P. M., & Trueb, B. (2015). Cell-cell fusion induced by the Ig3 domain of receptor FGFRL1 in CHO cells. *Biochimica et Biophysica Acta, 1853*(10 Pt A), 2273–2285.
Zimmer, Y., Givol, D., & Yayon, A. (1993). Multiple structural elements determine ligand binding of fibroblast growth factor receptors. Evidence that both Ig domain 2 and 3 define receptor specificity. *The Journal of Biological Chemistry, 268*(11), 7899–7903.

CHAPTER THREE

Klotho-Dependent Cellular Transport Regulation

M. Sopjani[1], M. Dërmaku-Sopjani[1]
University of Prishtina, Prishtinë, Republic of Kosova
[1]Corresponding authors: e-mail address: msopjani@yahoo.com; miribanedermaku@yahoo.com

Contents

Abstract

Klotho is a transmembrane protein that in humans is encoded by the *hKL* gene. This protein is known to have aging suppressor effects and is predominantly expressed in the distal convoluted tubule of the kidney, parathyroid glands, and choroid plexus of the brain. The Klotho protein exists in both full-length membrane form and a soluble secreted form, which exerts numerous distinct functions. The extracellular domain of Klotho can be enzymatically cleaved off and released into the systemic circulation where it functions as β-glucuronidase and a hormone. Soluble Klotho is a multifunction protein present in the biological fluids including blood, urine, and cerebrospinal fluid of mammals. Klotho deficiency leads to multiple organ failure accompanied by early appearance of multiple age-related disorders and early death, whereas overexpression of Klotho results in the opposite effects. Klotho, an enzyme and hormone, has been reported to participate in the regulation of cellular transport processes across the plasma membrane either indirectly through inhibiting calcitriol ($1,25(OH)_2D_3$) formation or other mechanism, or by directly affecting transporter proteins, including ion channels, cellular carriers, and Na^+/K^+-ATPase. Accordingly, Klotho protein serves as a powerful regulator of cellular transport across the plasma membrane. Importantly, Klotho-dependent cellular transport regulation implies stimulatory or inhibitory effects.

Vitamins and Hormones, Volume 101
ISSN 0083-6729
http://dx.doi.org/10.1016/bs.vh.2016.02.003

Klotho has been shown to play a key role in the regulation of multiple calcium and potassium ion channels, and various cellular carriers including the Na^+-coupled cotransporters such as NaPi-IIa, NaPi-IIb, EAAT3, and EAAT4, CreaT1 as well as Na^+/K^+-ATPase. These regulations are parts of the antiaging function of Klotho, which will be discussing throughout this chapter. Clearly, further experimental efforts are required to investigate the effect of Klotho on other transport proteins and underlying molecular mechanisms by which Klotho exerts its effect.

1. INTRODUCTION

1.1 Gene Structure and Klotho Protein Types

The gene encoding the antiaging protein of Klotho was initially cloned from a transgenic mouse model (Kuro-o et al., 1997) that manifested multiple age-related disorders. This gene was named *Klotho* (*Kl*) based on the Greek mythology goddess of Fate who spins the thread of life. One year later it has been isolated cDNA clones for human Klotho (Matsumura et al., 1998). The human *Klotho* gene contains five exons residing on chromosome 13q12 within a region longer than 50 kb of genomic DNA (Matsumura et al., 1998). Sequence analysis of the cDNAs further revealed that two different *Klotho* gene transcripts encoding membrane or secreted protein exist (Kuro-o et al., 1997; Matsumura et al., 1998; Nabeshima, 2006). As genomic analysis revealed, these transcripts have been reported to arise from a single *Kl* gene through alternative RNA splicing (Matsumura et al., 1998). In humans one transcript encodes a putative type I, 1014 amino acid (1012 amino acids in mice (Kuro-o et al., 1997)) single-pass transmembrane protein (130 kDa protein) with an N-terminal signal sequence (SS), a large extracellular domain (with two internal repeats) of 952 amino acids, a 21-amino acid-long transmembrane domain, and a short 11 amino acids intracellular carboxyl terminus. The expression of a single 5.2-kb transcript has been demonstrated, using Northern blot analysis, in kidney, placenta, small intestine, and prostate. The other Klotho transcript has a 50-bp insertion encoding a truncated putative secreted protein of 549 amino acids, with a molecular weight of approximately 65–70 kDa (Matsumura et al., 1998; Tohyama et al., 2004), which lacks the intracellular domain, the transmembrane, and the second internal repeat of extracellular domain. Importantly, the expression of secreted form of Klotho predominates over the membrane form in all tissues examined, i.e., kidney, prostate, brain, hippocampus, placenta, and small intestine (Kuro, 2012; Kuro-o, 2001; Matsumura et al., 1998; Tohyama et al., 2004; Fig. 1).

Fig. 1 Schematic representation of domain structure of the Klotho (Huang & Moe, 2011; Imura et al., 2007; Kuro-o, 2010). Human Klotho protein is composed of 1014 (1012 in mice) amino acids in length. At its N-terminus it possesses a putative signal sequence (SS), followed by two internal repeats (Kl1 and Kl2) of the extracellular domain that shares sequence homology to the β-glucosidase. A linker (L) is positioned in between two internal repeats. The next segment is a putative transmembrane domain (TM) with a short cytoplasmic domain (CD) at the C-terminus (not drawn to scale). (See the color plate.)

Additionally, the extracellular domain of membrane form of Klotho can be released into the extracellular space, i.e., circulatory system, after losing its short transmembrane domain and accordingly slightly lowering its molecular weight (Kuro, 2012; Kuro-o, 2001). This extracellular domain of membrane form of Klotho protein (designated soluble Klotho) is composed of two internal repeats, named KL1 and KL2, each of them sharing an amino acid sequence homology with family 1 glycosidases (Cha et al., 2009; Chang et al., 2005; Dermaku-Sopjani, Kolgeci, Abazi, & Sopjani, 2013; Kuro-o et al., 1997). Each of the KL1 and KL2 internal repeats has about 450 amino acids long with a similarity of 21% to each other. In other words, Klotho protein exists in two different forms: a membranous and a soluble secreted form, both of them belonging to the β-glycosidase family, but mainly have distinct functions (Dermaku-Sopjani et al., 2013).

1.2 Ectodomain Shedding and Some Functions Mediated by Klotho

The antiaging protein Klotho is a mammalian antisenescence protein mainly expressed in distal convoluted tubules (DCTs) of the kidney, parathyroid glands cells, and in the epithelial cells of the choroid plexus. As it is indicated above, the large extracellular domain of full-length membrane Klotho is cleaved off by the two membrane-anchored proteases, ADAM10 and ADAM17 (Disintegrin and Metalloproteinase *Domain*-containing Protein 10 and 17), and shed into the extracellular fluid and thereby elicits biological effects on target cells. The presence of the soluble Klotho protein in the biological fluids such as blood, urine, and cerebrospinal fluid (Dermaku-Sopjani et al., 2013; Kuro-o et al., 1997) suggests that it may function as an endocrine or paracrine humoral factor through not fully clarified mechanism. Klotho-deficient ($Klotho^{-/-}$) mice are characterized by accelerated aging phenotype and an imbalance of ion homeostasis, especially phosphate and calcium. Klotho may also act as an obligatory coreceptor for the fibroblast growth factor receptor (FGFR) to bind the circulating hormone fibroblast growth factor 23 (FGF23), and therefore Klotho may play an important role in mediating FGF23-dependent functions. The latter, FGF23, a phosphate- and calcium-regulating hormone, is a circulating factor secreted by osteocytes and its effects in the control of phosphate (Pi) and calcium (Ca) concentrations are mediated by Klotho (Dermaku-Sopjani et al., 2013, 2011; Kuro-o, 2006; Sopjani et al., 2011). Maintaining normal phosphate and calcium levels is crucial for normal physiological processes, whereas abnormalities in the mineral metabolism and homeostasis are considered to be important risk factors for the development of a broad spectrum of disorders, including vascular and renal calcification, osteoporosis, hyperphosphatemia, and other related disorders of mineral homeostasis dysregulation (Dermaku-Sopjani et al., 2013). However, still remains unclear whether circulating Klotho can similarly serve the role of coreceptor for FGFR as transmembrane Klotho.

Aging is known to be associated with increased incidence of many diseases, including hypertension (Satoh et al., 2012; Wang & Sun, 2009), diabetes (Mochly-Rosen, Das, & Grimes, 2012), neurological disorders (Dehvari et al., 2007), chronic kidney disease (CKD) with impairment of endothelial function (Maekawa et al., 2009; Rakugi et al., 2007), and enhanced risk of cardiovascular morbidity and mortality (Semba et al., 2011a). Importantly, *Klotho* gene expression is influenced by many physiological and pathological factors (Aizawa et al., 1998). Decreased plasma, urinary and renal Klotho levels and increased mortality risk are associated with normal aging (Semba et al., 2011a, 2011b), kidney disease (Satoh et al., 2012), animal models of diabetes, and hypertension diseases (Semba et al., 2011a). Conversely, overexpression of Klotho in animals leads to reversal of the aging process, hence to a substantial extension of lifespan, and additionally provides cardiovascular and renal protection (Wang & Sun, 2009), possibly by protecting tissues from oxidative stress damage and by inducing resistance to oxidative stress (Mitobe et al., 2005). Thus, Klotho could represent an important potential target for therapeutic intervention (Dermaku-Sopjani et al., 2013) of high-clinical relevance in age-related diseases.

1.3 Transport Proteins in the Different Types of Plasma Membrane Transport Systems

The transport of organic and inorganic molecules across plasma membranes is one of the main general roles of all cellular membranes. This function is essential for all forms of life, as it enables cells to maintain an off equilibrium condition, thus, preserving appropriate conditions of life in the internal environment in order to keep their high level of organization. Importantly, 15–39% of the human proteome is dedicated to be plasma membrane proteins (Ahram, Litou, Fang, & Al Tawallbeh, 2006). In the cell membrane transport processes are involved wide-varieties of membrane-bound and soluble proteins. Most of the biological molecules are unable to freely diffuse through the phospholipids bilayer of the cell membrane, which represent a barrier that blocks their free movements (Sepulveda, Pablo, Teulon, & Niemeyer, 2015; Wang, Yue, Sun, & Lin, 2010). Even though the cellular lipid bilayer is a formidable barrier for most charged and polar molecules, however, it allows the passage of hydrophobic organic compounds through passive diffusion. Transport proteins of the cell membrane mediate the selective passage of many different small molecules across the membrane, allowing the cell to pass into or out of the cell particular substances (molecules and ions) (Paula, Volkov, Van Hoek, Haines, & Deamer, 1996).

There are several different types of cell membrane transport systems, depending on the characteristics such as passive diffusion (including facilitated diffusion by channels), active transport (Na^+/K^+-ATPase), and secondary active transport (including various Na^+-coupled transporters). Facilitated dissuasion mediated by ion channels allows passage the ions across cell membrane down to their concentration gradient, while active and secondary active transport systems require a source of free energy to transport substances against a chemical gradient (e.g., import of nutrients). This free energy generated either by a coupled enzymatic reaction (mostly through hydrolyzing ATP)-active transport or is provided by the potential energy of an existing chemical gradient-secondary active transport (for further details, see the Transporter Classification Database [www.tcdb.org]) (Paula et al., 1996; Sepulveda et al., 2015). In this chapter, we will focus on the regulation of several reported ion channels, cellular carriers, and Na^+/K^+-ATPase by Klotho protein. Klotho-dependent cellular transport regulation participates in the aging suppressor activities of Klotho. The present chapter compiles the novel functions and mechanisms of the Klotho protein on various channels, carriers, and Na^+/K^+-ATPase. Examples are provided of how Klotho-sensitive cellular transports could affect longevity. The reader is kindly recommended to consult excellent earlier reviews on similar topics (Dermaku-Sopjani, Abazi, Faggio, Kolgeci, & Sopjani, 2014; Dermaku-Sopjani et al., 2013; Sopjani, Rinnerthaler, Almilaji, Ahmeti, & Dermaku-Sopjani, 2014; Sopjani, Rinnerthaler, Kruja, & Dermaku-Sopjani, 2015).

2. KLOTHO IN THE REGULATION OF CELLULAR TRANSPORT

Several reports indicated Klotho participation in the regulation of numerous transporting molecules. Even though those effects are reported, however, the detail mechanisms for most of them are unclear. This merits to be further studied in order to better understand the role of Klotho protein and its potential use in treating several conditions/diseases related to aging.

2.1 Klotho in the Regulation of Ion Channels

The first evidence that Klotho plays a fundamental role in the regulation of ion channels was reported for the case of TRPV5 (transient receptor potential vanilloid 5) (Chang et al., 2005), a calcium-permeable channel, mainly expressed in renal epithelial cells. This channel is located at the apical surface (lumen side) of cells of the renal DCT cells, where it plays an essential role in the reabsorption of Ca^{2+} (Dermaku-Sopjani et al., 2013). Of note, maintaining calcium within a narrow physiological range is critical for the body to properly function. Many physiological vital processes in the body are dependent on the precise maintenance of the calcium balance. Along with body stores, both kidney and intestine plays a critical role in regulating serum levels of calcium. This depends on gastrointestinal absorption (Lu, Boros, Chang, Bindels, & Hoenderop, 2008) and renal reabsorption (Boros, Bindels, & Hoenderop, 2009) of calcium. Calcium homeostasis is tightly under hormonal regulation (Boros et al., 2009; Lu et al., 2008). In this regard, the TRPV5 and TRPV6 channels are the most Ca^{2+}-selective members of the TRP superfamily and are commonly considered as the primary gatekeepers of Ca^{2+} influx across renal (TRPV5) and intestinal (TRPV6) epithelia (Imura et al., 2007). Klotho-sensitive stimulation of TRPV5 involves the β-glucuronidase of the Klotho by hydrolyzing extracellular N-linked oligosaccharides residues on the TRPV5 channel (Chang et al., 2005). This cause entrapping the channel in the cell membrane, thus, resulting in a prolonged presence of TRPV5 at the cell membrane, thereby increasing the membrane calcium permeability in kidney. In addition, a most recent paper demonstrated that experimental colitis induced in mice leads to a decreased expression of Klotho. Mice with colitis have also reduced renal TRPV5 channels associated with an enhanced renal calcium excretion (Radhakrishnan et al., 2013). However, numerous cell membrane proteins participate in renal calcium reabsorption. Another study (Oz et al., 2007) indicated that Klotho has a potential role in the regulation of the Na^+/Ca^+ exchanger (NCX1), an ubiquitously expressed membrane protein essential in calcium balance for many cells, and in this line important for the renal calcium reabsorption as well (Boros et al., 2009; Hoenderop, Nilius, & Bindels, 2002). Later on, a most recent paper demonstrated stimulation of Na^+/Ca^{2+} exchange in megakaryocytes of the Klotho-deficient mice (Schmid et al., 2015). Since plasma levels of the active form of vitamin D, $1,25(OH)_2D_3$, are excessive in Klotho-deficient mice and $1,25(OH)_2D_3$ is inhibited by FGF23, a function requiring presence of Klotho (Dermaku-Sopjani et al., 2013), these effects are due to absence of Klotho through its role in vitamin D. Importantly, $1,25(OH)_2D_3$ is an important regulator of cytosolic Ca^{2+}-concentration ($[Ca^{2+}]_i$) in a variety of cell types (Boros et al., 2009; Frick, Asplin, Favus, et al., 2013a; Frick, Asplin, Krieger, et al., 2013b; Torres et al., 2009, 2007), including upregulating the Na^+/Ca^{2+}-exchanger, which could, depending on plasma membrane potential and cytosolic sodium concentration, either increase or decrease $[Ca^{2+}]_i$.

Another previous study demonstrated Klotho involvement in platelets activation via calcium channels regulation (Borst et al., 2014). The deficiency of Klotho is followed by inhibition of platelets activation, a function at least in part due to blunted store-operated Ca^{2+} entry (SOCE) (Borst et al., 2014). Platelets are activated by enhanced $[Ca^{2+}]_i$ concentration following SOCE accomplished by calcium-release-activated calcium channel moiety Orai1, a pore forming ion channel unit, and its regulator STIM1. In other words, deficiency of Klotho downregulates platelet Ca^{2+} signaling and activation, an effect indicated at least partially to be due to $1,25(OH)_2D_3$-dependent inhibition of NF-κB activity and Orai1/STIM1 expression in megakaryocytes.

The renal Ca^{2+} reabsorption is also facilitated by the epithelial Ca^{2+} channel TRPV6. In addition to TRPV5 explained earlier, a stimulatory and similarly effect of Klotho has also been demonstrated for the renal TRPV6, but no effect has been observed for TRPV4 and TRPM6 (Lu et al., 2008), two other types of Ca^{2+}-permeable, nonselective cation channels—TRP cation channels—also found in renal tubules that are significant for ion transport in the kidney, including their role in the regulation of systemic osmotic pressure.

Both TRPV5 and TRPV6 are also present in the inner ear in mice (Takumida, Ishibashi, Hamamoto, Hirakawa, & Anniko, 2009). It is known that the Klotho protein content decreases with increased age in animals.

This, along with other effects, is accompanied by the inhibition of TRPV5 and TRPV6 that leads to an imbalance in Ca^{2+} levels in the inner ear, to a dysregulation of sensory transduction, and to cell damage, thus, causing hearing loss and/or vestibular disorders (Takumida et al., 2009).

Noteworthy, the exact sugar substrate and underlying mechanism for the regulation of the TRPV5 ion channel by Klotho are reported later (Cha, Ortega, et al., 2008). Based on that, the extracellular domain of the Klotho protein increases the surface abundance of TRPV5 through mediating removal of α2,6-, but not α2,3-linked sialic acids from the channel protein glycan chains (Cha, Ortega, et al., 2008). This leads to exposing underlying disaccharide *N*-acetyllactosamine (LacNAc), which is a high-affinity ubiquitous ligand for galectin-1, a family of galactoside-binding animal lectins. Binding of galectin-1 to LacNAc at the extracellular surface of the TRPV5 channel results to an accumulation of a functional channel variant at the plasma membrane and thereby promotes renal calcium uptake (Cha, Ortega, et al., 2008). Biochemical assays in HEK293 (Human Embryonic kidney) cells that are expressing TRPV channels revealed that in addition to a Klotho-dependent stimulatory effect on TRPV5 via the TRPV5 N-glycan, there exist another mechanism via sialidase as well. By inhibition, lipid raft-mediated internalization of TRPV5 this enzyme stimulates the TRPV5. The sialidase-dependent channel upregulation is independent of the N-glycosylation state of channel (Leunissen et al., 2013). In this study is also reported that galectin-3, but not galectin-1, is expressed in the DCT of the kidney. Therefore, after galectin-3 treatment authors detected an increase in TRPV5-mediated Ca^{2+} influx.

Klotho apparently have a broader role in the regulation of ion transport in the kidney. Treatment with the Klotho protein is followed by an increase of the cell membrane abundance of the renal outer medullary potassium channel 1 (ROMK1 or $K_{ir}1.1$) (Cha et al., 2009). This potassium channel is an ATP-dependent cell membrane channel that transports potassium out of cells and plays an important role in potassium recycling in the renal thick ascending limb and therefore potassium homeostasis (Siraskar et al., 2013). Similar to TRPV5, circulating Klotho protein increases the amount of ROMK1 protein at the cell membrane by removing α2,6-specific sialic acids from the channel N-glycans as demonstrated in HEK cells-expressing ROMK1 channel. This exposes underlying disaccharide galactose-*N*-acetylglucosamine that is a ligand for binding to galectin-1 at the extracellular surface. Binding to galectin-1 prevents clathrin-mediated endocytosis of channel and results in the accumulation of functional ROMK1 on the cell

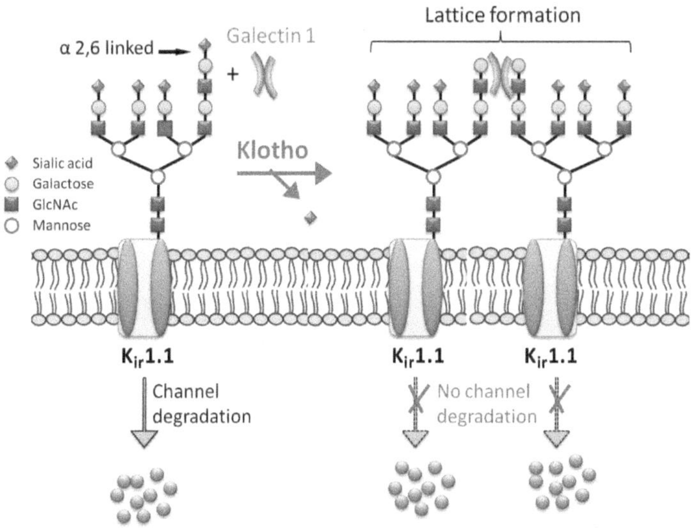

Fig. 2 Tentative model showing Klotho function in upregulating the plasma membrane abundance of ion channels $K_{ir}1.1$ (Cha et al., 2009; Sopjani et al., 2014). Soluble Klotho protein through its enzymatic activity removes the α2,6-specific sialic acids from $K_{ir}1.1$ (ROMK1) potassium channel N-glycans. By removing sialic acids, which are terminal acidic monosaccharides, it exposes underlying GlcNAc, a ligand-binding galectin-1 at the extracellular surface. This allows binding to galectin-1 and subsequent lattice formation at the extracellular surface between two adjacent ion channels that prevents endocytosis and diverting the channel to degradation. Entrapping of the functional channel on cell membrane leads to increasing $K_{ir}1.1$-mediated potassium transport on the plasma membrane. Similar model is reported for TRPV5 channel. See text for further details. (See the color plate.)

membrane (Fig. 2; Cha et al., 2009). Further, it has been shown that intravenous administration of soluble Klotho in rats increases both, the levels of Klotho and potassium in urine (Cha et al., 2009), thus, supporting the idea of a broader role of Klotho in the regulation of ion transport systems in the kidney.

Although, the underlying mechanism of Klotho-mediated regulation of TRPV5 (Cha, Ortega, et al., 2008) and ROMK1 (Cha et al., 2009) are similar, it should be noted that both channels appear to undergo endocytosis with subsequent degradation through different endocytic pathways, clathrin- and caveolae-mediated endocytosis, respectively. TRPV5 is internalized through a caveolae-mediated pathway that needs target proteins to signal to caveolin-1-containing lipid rafts (Cha, Wu, & Huang, 2008), whereas ROMK1 is internalized through clathrin-coated vesicles (Zeng et al., 2002), which needs target proteins to be recruited to clathrin-coated pits. Both ROMK1 and TRPV5 channels have only one N-glycan chain. However, each of those channels exist in a tetrameric form (Hebert, 1995; Hoenderop et al., 2002).

It should be mentioned that similar galectin-3-binding mechanisms have been reported to be essential for regulating the cell membrane abundance of cell surface glycoproteins including receptors for growth-stimulating and growth-inhibiting factors (Lau et al., 2007; Partridge et al., 2004), as well as glucose transporter 2 (Glut-2) (Ohtsubo et al., 2005). The latter is crucial for glucose-stimulated insulin secretion and lowering of blood glucose, thereby controlling blood glucose homeostasis in response to dietary intake. Accordingly, beta cell glucose transporter glycosylation mediates a relation between diet and insulin production and therefore suppresses the pathogenesis of type 2 diabetes. Thus, potentiating further beneficial role of Klotho, but additional research is needed to evaluate these functions.

Klotho plays an important role for endothelial function (Dermaku-Sopjani et al., 2013) and its deficiency deregulates endothelial integrity (Shimada et al., 2004), whereas Klotho expression recovery of endothelial integrity and function. Protein of Klotho has been shown to modify the activity of several other channels, including the TRP channel 1 (TRPC1). This is a nonspecific cation channel expressed in endothelial cells that can transport both sodium and calcium ions and participates in various functions, including endothelial Ca^{2+} entry mediated by vascular endothelial growth factor (VEGF) (Cheng, James, Foster, Hancox, & Bates, 2006), an important function to maintain endothelial integrity. The endothelial integrity is disrupted in *Klotho*$^{-/-}$ mice resulting from endothelial hyperpermeability induced by abnormal Ca^{2+} handling. Klotho is reported to bind directly to both VEGF receptor-2 (VEGFR-2) and TRPC-1 channel and strengthens the association between these two that results in a ternary complex formation of Klotho with VEGFR-2 and TRPC-1 and their cointernalization. This leads to a decrease of TRPC-1 protein abundance at the plasma membrane and consequent decrease of endothelial Ca^{2+} influx through store-operated cation channel TRPC1 in endothelial cells. The function of Klotho in VEGF-mediated Ca^{2+} signaling and thus in the maintenance of endothelial functions, is however, lost in the absence of Klotho (Kusaba et al., 2010). This eventually, leads to extensive vascular calcification that is associated with increased prevalence of cardiovascular and renal diseases. In this regard, Klotho may therefore prove to be an important therapeutic target in preventing cardiovascular and renal diseases. Interestingly, it has been shown (Kusaba et al., 2010) that Klotho is not directly binding to other TRPC channel subtypes.

Noteworthy, except its potential role in glucose-stimulated insulin secretion through Glut-2 regulation, Klotho has been identified to play an important role in the regulation of insulin secretion in mouse insulinoma β-cells (MIN6 β-cells) through regulation of TRPV2 (Lin & Sun, 2012). Klotho is also, but to a lesser extent, expressed in mouse pancreatic islets and in MIN6 β-cells. It was shown that overexpression of Klotho may increase glucose-induced insulin secretion through stimulating the cell membrane protein levels of TRPV2 (Lin & Sun, 2012), a calcium channel exclusively expressed in β-cells that determines the calcium influx in those cells (Hisanaga et al., 2009). The resulting increase in cytosolic-free Ca^{2+} levels stimulates insulin secretion (Lin & Sun, 2012). In contrast, silencing of Klotho attenuates these effects (Lin & Sun, 2012). The molecular mechanism of Klotho-dependent TRPV2 upregulation is not reported yet, however, having in mind that, similar to TRPV5, TRPV2 does have a potential N-glycosylation site (Hisanaga et al., 2009), the underlying mechanism could be similar as for TRPV5. Nonetheless, further studies are required to evaluate the hypothesis that Klotho is modulating N-linked glycan or sialic acid on the TRPV2 channel in MIN6 cells.

The expression level of the calcium-permeable nonselective TRPC6 channel is increased in hypertrophic hearts induced by various stressors and its inhibition is reported to protect against cardiac hypertrophy in mice (Koitabashi et al., 2010). Therefore, it is considered that TRPC6 may represent a significant modulator of cardiac hypertrophy and may be used as a potential target for therapy; however, the potential involvement of Klotho in TRPC6 regulation was not known. It has been demonstrated that Klotho exert some aging suppressor effects, while aging is also associated with an enhanced risk of heart failure. In this regard, a recent study demonstrated that soluble Klotho, independent of FGF23 and phosphate metabolism, inhibits TRPC6 in hearts and protects the heart against stress caused by pathological hypertrophy and remodeling (Xie et al., 2012). As confirmed in experiments using cardiomyocytes, cardioprotection function of Klotho in normal mice is mediated by inhibition of TRPC6 channels in the heart through blocking PI3K (phosphoinositide-3-kinase)-sensitive exocytosis of TRPC6 channels. Moreover, heart-specific overexpression of this channel in mice cause spontaneous cardiac hypertrophy and remodeling, whereas deletion of the gene encoding TRPC6 prevents stress-caused exaggerated cardiac remodeling in Klotho-deficient mice. These findings have shown

TRPC6 to play a key role in heart failure and Klotho as a powerful regulator of this channel. This opens new avenues for treatment of the heart failure.

This Klotho-dependent function may similarly have multiple other implications, especially in CKD, a relatively extensive human disease that is accompanied as high as 90%, with cardiac dysfunction. In addition, since TRPC6 channels have broader expression in tissues (Dietrich & Gudermann, 2011; Song et al., 2013), e.g., kidney and lung, the potential protective function of Klotho through TRPC6 channels regulation may also play important in other tissue/organ pathologies. Other members of the TRPC family (including TRPC1, 3, 4, and 5) are also expressed in the heart (Wu, Eder, Chang, & Molkentin, 2010). Increased expression of these calcium channels is also linked with cardiac hypertrophy. Thus, these channels may also represent an interesting potential target to be regulated by Klotho.

In the heart, Klotho insufficiency leads to cardiac arrhythmia and sudden cardiac death (Takeshita et al., 2004). Except its role in the regulation of calcium-permeable channels, cardioprotective effects of Klotho are mediated by the regulation of other channels such is the human ether-a-go-go (hERG) potassium channel, which is a key channel for completing cardiac repolarization (Vincent, 1998), essential for many congenital long QT syndrome-associated loss of faction mutations in hERG and accounts for disturbed excitation following cardiac hypertrophy (Wang et al., 2004). In a very recent paper it was shown that in *Xenopus* oocytes that are expressing the voltage-gated K^+ channel hERG Klotho protein significantly increases the hERG protein abundance in cell membrane. The same effect was reported when oocytes expressing this channel were treated with the recombinant Klotho protein (Munoz et al., 2013), an effect apparently requiring glucuronidase activity. The effect of Klotho on hERG channel protein abundance and subsequently the channel activity obviously influences the cardiac action potential, accelerating the repolarization phase of the contracted ventricular myocytes and is shortening the cardiac action potential duration, respectively. Furthermore, hERG channels are inhibited in cardiac hypertrophy (Hu, Yan, Lin, Liu, & Li, 2011); therefore, it might be of interest whether treatment with Klotho would reverse the inhibition of channel activity in cardiac hypertrophy.

Other potassium channels expressed in heart and stimulated by Klotho protein is the heterotetrameric K^+ channel KCNQ1/KCNE1 (Alesutan et al., 2011; Almilaji, Pakladok, et al., 2014; Wilmes et al., 2012) as well as voltage-gated potassium channel $Kv_{1.3}$ (KCNA3) (Almilaji, Honisch, et al., 2014). Importantly, the KCNQ1/KCNE1 complex is expressed in a wide variety of tissues, including the heart (Barhanin et al., 1996; Wilmes et al., 2012), trachea (Grahammer, Warth, Barhanin, Bleich, & Hug, 2001), skeletal muscle (Finsterer & Stollberger, 2004), gastric parietal cells (Grahammer, Herling, et al., 2001), intestinal epithelia (Grahammer, Herling, et al., 2001), stria vascularis of the inner ear (Wangemann, 2006), renal proximal tubule (Vallon et al., 2001), and hepatocytes (Lan, Wang, & Hill, 2006). Multiple abnormalities could be attributed to genetic defects of the K^+ channel subunits KCNE1 or KCNQ1, including cardiac arrhythmia and hearing loss (Barhanin et al., 1996; Neyroud et al., 1997). Additionally, KCNQ1 polymorphisms are associated with diabetes (Yasuda et al., 2008). Klotho-sensitive upregulation of this slowly activating K^+ channel, KCNQ1/KCNE1, is accomplished by enhancing the channel protein abundance on the cell membrane, an effect at least partially mediated by the β-glucuronidase activity of Klotho protein (Almilaji, Pakladok, et al., 2014). In the same way as the involvements of hERG channel regulation by Klotho, the Klotho-dependent KCNQ1/KCNE1 regulation is also in theory very important for normal cardiac cycle and accordingly loss of this function may contribute to the consequences of Klotho deficiency. Furthermore, Klotho-dependent potassium ion channel involves stimulation of the voltage-gated K^+ channel $Kv_{1.3}$ apparently by increasing channel protein abundance in the plasma membrane without significantly affecting other properties of the channel (Almilaji, Honisch, et al., 2014), an effect requiring β-glucuronidase activity of Klotho. The $Kv_{1.3}$ is a channel expressed in numerous tissues, including hearts, and involved in the regulation of a wide variety of cellular functions such as cellular excitability, cell proliferation, insulin sensitivity, cell volume regulation, immune response, apoptosis, and platelet function (Almilaji, Honisch, et al., 2014; Tyan et al., 2010). Klotho may therefore participate, directly or indirectly, in the regulation on those functions. This however requires further investigations and particularly underlying molecular mechanisms.

A summary of Klotho-sensitive cellular transport proteins are listed in Table 1.

2.2 Klotho in the Regulation of Carriers

Klotho may regulate a wide variety of cellular carriers, and through that it influences multiple cellular functions. An early study (Manabe et al., 2001) indicated that B-lymphoid lineage cells correlate with the pathophysiology of bone disorders through regulating osteoclastogenesis, including Klotho

Table 1 The Functions of Klotho in the Regulation of Various Transport Proteins and Implications

Cell Transporter	Klotho Function	Effective Through	Cell Type	Implications	References
Stimulatory function of Klotho					
TRPV5	Increase cell membrane protein abundance	Hydrolyzing extracellular N-linked oligosaccharides residues on channel (removal of α2,6-linked sialic acids from channel glycan chains)	Kidney cells	Maintenance of calcium balance	Chang et al. (2005) and Cha, Ortega, et al. (2008)
TRPV6	-//-	Similar mechanism	-//-	Maintenance of calcium balance	Lu et al. (2008)
TRPV2	-//-	Not known	MIN6 β-cells	Enhanced glucose-induced insulin secretion	Lin and Sun (2012)
Na^+/Ca^{2+} exchanger (NCX1)	Possibly stimulation	Klotho-dependent inhibition of $1,25(OH)_2D_3$	Mouse megakaryocytes and kidney membrane preparations	Megakaryocytic and renal calcium handling regulation	Oz et al. (2007) and Schmid et al. (2015)
ROMK1 ($K_{ir}1.1$)	Increase cell membrane protein abundance	Removing α2,6-specific sialic acids from channel N-glycans	HEK cells	Regulation of salt and potassium homeostasis	Cha et al. (2009)
hERG	-//-	At least partially mediated by the β-glucuronidase activity of Klotho	*Xenopus* oocytes	Recovering cardiac side effects	Munoz et al. (2013)
KCNQ1/KCNE1	-//-	-//-	-//-	Protecting from hearing loss and cardiac arrhythmia	Almilaji, Pakladok, et al. (2014)

Kv1.3	-//-	-//-	*Xenopus* oocytes, Jurkat and Jcam lymphoma cells	Kv1.3-dependent functions	Almilaji, Honisch, et al. (2014)
EAAT3 and EAAT4	-//-	-//-	*Xenopus* oocytes	Contributes to neurodegeneration during neuroinflammation	Almilaji et al. (2013)
CreaT (Slc6A8)	-//-	-//-	-//-	Proper function of brain and kidney	Almilaji, Sopjani, et al. (2014)
Na^+/K^+-ATPase	-//-	-//-	-//-	Maintains the chemical gradient for Na^+-coupled transport	Sopjani et al. (2011)
Inhibitory function of Klotho					
TRPC1	Reduction of cell membrane protein abundance	VEGFR-2-mediated channel internalization	Endothelial cells	Protecting role in cardiovascular diseases	Cheng et al. (2006) and Kusaba et al. (2010)
TRPC6	Channel inhibition	Blocking the PI3K-dependent channel exocytosis	Heart cells	Protecting effects in cardiomyopathies	Xie et al. (2012)
NaPi-IIa	Transporter downregulation	Deglycosylating extracellular sugar residues on carrier protein	*Xenopus laevis*, OK cells, BBM vesicle preparations	Negative phosphate balance	Dermaku-Sopjani et al. (2011) and Hu et al. (2010)
NaPi-IIb	Transporter downregulation	Not known	*Xenopus laevis*	Negative phosphate balance	Dermaku-Sopjani et al. (2011)

involvement in that process. Enhanced levels of Klotho and inhibition of insulin-like growth factor I (IGF-I) signaling has been reported in the bone marrow mesenchymal stem cells (BMMSCs) recipient mice (Yamaza et al., 2009). Therefore, revealing the erythropoietin receptor/Stat5 pathway participation to BMMSC-organized ectopic hematopoiesis, a proposed treatment paradigm of reversing age-related degeneration of multiple organs in mice (Yamaza et al., 2009).

There is emerging evidence of Klotho participation in the regulation of cellular transport through various ways such as by inhibiting the formation of calcitriol (1,25-$(OH)_2D_3$), an active form of Vitamin D, or by directly affecting channel and carrier proteins. The transmembrane form of Klotho regulates the activity of the Na^+-coupled excitatory amino acid transporters EAAT3 and EAAT4 (Almilaji et al., 2013). Both of these carriers are expressed in neurons and participate in the cellular uptake of glutamate and aspartate from the extracellular space and thus to the clearance of the excitatory transmitters from the extracellular space (Sopjani, Alesutan, Dermaku-Sopjani, et al., 2010). Experiments using EAAT3- and EAAT4-expressing *Xenopus* oocytes revealed higher carrier activity when coexpressed with Klotho. This effect has been demonstrated to be due to enhancing of maximal current and cell membrane protein abundance of both transporters, an effect similarly mimicked by treating those carriers-expressing *Xenopus* oocytes with recombinant human β-Klotho protein. However, the effect is abolished in the presence of β-glucuronidase inhibitor D-saccharic acid 1,4-lactone monohydrate (DSAL) (Almilaji et al., 2013). Therefore, clearly indicating that Klotho-sensitive EAAT3 and EAAT4 upregulation is at least partially effective through β-glucuronidase activity of Klotho.

In contrast to its effect on EAATs, Klotho inhibits a variety of Na^+-coupled transporters, such as the renal NaPi-IIa (*SLC34A1*) (Dermaku-Sopjani et al., 2011; Hu, Shi, et al., 2011) and intestinal NaPi-IIb (*SLC34A2*) Na^+-coupled phosphate transporters (Dermaku-Sopjani et al., 2011). The NaPi-IIb is predominantly expressed at the apical membrane of intestinal epithelial cells where it mediates phosphate uptake from the small intestine (Dermaku-Sopjani et al., 2013). On the other hand, the renal proximal tubule plays a critical role in solute reabsorption from the glomerular filtrate in the kidney, mediated by various apical membrane phosphate transporters NaPi-IIa, NaPi-IIc, and an isoform of NaPi-3 named Pit-2 (Dermaku-Sopjani et al., 2013, 2011; Segawa et al., 2007; Villa-Bellosta et al., 2009). However, the most important carrier accomplishing renal phosphate is NaPi-IIa (Foller et al., 2011; Kempe et al., 2010). Importantly, it has previously been reported that *Klotho*$^{-/-}$ mice exhibit enhanced NaPi-IIa and NaPi-IIc carriers activity and protein abundance levels compared to their wild-type littermates (Segawa et al., 2007), thus, suggesting that the hyperphosphatemia observed in *Klotho*$^{-/-}$ mice, at least partly, is mediated via regulation of NaPi cotransporters, and in this line potentiating Klotho as an important regulator in this process. Noteworthy, Klotho protein can act as an endocrine, paracrine, or autocrine substance (Dermaku-Sopjani et al., 2013). Hu et al. using both in vivo and in vitro models demonstrated that Klotho phosphaturic function could be either FGF23-dependent or independent from FGF23 through its enzymatic action on renal NaPi-IIa involving glucuronidase activity. This leads to the carrier activity inhibition, proteolytic degradation, perhaps through carrier protein internalization. Importantly, NaPi-IIa cotransporter contains N-glycosylation sites at extracellular loop 2 (Hayes et al., 1994). Hu et al. (2010) observed no sialidase effect on NaPi-IIa, therefore this carrier regulation by Klotho does not proceed through modulation of sialic acid residues. However, glucuronidase and Klotho inhibit NaPi-IIa (Dermaku-Sopjani et al., 2011; Hu et al., 2010), whereas Klotho inhibitor DSAL (a glucuronidase inhibitor) blocked the inhibition of NaPi-IIa (Hu et al., 2010). Thus suggesting that Klotho downregulates NaPi-IIa through its β-glucuronidase activity. Specifically, Klotho removes glucuronate (deglycosylate) that renders NaPi-IIa protein in the plasma membrane. This makes it more susceptible to still not defined proteases residing in the renal brush border membrane (BBM) (Hu et al., 2010). Either the sugar modification of carrier, the proteolysis, or both processes is considered to impair the ability of NaPi-IIa protein to remain on the cell membrane and exerting its function, and therefore results in lowering the renal phosphate reabsorption levels (Fig. 3). Except its effects on NaPi-IIa and NaPi-IIb transporters regulation, Klotho may also regulate other phosphate transporters in the kidney, such as NaPi-IIc (Segawa et al., 2007; Villa-Bellosta et al., 2009) and Pit-2 (Moe, 2009; Villa-Bellosta et al., 2009). Whether Klotho regulates other phosphate transporters via the same mechanism remains to be determined. Indeed, Klotho has an important role in the phosphate metabolism (Dermaku-Sopjani et al., 2013). NaPi-II transporters are known to be regulated by multiple regulators (Bhandaru et al., 2011; Dermaku-Sopjani et al., 2011; Foller et al., 2011; Kempe et al., 2010). The role of Klotho in increasing the renal phosphate excretion and reducing intestinal phosphate uptake through inhibiting the NaPi transporters represent one of the main mechanisms of

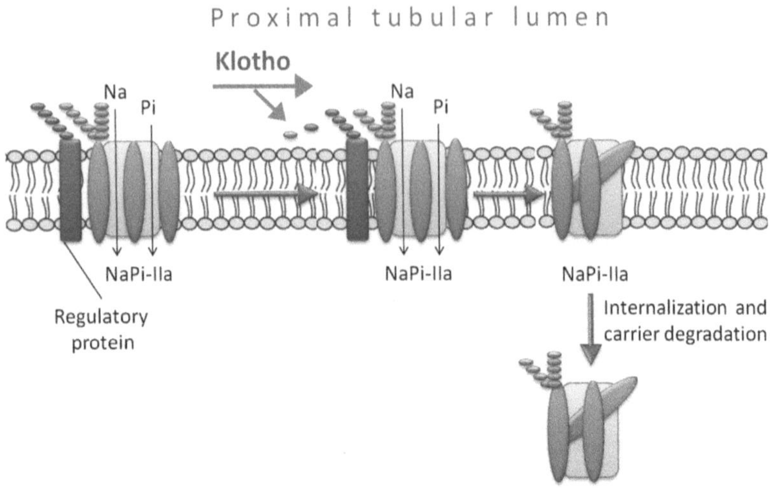

Fig. 3 A simple model showing the enzymatic activity of Klotho-dependent regulation of NaPi-IIa carrier protein abundance (Hu et al., 2010; Sopjani et al., 2014) at the plasma membrane. The circulating Klotho inhibits Na^+-coupled phosphate transporter NaPi-IIa activity in the apical membrane of the renal proximal tubules. Klotho-sensitive NaPi-IIa inhibition is mediated by the enzymatic activity of Klotho and is realized by deglycosylating extracellular sugar residues on the cell membrane proteins of NaPi-IIa and/or a regulatory protein. This makes NaPi-IIa protein more sensitive to proteases in brush border membrane (BBM) that results in proteolitic degradation in lysosomes. This leads to phosphaturia as a consequence of reduced phosphate transport mediated by NaPi-IIa due to carrier protein internalization from the apical membrane into intracellular pool. (See the color plate.)

Klotho functions in suppressing aging-induced processes. Another mechanism is attributed to an inhibitory effect of Klotho, as an obligate coreceptor to bind and activate FGFR on the cell membrane, on the calcitriol formation, either through inhibition of the renal enzyme 25-hydroxyvitamin D 1-α-hydroxylase (*Cyp27b1*) participating in the synthesis of 1,25(OH)2D, or stimulation of the enzyme 1,25-dihydroxyvitamin D 24-hydroxylase (*Cyp24a1*) catalyzing the catabolism of 1,25(OH)2D into inactive calcitroic acid in renal tubules (Perwad, Zhang, Tenenhouse, & Portale, 2007). Inhibition of the phosphate transporters decreases the Na^+ consignment for the Na^+/K^+-ATPase and thus energy consumption.

Among the cellular carriers expressed in both, brain and kidney cells, widely expressed is the Na^+,Cl^--coupled creatine transporter CreaT (*Slc6A8*). This carrier is very important for proper function of the brain, as genetic defects of CreaT may cause mental retardation with seizures (Baroncelli et al., 2014). A most recent study demonstrated Klotho participation in the regulation of CreaT (Almilaji, Sopjani, et al., 2014). As determined by utilizing dual electrode voltage clamp and using *Xenopus* oocytes, Klotho-dependent upregulation of CreaT enhances the maximal transport rate of the carrier presumably by stabilizing the CreaT protein in the plasma membrane, an effect requiring the presence of the β-glucuronidase activity of Klotho.

Klotho may have even a broader function in the regulation of cellular transport in the kidney. Intriguing, in an attempt to validate the role of *N*-acetylcysteine (NAC) on renal function, a potential role of antisenescence protein Klotho has been recently suggested in the regulation of the Na-K-Cl cotransporter (NKCC2) and urea transporter A1 (UT-A1), as well as aquaporin 2 (AQP2) (Shimizu et al., 2013), as indicated in experiments using kidney of rats.

2.3 Klotho in the Regulation of Na^+/K^+-ATPase

The Na^+,K^+-ATPase actively pumps Na^+ out of cells, while pumping K^+ into cells therefore known as the Na^+–K^+ pump. This pump maintains the electrochemical gradients of Na^+ and K^+ across the cell membrane of all living animal cells (Sopjani, Alesutan, Wilmes, et al., 2010). In an attempt to reveal fundamental roles of Klotho in the regulation of calcium metabolism, it has been reported that the molecular association of alpha-Klotho and Na^+,K^+-ATPase, and that the deficiency of *alpha-Klotho* gene impacts Na^+/K^+ ATPase activity at the cell membrane of parathyroid glands (Imura et al., 2007). Consistent with this finding another study further explored the effect of Klotho on the renal Na^+/K^+-ATPase activity and demonstrated enhanced Na^+/K^+-ATPase activity in *Xenopus* oocytes following coexpression with Klotho or treatment with the Klotho protein, an effect reversed by a β-glucuronidase inhibitor (Sopjani et al., 2011), thus indicating β-glucuronidase activity of Klotho in the regulation of Na^+/K^+-ATPase. Additionally, immunohistochemistry analysis and confocal microscopy experiments have shown that the abundance of Na^+/K^+-ATPase at least in collecting ducts is significantly lower in Klotho-hypomorphic (Klotho (hm)) mice as compared to their wild-type littermates (Klotho$^{+/+}$) (Sopjani et al., 2011). This decrease in Klotho (hm) mice may be a consequence of accelerated degradation of the Na^+/K^+-ATPase. Klotho has been demonstrated to be effective as an enzyme modifying the plasma membrane glycoproteins. Depending on the plasma membrane proteins, the effect can be either acceleration (Huang & Moe, 2011) or delay

(Cha, Ortega, et al., 2008) of the degradation of membrane proteins such as transport proteins. The β-glucuronidase activity of Klotho protein increases membrane abundance and activity of the Na^+/K^+-ATPase by stabilizing the pump in the cell membrane.

3. CONCLUSION AND FUTURE DIRECTIONS

The aging suppressor protein of Klotho, a hormone and enzyme, is a powerful regulator of a wide variety of cellular transport systems, including ion channels, carriers, and the Na^+/K^+-ATPase pump. Depending on the kind of cellular transport protein regulation, the effect can be either stimulatory or inhibitory. Klotho-dependent functions on transport proteins may be either direct or indirect. Stimulatory functions through enhancing transport protein plasma membrane abundance are fully or at least partially mediated by the β-glucuronidase activity of Klotho, whereas inhibitory functions are mediated by various mechanisms such as the interference with VEGFR-2-mediated channel internalization, blocking the PI3K-dependent channel exocytosis, or by still unknown mechanism or perhaps several unknown mechanisms. It should be emphasized that Klotho-mediated regulation of transport proteins is however an integral part of the antiaging effects of Klotho. Clearly, tremendous additional studies will be required, however, to find other potential cellular transport proteins regulated by Klotho, and to define the underlying mechanisms involved in particular transport proteins regulated by Klotho under physiological and pathophysiological conditions.

ACKNOWLEDGMENT

This work was supported by the University of Prishtina of the Republic of Kosova as well as by the MEST (Ministry of Education, Science, and Technology) small research grant project, no. 2-3214/2/2015 (M. Sopjani, 2015).

Conflict of Interest: None declared.

REFERENCES

Ahram, M., Litou, Z. I., Fang, R., & Al Tawallbeh, G. (2006). Estimation of membrane proteins in the human proteome. *In Silico Biology*, *6*, 379–386.

Aizawa, H., Saito, Y., Nakamura, T., Inoue, M., Imanari, T., Ohyama, Y., et al. (1998). Downregulation of the Klotho gene in the kidney under sustained circulatory stress in rats. *Biochemical and Biophysical Research Communications*, *249*, 865–871.

Alesutan, I., Foller, M., Sopjani, M., Dermaku-Sopjani, M., Zelenak, C., Frohlich, H., et al. (2011). Inhibition of the heterotetrameric K+ channel KCNQ1/KCNE1 by the AMP-activated protein kinase. *Molecular Membrane Biology*, *28*, 79–89.

Almilaji, A., Honisch, S., Liu, G., Elvira, B., Ajay, S. S., Hosseinzadeh, Z., et al. (2014). Regulation of the voltage gated K channel Kv1.3 by recombinant human klotho protein. *Kidney & Blood Pressure Research*, *39*, 609–622.

Almilaji, A., Munoz, C., Pakladok, T., Alesutan, I., Feger, M., Foller, M., et al. (2013). Klotho sensitivity of the neuronal excitatory amino acid transporters EAAT3 and EAAT4. *PLoS One*, *8*, e70988.

Almilaji, A., Pakladok, T., Munoz, C., Elvira, B., Sopjani, M., & Lang, F. (2014). Upregulation of KCNQ1/KCNE1 K channels by Klotho. *Channels (Austin, Tex.)*, *8*, 222–229.

Almilaji, A., Sopjani, M., Elvira, B., Borras, J., Dermaku-Sopjani, M., Munoz, C., et al. (2014). Upregulation of the creatine transporter Slc6A8 by Klotho. *Kidney & Blood Pressure Research*, *39*, 516–525.

Barhanin, J., Lesage, F., Guillemare, E., Fink, M., Lazdunski, M., & Romey, G. (1996). K(V) LQT1 and lsK (minK) proteins associate to form the I(Ks) cardiac potassium current. *Nature*, *384*, 78–80.

Baroncelli, L., Alessandri, M. G., Tola, J., Putignano, E., Migliore, M., Amendola, E., et al. (2014). A novel mouse model of creatine transporter deficiency. *F1000Research*, *3*, 228.

Bhandaru, M., Kempe, D. S., Rotte, A., Capuano, P., Pathare, G., Sopjani, M., et al. (2011). Decreased bone density and increased phosphaturia in gene-targeted mice lacking functional serum- and glucocorticoid-inducible kinase 3. *Kidney International*, *80*, 61–67.

Boros, S., Bindels, R. J., & Hoenderop, J. G. (2009). Active Ca(2+) reabsorption in the connecting tubule. *Pflügers Archiv*, *458*, 99–109.

Borst, O., Munzer, P., Schmid, E., Schmidt, E. M., Russo, A., Walker, B., et al. (2014). 1,25 (OH)2 vitamin D3-dependent inhibition of platelet Ca2+ signaling and thrombus formation in klotho-deficient mice. *The FASEB Journal*, *28*, 2108–2119.

Cha, S. K., Hu, M. C., Kurosu, H., Kuro-o, M., Moe, O., & Huang, C. L. (2009). Regulation of renal outer medullary potassium channel and renal K(+) excretion by Klotho. *Molecular Pharmacology*, *76*, 38–46.

Cha, S. K., Ortega, B., Kurosu, H., Rosenblatt, K. P., Kuro, O., & Huang, C. L. (2008). Removal of sialic acid involving Klotho causes cell-surface retention of TRPV5 channel via binding to galectin-1. *Proceedings of the National Academy of Sciences of the United States of America*, *105*, 9805–9810.

Cha, S. K., Wu, T., & Huang, C. L. (2008). Protein kinase C inhibits caveolae-mediated endocytosis of TRPV5. *American Journal of Physiology. Renal Physiology*, *294*, F1212–F1221.

Chang, Q., Hoefs, S., van der Kemp, A. W., Topala, C. N., Bindels, R. J., & Hoenderop, J. G. (2005). The beta-glucuronidase klotho hydrolyzes and activates the TRPV5 channel. *Science*, *310*, 490–493.

Cheng, H. W., James, A. F., Foster, R. R., Hancox, J. C., & Bates, D. O. (2006). VEGF activates receptor-operated cation channels in human microvascular endothelial cells. *Arteriosclerosis, Thrombosis, and Vascular Biology*, *26*, 1768–1776.

Dehvari, N., Cedazo-Minguez, A., Isacsson, O., Nilsson, T., Winblad, B., Karlstrom, H., et al. (2007). Presenilin dependence of phospholipase C and protein kinase C signaling. *Journal of Neurochemistry*, *102*, 848–857.

Dermaku-Sopjani, M., Abazi, S., Faggio, C., Kolgeci, J., & Sopjani, M. (2014). AMPK-sensitive cellular transport. *Journal of Biochemistry*, *155*, 147–158.

Dermaku-Sopjani, M., Kolgeci, S., Abazi, S., & Sopjani, M. (2013). Significance of the anti-aging protein Klotho. *Molecular Membrane Biology*, *30*(8), 369–385.

Dermaku-Sopjani, M., Sopjani, M., Saxena, A., Shojaiefard, M., Bogatikov, E., Alesutan, I., et al. (2011). Downregulation of NaPi-IIa and NaPi-IIb Na-coupled phosphate transporters by coexpression of Klotho. *Cellular Physiology and Biochemistry*, *28*, 251–258.

Dietrich, A., & Gudermann, T. (2011). TRP channels in the cardiopulmonary vasculature. *Advances in Experimental Medicine and Biology*, *704*, 781–810.

Finsterer, J., & Stollberger, C. (2004). Skeletal muscle involvement in congenital long QT syndrome. *Neurological Sciences*, *25*, 238–240.

Foller, M., Kempe, D. S., Boini, K. M., Pathare, G., Siraskar, B., Capuano, P., et al. (2011). PKB/SGK-resistant GSK3 enhances phosphaturia and calciuria. *Journal of the American Society of Nephrology, 22*, 873–880.

Frick, K. K., Asplin, J. R., Favus, M. J., Culbertson, C., Krieger, N. S., & Bushinsky, D. A. (2013a). Increased biological response to 1,25(OH)(2)D(3) in genetic hypercalciuric stone-forming rats. *American Journal of Physiology. Renal Physiology, 304*, F718–F726.

Frick, K. K., Asplin, J. R., Krieger, N. S., Culbertson, C. D., Asplin, D. M., & Bushinsky, D. A. (2013b). 1,25(OH)(2)D(3)-enhanced hypercalciuria in genetic hypercalciuric stone-forming rats fed a low-calcium diet. *American Journal of Physiology. Renal Physiology, 305*, F1132–F1138.

Grahammer, F., Herling, A. W., Lang, H. J., Schmitt-Graff, A., Wittekindt, O. H., Nitschke, R., et al. (2001). The cardiac K+ channel KCNQ1 is essential for gastric acid secretion. *Gastroenterology, 120*, 1363–1371.

Grahammer, F., Warth, R., Barhanin, J., Bleich, M., & Hug, M. J. (2001). The small conductance K+ channel, KCNQ1: Expression, function, and subunit composition in murine trachea. *The Journal of Biological Chemistry, 276*, 42268–42275.

Hayes, G., Busch, A., Lotscher, M., Waldegger, S., Lang, F., Verrey, F., et al. (1994). Role of N-linked glycosylation in rat renal Na/Pi-cotransport. *The Journal of Biological Chemistry, 269*, 24143–24149.

Hebert, S. C. (1995). An ATP-regulated, inwardly rectifying potassium channel from rat kidney (ROMK). *Kidney International, 48*, 1010–1016.

Hisanaga, E., Nagasawa, M., Ueki, K., Kulkarni, R. N., Mori, M., & Kojima, I. (2009). Regulation of calcium-permeable TRPV2 channel by insulin in pancreatic beta-cells. *Diabetes, 58*, 174–184.

Hoenderop, J. G., Nilius, B., & Bindels, R. J. (2002). Molecular mechanism of active Ca2+ reabsorption in the distal nephron. *Annual Review of Physiology, 64*, 529–549.

Hu, M. C., Shi, M., Zhang, J., Pastor, J., Nakatani, T., Lanske, B., et al. (2010). Klotho: A novel phosphaturic substance acting as an autocrine enzyme in the renal proximal tubule. *The FASEB Journal, 24*, 3438–3450.

Hu, M. C., Shi, M., Zhang, J., Quinones, H., Griffith, C., Kuro-o, M., et al. (2011). Klotho deficiency causes vascular calcification in chronic kidney disease. *Journal of the American Society of Nephrology, 22*, 124–136.

Hu, C., Yan, C., Lin, J., Liu, S., & Li, Y. (2011). Down-regulation of the human ether-a-go-go-related gene in rat cardiac hypertrophy. *The American Journal of the Medical Sciences, 341*, 119–125.

Huang, C. L., & Moe, O. W. (2011). Klotho: A novel regulator of calcium and phosphorus homeostasis. *Pflügers Archiv, 462*, 185–193.

Imura, A., Tsuji, Y., Murata, M., Maeda, R., Kubota, K., Iwano, A., et al. (2007). alpha-Klotho as a regulator of calcium homeostasis. *Science, 316*, 1615–1618.

Kempe, D. S., Dermaku-Sopjani, M., Frohlich, H., Sopjani, M., Umbach, A., Puchchakayala, G., et al. (2010). Rapamycin-induced phosphaturia. *Nephrology, Dialysis, Transplantation, 25*, 2938–2944.

Koitabashi, N., Aiba, T., Hesketh, G. G., Rowell, J., Zhang, M., Takimoto, E., et al. (2010). Cyclic GMP/PKG-dependent inhibition of TRPC6 channel activity and expression negatively regulates cardiomyocyte NFAT activation Novel mechanism of cardiac stress modulation by PDE5 inhibition. *Journal of Molecular and Cellular Cardiology, 48*, 713–724.

Kuro, O. (2012). Klotho in health and disease. *Current Opinion in Nephrology and Hypertension, 21*, 362–368.

Kuro-o, M. (2001). Disease model: Human aging. *Trends in Molecular Medicine, 7*, 179–181.

Kuro-o, M. (2006). Klotho as a regulator of fibroblast growth factor signaling and phosphate/calcium metabolism. *Current Opinion in Nephrology and Hypertension, 15*, 437–441.

Kuro-o, M. (2010). Klotho. *Pflügers Archiv, 459*, 333–343.

Kuro-o, M., Matsumura, Y., Aizawa, H., Kawaguchi, H., Suga, T., Utsugi, T., et al. (1997). Mutation of the mouse klotho gene leads to a syndrome resembling ageing. *Nature, 390*, 45–51.

Kusaba, T., Okigaki, M., Matui, A., Murakami, M., Ishikawa, K., Kimura, T., et al. (2010). Klotho is associated with VEGF receptor-2 and the transient receptor potential canonical-1 Ca2+ channel to maintain endothelial integrity. *Proceedings of the National Academy of Sciences of the United States of America, 107*, 19308–19313.

Lan, W. Z., Wang, P. Y., & Hill, C. E. (2006). Modulation of hepatocellular swelling-activated K+ currents by phosphoinositide pathway-dependent protein kinase C. *American Journal of Physiology. Cell Physiology, 291*, C93–C103.

Lau, K. S., Partridge, E. A., Grigorian, A., Silvescu, C. I., Reinhold, V. N., Demetriou, M., et al. (2007). Complex N-glycan number and degree of branching cooperate to regulate cell proliferation and differentiation. *Cell, 129*, 123–134.

Leunissen, E. H., Nair, A. V., Bull, C., Lefeber, D. J., van Delft, F. L., Bindels, R. J., et al. (2013). The epithelial calcium channel TRPV5 is regulated differentially by klotho and sialidase. *The Journal of Biological Chemistry, 288*, 29238–29246.

Lin, Y., & Sun, Z. (2012). Antiaging gene Klotho enhances glucose-induced insulin secretion by up-regulating plasma membrane levels of TRPV2 in MIN6 beta-cells. *Endocrinology, 153*, 3029–3039.

Lu, P., Boros, S., Chang, Q., Bindels, R. J., & Hoenderop, J. G. (2008). The beta-glucuronidase klotho exclusively activates the epithelial Ca2+ channels TRPV5 and TRPV6. *Nephrology, Dialysis, Transplantation, 23*, 3397–3402.

Maekawa, Y., Ishikawa, K., Yasuda, O., Oguro, R., Hanasaki, H., Kida, I., et al. (2009). Klotho suppresses TNF-alpha-induced expression of adhesion molecules in the endothelium and attenuates NF-kappaB activation. *Endocrine, 35*, 341–346.

Manabe, N., Kawaguchi, H., Chikuda, H., Miyaura, C., Inada, M., Nagai, R., et al. (2001). Connection between B lymphocyte and osteoclast differentiation pathways. *Journal of Immunology, 167*, 2625–2631.

Matsumura, Y., Aizawa, H., Shiraki-Iida, T., Nagai, R., Kuro-o, M., & Nabeshima, Y. (1998). Identification of the human klotho gene and its two transcripts encoding membrane and secreted klotho protein. *Biochemical and Biophysical Research Communications, 242*, 626–630.

Mitobe, M., Yoshida, T., Sugiura, H., Shirota, S., Tsuchiya, K., & Nihei, H. (2005). Oxidative stress decreases klotho expression in a mouse kidney cell line. *Nephron. Experimental Nephrology, 101*, e67–e74.

Mochly-Rosen, D., Das, K., & Grimes, K. V. (2012). Protein kinase C, an elusive therapeutic target? *Nature Reviews. Drug Discovery, 11*, 937–957.

Moe, O. W. (2009). PiT-2 coming out of the pits. *American Journal of Physiology. Renal Physiology, 296*, F689–F690.

Munoz, C., Pakladok, T., Almilaji, A., Elvira, B., Seebohm, G., Voelkl, J., et al. (2013). Klotho sensitivity of the hERG channel. *FEBS Letters, 587*, 1663–1668.

Nabeshima, Y. (2006). Toward a better understanding of Klotho. *Science of Aging Knowledge Environment, 2006*, e11.

Neyroud, N., Tesson, F., Denjoy, I., Leibovici, M., Donger, C., Barhanin, J., et al. (1997). A novel mutation in the potassium channel gene KVLQT1 causes the Jervell and Lange-Nielsen cardioauditory syndrome. *Nature Genetics, 15*, 186–189.

Ohtsubo, K., Takamatsu, S., Minowa, M. T., Yoshida, A., Takeuchi, M., & Marth, J. D. (2005). Dietary and genetic control of glucose transporter 2 glycosylation promotes insulin secretion in suppressing diabetes. *Cell, 123*, 1307–1321.

Oz, O. K., Hajibeigi, A., Howard, K., Cummins, C. L., van Abel, M., Bindels, R. J., et al. (2007). Aromatase deficiency causes altered expression of molecules critical for calcium reabsorption in the kidneys of female mice ⋆. *Journal of Bone and Mineral Research, 22*, 1893–1902.

Partridge, E. A., Le Roy, C., Di Guglielmo, G. M., Pawling, J., Cheung, P., Granovsky, M., et al. (2004). Regulation of cytokine receptors by Golgi N-glycan processing and endocytosis. *Science*, *306*, 120–124.

Paula, S., Volkov, A. G., Van Hoek, A. N., Haines, T. H., & Deamer, D. W. (1996). Permeation of protons, potassium ions, and small polar molecules through phospholipid bilayers as a function of membrane thickness. *Biophysical Journal*, *70*, 339–348.

Perwad, F., Zhang, M. Y., Tenenhouse, H. S., & Portale, A. A. (2007). Fibroblast growth factor 23 impairs phosphorus and vitamin D metabolism in vivo and suppresses 25-hydroxyvitamin D-1alpha-hydroxylase expression in vitro. *American Journal of Physiology. Renal Physiology*, *293*, F1577–F1583.

Radhakrishnan, V. M., Ramalingam, R., Larmonier, C. B., Thurston, R. D., Laubitz, D., Midura-Kiela, M. T., et al. (2013). Post-translational loss of renal TRPV5 calcium channel expression, Ca(2+) wasting, and bone loss in experimental colitis. *Gastroenterology*, *145*, 613–624.

Rakugi, H., Matsukawa, N., Ishikawa, K., Yang, J., Imai, M., Ikushima, M., et al. (2007). Anti-oxidative effect of Klotho on endothelial cells through cAMP activation. *Endocrine*, *31*, 82–87.

Satoh, M., Nagasu, H., Morita, Y., Yamaguchi, T. P., Kanwar, Y. S., & Kashihara, N. (2012). Klotho protects against mouse renal fibrosis by inhibiting Wnt signaling. *American Journal of Physiology. Renal Physiology*, *303*, F1641–F1651.

Schmid, E., Yan, J., Hosseinzadeh, Z., Almilaji, A., Shumilina, E., Kuro, O., et al. (2015). Up-regulation of megakaryocytic Na/Ca exchange in klotho-deficient mice. *Biochemical and Biophysical Research Communication*, *460*, 177–182.

Segawa, H., Yamanaka, S., Ohno, Y., Onitsuka, A., Shiozawa, K., Aranami, F., et al. (2007). Correlation between hyperphosphatemia and type II Na-Pi cotransporter activity in klotho mice. *American Journal of Physiology. Renal Physiology*, *292*, F769–F779.

Semba, R. D., Cappola, A. R., Sun, K., Bandinelli, S., Dalal, M., Crasto, C., et al. (2011a). Plasma klotho and cardiovascular disease in adults. *Journal of the American Geriatrics Society*, *59*, 1596–1601.

Semba, R. D., Cappola, A. R., Sun, K., Bandinelli, S., Dalal, M., Crasto, C., et al. (2011b). Plasma klotho and mortality risk in older community-dwelling adults. *The Journals of Gerontology. Series A, Biological Sciences and Medical Sciences*, *66*, 794–800.

Sepulveda, F. V., Pablo, C. L., Teulon, J., & Niemeyer, M. I. (2015). Molecular aspects of structure, gating, and physiology of pH-sensitive background K2P and Kir K+-transport channels. *Physiological Reviews*, *95*, 179–217.

Shimada, T., Takeshita, Y., Murohara, T., Sasaki, K., Egami, K., Shintani, S., et al. (2004). Angiogenesis and vasculogenesis are impaired in the precocious-aging klotho mouse. *Circulation*, *110*, 1148–1155.

Shimizu, M. H., Volpini, R. A., de Braganca, A. C., Campos, R., Canale, D., Sanches, T. R., et al. (2013). N-acetylcysteine attenuates renal alterations induced by senescence in the rat. *Experimental Gerontology*, *48*, 298–303.

Siraskar, B., Huang, D. Y., Pakladok, T., Siraskar, G., Sopjani, M., Alesutan, I., et al. (2013). Downregulation of the renal outer medullary K(+) channel ROMK by the AMP-activated protein kinase. *Pflügers Archiv*, *465*, 233–245.

Song, J., Wang, Y., Li, X., Shen, Y., Yin, M., Guo, Y., et al. (2013). Critical role of TRPC6 channels in the development of human renal cell carcinoma. *Molecular Biology Reports*, *40*, 5115–5122.

Sopjani, M., Alesutan, I., Dermaku-Sopjani, M., Fraser, S., Kemp, B. E., Foller, M., et al. (2010a). Down-regulation of Na+-coupled glutamate transporter EAAT3 and EAAT4 by AMP-activated protein kinase. *Journal of Neurochemistry*, *113*, 1426–1435.

Sopjani, M., Alesutan, I., Dermaku-Sopjani, M., Gu, S., Zelenak, C., Munoz, C., et al. (2011). Regulation of the Na+/K+ ATPase by Klotho. *FEBS Letters*, *585*, 1759–1764.

Sopjani, M., Alesutan, I., Wilmes, J., Dermaku-Sopjani, M., Lam, R. S., Koutsouki, E., et al. (2010b). Stimulation of Na+/K+ ATPase activity and Na+coupled glucose transport by beta-catenin. *Biochemical and Biophysical Research Communications, 402*, 467–470.

Sopjani, M., Rinnerthaler, M., Almilaji, A., Ahmeti, S., & Dermaku-Sopjani, M. (2014). Regulation of cellular transport by klotho protein. *Current Protein & Peptide Science, 15*, 828–835.

Sopjani, M., Rinnerthaler, M., Kruja, J., & Dermaku-Sopjani, M. (2015). Intracellular signaling of the aging suppressor protein Klotho. *Current Molecular Medicine, 15*, 27–37.

Takeshita, K., Fujimori, T., Kurotaki, Y., Honjo, H., Tsujikawa, H., Yasui, K., et al. (2004). Sinoatrial node dysfunction and early unexpected death of mice with a defect of klotho gene expression. *Circulation, 109*, 1776–1782.

Takumida, M., Ishibashi, T., Hamamoto, T., Hirakawa, K., & Anniko, M. (2009). Age-dependent changes in the expression of klotho protein, TRPV5 and TRPV6 in mouse inner ear. *Acta Oto-Laryngologica, 129*, 1340–1350.

Tohyama, O., Imura, A., Iwano, A., Freund, J. N., Henrissat, B., Fujimori, T., et al. (2004). Klotho is a novel beta-glucuronidase capable of hydrolyzing steroid beta-glucuronides. *The Journal of Biological Chemistry, 279*, 9777–9784.

Torres, P. U., Prie, D., Beck, L., De Brauwere, D., Leroy, C., & Friedlander, G. (2009). Klotho gene, phosphocalcic metabolism, and survival in dialysis. *Journal of Renal Nutrition, 19*, 50–56.

Torres, P. U., Prie, D., Molina-Bletry, V., Beck, L., Silve, C., & Friedlander, G. (2007). Klotho: An antiaging protein involved in mineral and vitamin D metabolism. *Kidney International, 71*, 730–737.

Tyan, L., Sopjani, M., Dermaku-Sopjani, M., Schmid, E., Yang, W., Xuan, N. T., et al. (2010). Inhibition of voltage-gated K+ channels in dendritic cells by rapamycin. *American Journal of Physiology. Cell Physiology, 299*, C1379–C1385.

Vallon, V., Grahammer, F., Richter, K., Bleich, M., Lang, F., Barhanin, J., et al. (2001). Role of KCNE1-dependent K+ fluxes in mouse proximal tubule. *Journal of the American Society of Nephrology, 12*, 2003–2011.

Villa-Bellosta, R., Ravera, S., Sorribas, V., Stange, G., Levi, M., Murer, H., et al. (2009). The Na+-Pi cotransporter PiT-2 (SLC20A2) is expressed in the apical membrane of rat renal proximal tubules and regulated by dietary Pi. *American Journal of Physiology. Renal Physiology, 296*, F691–F699.

Vincent, G. M. (1998). The molecular genetics of the long QT syndrome: Genes causing fainting and sudden death. *Annual Review of Medicine, 49*, 263–274.

Wang, Y., & Sun, Z. (2009). Klotho gene delivery prevents the progression of spontaneous hypertension and renal damage. *Hypertension, 54*, 810–817.

Wang, J., Wang, H., Zhang, Y., Gao, H., Nattel, S., & Wang, Z. (2004). Impairment of HERG K(+) channel function by tumor necrosis factor-alpha: Role of reactive oxygen species as a mediator. *The Journal of Biological Chemistry, 279*, 13289–13292.

Wang, W. H., Yue, P., Sun, P., & Lin, D. H. (2010). Regulation and function of potassium channels in aldosterone-sensitive distal nephron. *Current Opinion in Nephrology and Hypertension, 19*, 463–470.

Wangemann, P. (2006). Supporting sensory transduction: Cochlear fluid homeostasis and the endocochlear potential. *The Journal of Physiology, 576*, 11–21.

Wilmes, J., Haddad-Tovolli, R., Alesutan, I., Munoz, C., Sopjani, M., Pelzl, L., et al. (2012). Regulation of KCNQ1/KCNE1 by beta-catenin. *Molecular Membrane Biology, 29*, 87–94.

Wu, X., Eder, P., Chang, B., & Molkentin, J. D. (2010). TRPC channels are necessary mediators of pathologic cardiac hypertrophy. *Proceedings of the National Academy of Sciences of the United States of America, 107*, 7000–7005.

Xie, J., Cha, S. K., An, S. W., Kuro, O., Birnbaumer, L., & Huang, C. L. (2012). Cardioprotection by Klotho through downregulation of TRPC6 channels in the mouse heart. *Nature Communications*, *3*, 1238.

Yamaza, T., Miura, Y., Akiyama, K., Bi, Y., Sonoyama, W., Gronthos, S., et al. (2009). Mesenchymal stem cell-mediated ectopic hematopoiesis alleviates aging-related phenotype in immunocompromised mice. *Blood*, *113*, 2595–2604.

Yasuda, K., Miyake, K., Horikawa, Y., Hara, K., Osawa, H., Furuta, H., et al. (2008). Variants in KCNQ1 are associated with susceptibility to type 2 diabetes mellitus. *Nature Genetics*, *40*, 1092–1097.

Zeng, W. Z., Babich, V., Ortega, B., Quigley, R., White, S. J., Welling, P. A., et al. (2002). Evidence for endocytosis of ROMK potassium channel via clathrin-coated vesicles. *American Journal of Physiology. Renal Physiology*, *283*, F630–F639.

CHAPTER FOUR

Klotho and the Growth Hormone/ Insulin-Like Growth Factor 1 Axis: Novel Insights into Complex Interactions

T. Rubinek*[,1], D. Modan-Moses[†,‡]

*Institute of Oncology, Tel Aviv Sourasky Medical Center, Tel Aviv, Israel
[†]The Edmond and Lily Safra Children's Hospital, Chaim Sheba Medical Center, Tel-Hashomer, Ramat-Gan, Israel
[‡]Tel Aviv University, Tel Aviv, Israel
[1]Corresponding author: e-mail address: rubinekt@gmail.com

Contents

Abstract

The growth hormone (GH)/insulin-like growth factor (IGF)-1 axis is pivotal for many metabolic functions, including proper development and growth of bones, skeletal muscles, and adipose tissue. Defects in the axis' activity during childhood result in growth abnormalities, while increased secretion of GH from the pituitary results in acromegaly. In order to keep narrow physiologic concentration, GH and IGF-1 secretion and activity are tightly regulated by hypothalamic, pituitary, endocrine, paracrine, and autocrine factors.

Vitamins and Hormones, Volume 101
ISSN 0083-6729
http://dx.doi.org/10.1016/bs.vh.2016.02.009

Klotho was first discovered as an aging-suppressor gene. Mice that do not express klotho die prematurely with multiple symptoms of aging, several of them are also characteristic of decreased GH/IGF-1 axis activity. Klotho is highly expressed in the brain, the kidney, and parathyroid and pituitary glands, but can also serve as a circulating hormone by its shedding, forming soluble klotho that can be detected in blood, cerebrospinal fluid, and urine. Several lines of evidence suggest an association between klotho levels and activity of the GH/IGF-1 axis: the GH-secreting cells in the anterior pituitary of klotho-deficient mice are hypotrophic; klotho levels are altered in subjects with pathologies of the GH/IGF-1 axis; and accumulating data indicate that klotho is a direct regulator of GH secretion. Thus, klotho seems to be a new player in the intricate regulation of the GH/IGF-1 axis.

1. THE PITUITARY GLAND AND THE GH/IGF-1 AXIS

The pituitary gland is located at the base of the brain. It is in the size of 5 mm and is comprised of the anterior, posterior, and vestigial lobes. The mature anterior pituitary lobe consists of five distinct hormone producing cells: corticotroph cells secreting proopiomelanocortin (POMC) peptides including adenocorticotropic hormone; somatotroph cells which secrete growth hormone (GH); thyrotroph cells, which secrete glycoprotein α-subunit (αGSU) and the specific β-subunit of thyroid-stimulating hormone; gonadotrophs which secrete the α- and β-subunits of follicle-stimulating hormone (FSH) and luteinizing hormone (LH); and lactotrophs which secrete prolactin (PRL) (reviewed in detail in Keiser & Ho, 2011). The pituitary blood supply is unique. Arterial blood is supplied by the superior hypophyseal arteries, which arise from the internal carotid arteries and branch toward the hypothalamus where they form a capillary network that is external to the brain–blood barrier. In addition, the hypophyseal portal system connects the hypothalamus with the anterior pituitary, facilitates transport and exchange of hormones from the hypothalamus to the pituitary and allows fast communication between both glands.

1.1 Regulation of GH Transcription and Secretion

The GH-secreting cells, the somatotrophs, are located mainly at the lateral wings of the anterior pituitary, and comprise 35–45% of pituitary cells. They express abundant GH, up to 15 mg per pituitary, which is mostly located in 250–700 μm secretory granules (Keiser & Ho, 2011).

The human GH is encoded by the *GH1* gene, located at the GH locus on chromosome 17 (Martial, Hallewell, Baxter, & Goodman, 1979), along with four other related genes (Miller & Eberhardt, 1983). The five genes share a remarkably high degree of sequence identity. While GH1 is expressed in the pituitary, the other family members are expressed in the placental tissue (Theill & Karin, 1993). The promoter of GH1 contains *cis* elements that control GH expression and the two major transcription factors are POU1F1 (Pit-1) that confers tissue specificity and restricts GH expression to the somatotrophs, and the ubiquitous Sp1 (Bodner & Karin, 1987; Theill & Karin, 1993).

GH, the gene product of GH1, is a 22-kDa (191-amino acid) protein (Lewis, Singh, Bonewald, Lewis, & Vanderlaan, 1979). Several forms of GH appear in the circulation and the three most common are a 22-kDa monomer, which accounts for 75% of secreted GH, a 20-kDa monomer, a splice variant with a higher stability, which accounts for 10% of pituitary GH and an acetylated 22 kDa form (Lewis et al., 1979).

Normal GH secretion is pulsatile, and two-thirds of its secretion occurs at night. GH secretion is high at midgestation and declines at the neonatal stage, childhood, and prepuberty, where it is 200–600 μg/day. Its secretion peaks during puberty to 2 mg/day, then declines back to levels as low as 20 μg/day at older age (Keiser & Ho, 2011).

Three tiers participate in the regulation of GH secretion from somatotrophs:

1. Hypothalamic regulation through secretion of hormones into the portal system.
2. Peripheral regulation through the arterial blood supply.
3. Autocrine and paracrine regulation of the anterior pituitary itself.

While the two most prominent players are the positive regulator GH-releasing hormone (GHRH) and the negative regulator somatostatin, additional players contribute to the fine-tuning of the process.

1.1.1 Hypothalamic Regulation of GH Secretion

Positive regulators: GHRH is a 44-amino acid peptide hormone produced in the arcuate nucleus of the hypothalamus (Mayo, 1992; Murray, Higham, & Clayton, 2015). It binds a G-protein-coupled receptor located in the somatotrophs and triggers cAMP production and an increase in Ca^{2+} cellular levels. cAMP production promotes the transcriptional activity of cAMP-responsive elements binding protein (CREB), which serves as a cofactor for POU1F1, thus inducing GH transcription (Keiser & Ho, 2011).

Ghrelin is a peptide produced in the hypothalamus, as well as in ghrelinergic cells in the gastrointestinal tract. Although data are somewhat conflicting, several studies showed that ghrelin administration provokes both GH secretion and food intake (Carreira, Crujeiras, Andrade, Monteiro, & Casanueva, 2013). Current data suggest that ghrelin GH-releasing activity is mediated mostly by the hypothalamus-derived peptide, which operates through GH-secretagogue receptors located on somatotrophs (Reichenbach, Steyn, Sleeman, & Andrews, 2012).

Negative regulator: Somatostatin is a peptide hormone widely distributed in mammals, which can act as a peripheral hormone, a hypothalamic-secreted factor, a paracrine factor, and even as an autocrine factor (Brazeau et al., 1973). The two different forms of mammalian somatostatin (SRIF-14 and SRIF-28) are derived from a single precursor, preprosomatostatin I (PPSS-I) (Siler et al., 1973). Five somatostatin receptor subtypes (SSTR1-5) are recognized, where subtypes 2 and 4 are most common in somatotrophs (Miller et al., 1995; Yamada et al., 1992). Somatostatin binding to SSTR on the somatotrophs results in a suppression of cAMP accumulation, which leads to inhibition of GH secretion but does not affect GH transcription (Eigler & Ben-Shlomo, 2014).

A complex interplay exists between the positive and negative regulators of GH secretion. For example, while GHRH and ghrelin act on different receptors, which activate distinct signaling pathways, GHRH can also act as an allosteric coagonist for ghrelin receptor. GHRH and somatostatin can also interact with each other. Somatostatin-producing neurons project to GHRH neurons and inhibit GHRH release. Another layer of regulation is by GH itself, which exerts stimulatory influence on somatostatin release, thus decreasing GHRH release. The relative contributions of GHRH and somatostatin in regulating the pulsatile pattern of GH release depend on multiple factors, including age and sex.

1.1.2 Peripheral Regulation of GH Secretion

IGF-1 is the major growth factor secreted in response to GH. It is a highly conserved peptide, expressed in a wide range of tissues, with the highest secretion observed in the liver in response to GH (LeRoith & Yakar, 2007). Circulating IGF-1 exerts a long-loop negative feedback on GH transcription and secretion by acting at both the hypothalamus and pituitary level, resulting in 50–80% GH-secretion inhibition. In the somatotrophs, IGF-1 inhibits GH transcription by disrupting of the POU1F1/CBP complex (Romero et al., 2012). In the hypothalamus, IGF-1 induces somatostatin release, which in turn inhibits GH secretion from the somatotrophs (Berelowitz et al., 1981).

Additional peripheral factors that positively regulate GH secretion include sex-steroid hormones, which augment GH secretory bursts, thyroid and adrenal hormones, nutrients, and metabolites. On the other hand, aging, obesity, and lack of physical activity reduce the amplitude of pulsatile GH output (Keiser & Ho, 2011).

1.1.3 Autocrine and Paracrine Regulation of GH Secretion

GH synthesis and secretion are affected by paracrine factors secreted by neighboring cells, both hormone-secreting and nonsecreting cells like folliculo-satellite (FS) cells. These may include members of the fibroblast growth factor (FGF) family, ie, bFGF and aFGF (Atkin et al., 1993; Black, Logan, Davis, & Sheppard, 1990; Liu, Luo, Chen, & Shang, 2013; Shahmoon et al., 2014), as well as other growth and angiogenic factors, including epidermal growth factor, vascular endothelial growth factor, and transforming growth factor (TGF) β. GH secretion is also affected by hormones secreted from neighboring cells, including GnRH and TRH (Denef, 2008). Additionally, nitric oxide (NO) produced in neighboring cells may also affect GH secretion (Denef, 2008; Rubinek, Rubinfeld, Hadani, Barkai, & Shimon, 2005). Cell–cell contact mediated by adhesion molecules like N-cadherin and N-CAM was shown to stimulate GH secretion (Rubinek et al., 2003). Factors produced by somatotrophs, like gherlin, TGFα, and leptin, can also affect GH secretion (Denef, 2008).

1.2 The Activities of GH and IGF-1

GH operates through GH receptor (GHR) that triggers activation and phosphorylation of the JAK/STAT pathway. The GHR is abundantly expressed in liver cells but is also expressed in moderate levels in other tissues, including muscle and fat. Activation of the receptor is initiated by binding of GH to a dimerized GHR, leading to JAK2 tyrosine kinase activation, which in turn phosphorylates and activates STAT1, 3, and 5. The activated STATs molecules translocate to the nucleus, where they elicit GH-specific target gene expression. Activation of JAK2 by GH can activate the PI3K pathway by phosphorylation of the insulin receptor substrate-1 (IRS-1) and the mitogen-activated protein kinase (MAPK) pathway by phosphorylation of SHC proteins.

GH stimulates linear growth in children both directly and indirectly, via IGF-1, by acting on the epiphyseal plates of long bones and regulating bone turnover and chondrocyte proliferation. GH also plays a role in the regulation of diverse metabolic and endocrine activities. It participates in the regulation of lipid and carbohydrate metabolism, particularly by liver and adipose tissue, partitioning nutrients into muscle and away from fat deposition. It is associated with insulin resistance, yet promotes insulin production. Together with IGF-1, it is necessary for normal reproductive function, particularly in females and has beneficial physiologic actions on the cardiovascular system and the epidermis. Additionally, it is a modulator of gut function, enhancing the uptake of macro- and micronutrients, and influences immune function, notably of B cells (Bartke, Sun, & Longo, 2013; Lichanska & Waters, 2008; Waters & Brooks, 2011).

While anabolic actions of GH are mediated through the IGF peptides, GH is capable of stimulating a variety of effects that are independent of IGF-1 activity. Indeed, the effects of GH and IGF are on occasion contradictory. Among the established independent effects of GH are gluconeogenesis and lipolysis, both insulin-antagonistic effects. On the other hand, IGF-1 exerts insulin-like effects of increased tissue glucose uptake, inhibition of gluconeogenesis, and enhanced adipogenesis (Kaplan & Cohen, 2007).

GH activity is regulated also by a high-affinity and a low-affinity, circulating GH-binding proteins (GHBPs). In humans, the high-affinity GHBP is generated by proteolytic cleavage of the extracellular domain of the GHR by metalloprotease A disintegrin and metalloprotease domain (ADAM) 17 in response to protein kinase C and binds about 50% of the 22-kDa GH (Keiser & Ho, 2011; Zhang, Jiang, Black, Baumann, & Frank, 2000). The low-affinity GHBP binds preferentially the 20-kDa GH and is unrelated to the GHR. The GHBPs dampen the acute oscillations in serum GH levels associated with pulsatile pituitary GH secretion, but increase the plasma GH half-life due to the decreased renal GH clearance of bound GH. The high-affinity-binding protein also competes with GH for binding to surface GHRs and as such alters GH pharmacokinetics and distribution.

2. KLOTHO: HISTORY, STRUCTURE, FUNCTIONS

The α-klotho gene (*kl*, referred from here on as klotho) was discovered at 1997 as an aging-suppressor gene (Kuro-o et al., 1997). Mice that do not express klotho exhibit a syndrome resembling human premature aging that includes soft tissue calcification, arteriosclerosis, skin atrophy, osteoporosis, emphysema, and a significantly shorter lifespan of 3–4 months. On the other hand, overexpression of the klotho gene in mice extends life span and is associated with inhibition of the insulin and IGF-1 activity (Kurosu et al., 2005). The klotho gene is highly conserved in humans, mice, and rats, and is also found in Zebra fish and nematodes. Human klotho is a 1012-amino acid single-pass transmembrane protein. It is composed of short intracellular and transmembrane domains, and a 980-amino acid long extracellular domain. The extracellular domain of klotho is composed of two internal repeats, KL1 and KL2. While the main product of the klotho gene is the full-length protein, a splice variant that translates only the KL1 domain, as a soluble, secreted protein, was also described (Matsumura et al., 1998; Shiraki-Iida et al., 1998). KL1 and KL2 share amino acid sequence homology with β-glucosidase but lack glutamic acid residues considered to be essential for enzymatic activity (Ligumsky et al., 2015). However, glucoronidase and sialidase activities were attributed to klotho (Cha et al., 2008; Chang et al., 2005; Tohyama et al., 2004).

Klotho is expressed in a wide array of tissues with highest expression in the distal convoluted tubules of the kidney, the choroid plexus in the brain, the pituitary, and the parathyroid gland (Kuro-o et al., 1997). The extracellular domain of klotho can also be cleaved and shed to act as a circulating hormone. The main proteases that participate in klotho's shedding are ADAM 10, ADAM 17, and γ-secretases (Chen et al., 2014; Chen, Podvin, Gillespie, Leeman, & Abraham, 2007). The activity of these sheddases yield secreted klotho that is composed of the entire extracellular domain, containing both KL1 and KL2, termed soluble klotho (sKlotho). sKlotho can be detected in blood, cerebrospinal fluid, and urine. Smaller cleaved products of KL1 and KL2 can also be detected as separate polypeptides, yet their abundance in the circulation was not elucidated.

Major physiologic activities attributed to klotho include regulation of phosphate and calcium homeostasis, aging, adipogenesis, glucose metabolism, antioxidant activity, and tumor suppressor activities (Xu & Sun, 2015). These activities are mediated through diverse mechanisms, including the following:

1. Regulation of FGF23 signaling. Klotho is a mandatory cofactor for the interaction of the endocrine FGF23 with FGF receptors (FGFR) 1, 3, or 4 (Kurosu et al., 2006; Urakawa et al., 2006). Indeed, genetically engineered mice lacking klotho share similar phenotype with mice lacking FGF23 (Shimada et al., 2004). Klotho mediates this effect by directly interacting with both the factor (FGF23) and its cognate receptor. This can only be mediated by the full-length membrane-bound protein.
2. Inhibition of insulin and IGF-1 signaling pathways (Abramovitz et al., 2011; Kurosu et al., 2005; Wolf et al., 2008). Inhibition of these pathways has been demonstrated in mice overexpressing klotho and may be

associated with the ability of klotho to extend life (Kurosu et al., 2005). Recent data also indicated klotho as a regulator of insulin secretion in response to glucose levels in isolated Langerhans islets and in MIN6 cells (Lin & Sun, 2012). Thus, klotho improved β cell function and attenuated the development of type 2 diabetes mellitus (T2DM) (Lin & Sun, 2015). This activity can be mediated by the full-length membrane-bound protein or the cleavage product sKlotho of KL1 (Ligumsky et al., 2015). Inhibition of the insulin and IGF-1 pathways was also noted in various cancer cells (Abramovitz et al., 2011; Li et al., 2014; Wolf et al., 2008). While klotho can directly bind the IGF-1R, it probably does not alter ligand binding and the exact mechanism of inhibition remains to be elucidated (Kurosu et al., 2005; Ligumsky et al., 2015; Wolf et al., 2008).

3. Activation of the transient receptor potential cation channel subfamily V (TRPV) 5 and 6 (Cha et al., 2008; Chang et al., 2005; Lu, Boros, Chang, Bindels, & Hoenderop, 2008). Klotho, either directly or via other proteins, promotes hydrolysis of α-2,6-linked sialic acids from the N-glycosylation sites of the TRPV proteins, which in turn leads to their recruitment to the plasma membrane. This activity has been associated with the ability of klotho to promote calcium shuttling in the kidney, as well as in the intestine (Wolf, An, Nie, Bal, & Huang, 2014).
4. Downregulation of Na-coupled phosphate transporters (Na-Pi) (Hu et al., 2010). Overexpression of klotho reduces plasma phosphate concentrations, and these are reduced in *kl/kl* mice. Na-Pi IIa and Na-Pi IIb transporters participate in phosphate reabsorption in the kidney and intestine, respectively, thus participate in regulating plasma phosphate concentrations. It was shown that klotho downregulates the expression and activity of both transporters (Dermaku-Sopjani et al., 2011).
5. Regulation of paracrine FGF signaling and binding. Klotho regulates bFGF signaling in various malignancies, yet its role in the physiologic activity of klotho remains to be elucidated (Abramovitz et al., 2011; Kurosu et al., 2006; Shahmoon et al., 2014; Wolf et al., 2008).
6. Inhibition of Wnt signaling. This activity was noted in the stem cell niche in the skin, as well as in several cancer types (Camilli et al., 2011; Lee et al., 2010; Liu et al., 2007).
7. Inhibition of TGFβ signaling. This activity was observed in the kidney and in lung cancer cells (Doi et al., 2011).

No "klotho receptor" has been identified and the common link between these diverse activities remains to be deciphered.

3. KLOTHO AS A CIRCULATING HORMONE

3.1 Available Tests

The research of klotho blood levels was limited until reliable methods to detect sKlotho were developed. While Western blotting and home-made ELISAs were used earlier, in recent years commercially available ELISA kits revolutionized the field and numerous studies have been published regarding klotho levels in various physiologic and pathologic states (Yamazaki et al., 2010).

Several vendors provide commercial kits and a comparison between the different kits was conducted recently. The authors concluded that the kit developed by Nabeshima's group is the most reliable one (Heijboer et al., 2013).

3.2 Factor Affecting Circulating Klotho Levels

Klotho can be detected in the blood, as well as in other body fluids in pictogram per milliliter concentrations. Klotho levels are higher in childhood and lower among elderly (Yamazaki et al., 2010). Lower levels are associated with shorter lifespan, frailty in elderly, chronic kidney disease (CKD), and aortic calcification (Kitagawa et al., 2013; Rotondi et al., 2015; Semba et al., 2011a, 2011b).

The major source of circulating klotho is the cleaved product of membrane-bound full-length klotho (Lindberg et al., 2014). A splice variant comprises of mostly KL1 (sKL1), which may be secreted directly from the cells to the circulation, has also been described, but little is known regarding its expression pattern or its circulating levels. To our knowledge, the majority of circulating klotho is derived from the kidneys. Thus, deletion of klotho in mice throughout the nephron (Six2-KL2/2) resulted in 80% reduction of serum levels compared with wild-type littermates (Lindberg et al., 2014). Studies in humans with kidney diseases support these observations and noted reduced klotho blood levels in patients with CKDs (Lee et al., 2014; Rotondi et al., 2015; Scholze et al., 2014; Wu et al., 2014). Klotho circulating levels can be regulated by its expression levels in the kidneys, as well as by its cleavage. Not much is known regarding factors regulating klotho expression. Epigenetic mechanisms, and in particular CpG hypermethylation of klotho's promoter, reduce klotho expression in aging and cancer (King, Rosene, & Abraham, 2012; Lee et al., 2010; Rubinek et al., 2011). Hypermethylation may also play a role in regulating klotho expression in the kidneys. Thus, fibrosis of the kidney in mice is associated with hypermethylation of the klotho gene promoter and decreased klotho expression (Sun, Chang, & Wu, 2012). In patients with CKD, renal klotho promoter methylation level positively correlated with the severity of the disease and both klotho mRNA and protein were reduced (Chen, Zhang, et al., 2013). Klotho levels in the kidneys are also regulated by Ca^{2+} levels. Thus, low Ca^{2+} levels induce klotho expression, whereas elevated Ca^{2+} levels decrease its expression (Imura et al., 2007). Ex vivo experiments demonstrated that exposure of the kidney and the parathyroid gland to high Ca^{2+} levels reduced both klotho's expression and secretion.

3.3 Regulation of Klotho Shedding

Klotho's cleavage from the kidneys can be induced by insulin. Thus, treatment of the kidney-derived COS-7 cells with insulin induced klotho cleavage by phosphoinositide 3 kinase (PI3K)-dependent activation of the metalloproteases ADAM 10 and ADAM 17 (Chen et al., 2007). Another study determined the specific site for the protease activity in klotho polypeptide and further demonstrated that klotho shedding is mediated primarily by ADAM 17 (Chen et al., 2014). As discussed earlier, ADAM17 also cleaves the transmembrane region of the GHR, yielding a soluble GHR which functions as a GHBP (Zhang et al., 2000). Thus klotho circulating levels and GH activity are regulated by a common mechanism.

4. KLOTHO AND THE PITUITARY: LESSONS FROM MICE AND HUMANS

The first clue for the role klotho plays in growth was observed in the phenotype of klotho hypomorphic mice (*kl/kl*) (Kuro-o et al., 1997). These mice are significantly smaller and show reduced mass of most tissues affected by the GH/IGF-1 axis (ie, bone, muscle, and adipose tissue). Moreover, they represent a phenotype resembling growth hormone deficiency (GHD), including growth retardation, reduced bone mass, reduced thymus weight, and reduced skin thickness. Indeed, a closer look at the pituitary of these mice revealed several abnormalities. Their somatotrophs are smaller and the number of secretory granules in these cells is significantly reduced compared to control mice. LH- and FSH-producing cells also seemed slightly atrophic (Kuro-o et al., 1997). While no changes in GH levels were observed in these mice (Utsugi et al., 2000), it is important to note that direct

GH measurement in mice or humans, is not accurate and is subjected to various biases (Steyn et al., 2011). Therefore, additional studies are required in order to reflect GH levels in these mice.

Further studies in mice and humans indicated high expression of klotho in the anterior pituitary. Thus, high mRNA levels of klotho are observed in mice pituitary (Kuro-o et al., 1997). In humans, both klotho mRNA and protein were detected (Neidert et al., 2013; Shahmoon et al., 2014), in the normal pituitary as well as in GH and PRL secreting pituitary adenomas and in nonfunctioning adenomas.

5. KLOTHO REGULATES GH SECRETION

The abnormalities of somatotrophs histology of *kl/kl* mice, together with the distinct growth-retarded phenotype of these mice, prompted our group to study the role klotho plays in GH secretion. The direct effect of klotho on GH secretion was examined in three different settings: rat pituitary adenomacarcinoma GH3 cell line, cultured rat pituitaries, and cultured human GH-secreting adenomas (Shahmoon et al., 2014). In all three models klotho treatment increased GH secretion by about twofold. This increase was achieved at relatively low concentrations (0.1 ng/mL) and at 24–48 h. Importantly, the majority of acromegaly cultures responded to klotho, and only two of seven samples were insensitive to it (Fig. 1). These results were strengthened in an in vivo model, in which prolonged treatment of mice with intraperitoneal sKlotho injections increased mRNA levels of IGF-1 and IGF-1-binding protein-3 mRNA in the liver, reflecting elevated serum GH levels. Our observations were strongly supported by a study aiming to reveal the effect of klotho treatment on *kl/kl* mice. In this study, treatment with sKlotho extended the life span of *kl/kl* mice by 17% and also ameliorated premature aging characteristics, including growth retardation, premature thymus involution, and vascular calcification. Importantly, treatment significantly increased expression of the GH target gene, acid labile subunit in the liver (Chen, Kuro, et al., 2013). Taken together, these data indicate klotho as a positive direct regulator of GH secretion from somatotrophs.

5.1 Mechanisms of Klotho Activity on GH Secretion

The IGF-1 pathway is an important negative regulator of GH secretion and klotho is a potent inhibitor of the IGF-1 pathway (Keiser & Ho, 2011; Kurosu et al., 2005). Indeed, in accordance with this activity, klotho partially

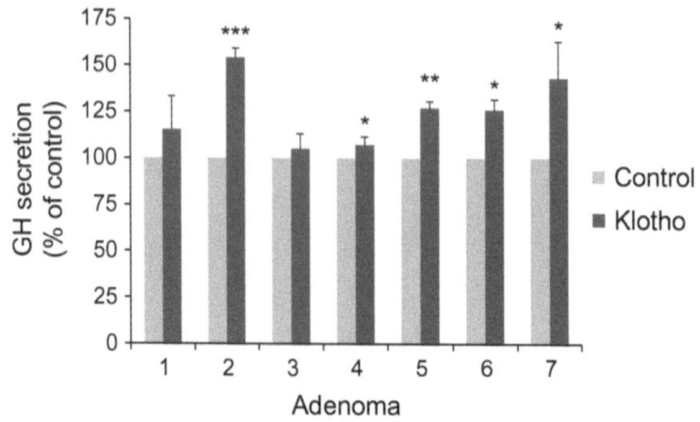

Fig. 1 Klotho increase GH secretion from cultured pituitary adenomas. Seven human GH adenomas were harvested and cultured. Cultures were treated with klotho (0.1 ng/mL) or control vehicle and 24 h later aliquots were removed for GH measurements. Each bar represents GH secretion relative to control treated cells. $^{*}p < 0.05$, $^{**}p < 0.005$, treatment vs vehicle-treated control. *Shahmoon, S., Rubinfeld, H., Wolf, I., Cohen, Z. R., Hadani, M., Shimon, I., & Rubinek, T. (2014). The aging suppressor klotho: A potential regulator of growth hormone secretion.* American Journal of Physiology. Endocrinology and Metabolism, 307(3), *17, copyright (2014).* (See the color plate.)

abrogated the inhibitory effect of IGF-1 on GH secretion in GH-secreting adenoma cultures, leading to increased GH secretion (Shahmoon et al., 2014) (Fig. 2). Thus, at least part of the direct role klotho exhibits on GH secretion from somatotrophs can be attributed to its inhibitory effect on the IGF-1 pathway.

Klotho is also a regulator of bFGF activity (Kurosu et al., 2006; Urakawa et al., 2006; Wolf et al., 2008). As the precise role bFGF in regulating GH secretion is not fully elucidated (Atkin et al., 1993; Black et al., 1990; Liu et al., 2013), we first evaluated the effect of bFGF on GH secretion from normal rat pituitary cultures and from human GH-secreting adenoma cultures, and discovered that bFGF induced a modest increase in GH secretion. However, cotreatment with bFGF and klotho caused a further increase in GH secretion (Shahmoon et al., 2014). The interaction of klotho and bFGF is cell type dependent. While klotho augments bFGF signaling in breast cancer cells (Wolf et al., 2008), it inhibits bFGF signaling in pancreatic cancer cells and HEK293 cells (Abramovitz et al., 2011; Kurosu & Kuro-o, 2008). In the rat pituitary adenoma GH3 cell line, treatment with either klotho or bFGF activated the ERK cascade, the downstream effector of FGF signaling (Shahmoon et al., 2014). In line with the increased GH secretion observed upon cotreatment with both agents, cotreatment further increased ERK activation. Moreover, specific inhibition of the ERK cascade abolished

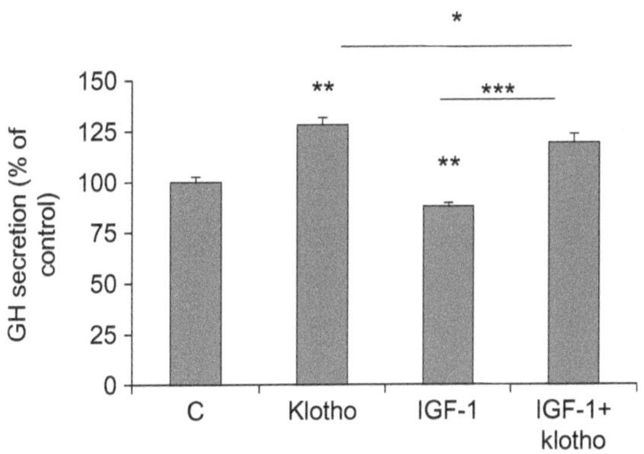

Fig. 2 Klotho partially overrides the inhibitory effect of IGF-1 on GH secretion. Human GH adenoma was harvested, cells were cultured and 48 h later cultures were treated with IGF-1 (50 ng/mL), klotho (0.1 ng/mL), their combination, or treated with control vehicle. Twenty-four hours later aliquots were removed for GH measurements. $^{*}p<0.05$, $^{**}p<0.005$, $^{***}p<0.0005$. *Shahmoon, S., Rubinfeld, H., Wolf, I., Cohen, Z. R., Hadani, M., Shimon, I., & Rubinek, T. (2014). The aging suppressor klotho: A potential regulator of growth hormone secretion.* American Journal of Physiology. Endocrinology and Metabolism, 307(3), *17, copyright (2014).* (See the color plate.)

the promoting effect of klotho and bFGF on GH secretion in normal rat pituitary cultures and in human GH-secreting adenoma cultures.

In addition to its direct effect on the somatotrophs, IGF-1 may also act through the hypothalamus where IGF-1 induces SRIF secretion leading to decreased GH secretion (Keiser & Ho, 2011). Possibly, klotho may inhibit the IGF-1 signal in the hypothalamus, thus leading to further increase in GH secretion.

6. KLOTHO AND THE GH/IGF-1 AXIS IN HUMANS: HEALTH AND DISEASE STATES

A limited number of studies assessed the interaction between klotho and the GH/IGF-1 axis in healthy subjects and in clinical settings of GH/IGF-1 axis alterations.

6.1 Healthy Subjects

In a study comprising 159 healthy children (aged 8.8 ± 3.5 years, 82 males), median klotho levels were 1945 pg/mL with a range of 372–5866 (Gkentzi, Efthymiadou, Kritikou, & Chrysis, 2014). Klotho levels were not associated with age. However, pubertal children had higher klotho levels compared to prepubertal children, and females had higher klotho levels compared to males. Klotho was positively and strongly correlated with IGF-1, and was also correlated with serum phosphate and 1,25-hydroxy vitamin D (Gkentzi et al., 2014). In accordance, we recently noted similar klotho levels (1645 ± 778 pg/mL) and a positive correlation between klotho and IGF-1 in 59 GH sufficient pediatric patients with short stature (Wolf, Shahmoon, et al., 2014). We are not aware of any studies that investigated the association between klotho and the GH/IGF-1 axis in healthy adults.

6.2 GH/IGF-1 Deficiency

6.2.1 GH Deficiency

GHD is a rare disorder with a prevalence of approximately 1 in 4000 during childhood (Murray, Dattani, & Clayton, 2015). Accurate diagnosis allows efficacious treatment with GH in these patients, while a false-positive diagnosis may result in many years of daily subcutaneous injections, significant superfluous expenditure, and unnecessary exposure to potential adverse effects (Murray, Dattani, et al., 2015). Currently, there is no "gold-standard" for the biochemical diagnosis of GHD in childhood; the diagnosis is a multifaceted process requiring comprehensive clinical and auxological assessment, combined with biochemical tests of the GH–IGF axis and imaging of the hypothalamo-pituitary area (Hokken-Koelega, 2011; Murray, Dattani, et al., 2015). As GH is secreted at intervals, random GH measurements are useless, and the secretion is assessed using stimulation tests with insulin, glucagon, clonidine, arginine, or L-dopa (GH Research Society, 2000). There is a considerable overlap in peak GH concentrations between children with normal GH secretion and GHD (GH Research Society, 2000). Only 15–36% of patients with childhood idiopathic GHD retest as GH deficient after discontinuation of GH treatment (Quigley et al., 2013; Tauber, Moulin, Pienkowski, Jouret, & Rochiccioli, 1997; Thomas et al., 2003; Toogood, Beardwell, & Shalet, 1994), and as many as 20% of children with normal growth and stature may test "deficient" if a single stimulation test is used. Consequently, the performance of two stimulation tests is recommended in order to establish the diagnosis of GHD (Bright, Julius, Lima, & Blethen, 1999). Several biomarkers (eg, IGF-1, IGFBP-3) have been tested to date, but all failed to accurately distinguish between GH sufficient and deficient children (Hokken-Koelega, 2011).

We hypothesized that klotho levels may be associated with pituitary function and GH secretion and (Wolf, Shahmoon, et al., 2014) investigated this question prospectively in children with short stature and growth

Table 1 Laboratory Measurements According to GH Status

	GH Sufficient (n=59)	Idiopathic GHD (n=29)	Organic GHD (n=11)	p Value
Peak GH (μg/L)	12.1±4.8	5.8±1.6	3.5±2.6	<0.001*
IGF-1 SDS	−1.4±1.5	−2.0±2.8	−3.7±2.5	0.004**
Klotho (pg/mL)	1645±778	1497±754	727±273	<0.001**
Glucose (mg/dL)	74±8	75±14	72±9	NS
Calcium (mg/dL)	10±0.3	9.9±0.5	10.5±0.5	0.02**
PO_4 (mg/dL)	5.0±0.6	4.9±0.6	4.1±0.3	0.03**

*GH sufficient compared to organic and idiopathic GHD.
**Organic GHD compared to both other groups.
GH, growth hormone; *GHD*, growth hormone deficiency.

impairment. Ninety-nine children and adolescents (aged 9.0±3.7 years, 49 male) undergoing GH stimulation tests for short stature (height-SDS-=−2.1±0.6) were included. Based on the results of GH stimulation tests, 59 were classified as GH sufficient and 40 were classified as GHD. Of these 40 patients, 29 had idiopathic GHD and 11 patients had organic GHD. Causes of organic GHD were brain tumors with cranial irradiation, craniopharyingioma with panhypopituitarism, pituitary hypoplasia, empty sella syndrome, and CNS bleeding. Significantly lower klotho levels were observed among children with organic GHD (n=11, 727±273 pg/mL) compared to both GH sufficient participants (n=59, 1645±778 pg/mL) and those with idiopathic GHD (n=29, 1497±754 pg/mL, $p<0.001$ for the comparison between the groups) (Table 1). Importantly, klotho levels were positively and significantly correlated with IGF-1 standard deviation scores (SDS) (r=0.45, $p<0.001$) (Fig. 3). This correlation remained significant even when patients with organic GHD were removed from the analysis (r=0.27, p=0.01). There was no correlation between klotho levels and gender, pubertal status, age, height-SDS, weight-SDS, and BMI-SDS, suggesting that low klotho levels were specifically associated with GH/IGF-1 activity and not with short stature per se.

Klotho levels of children with idiopathic GHD were lower compared to those of GH sufficient children, but the difference did not reach statistical

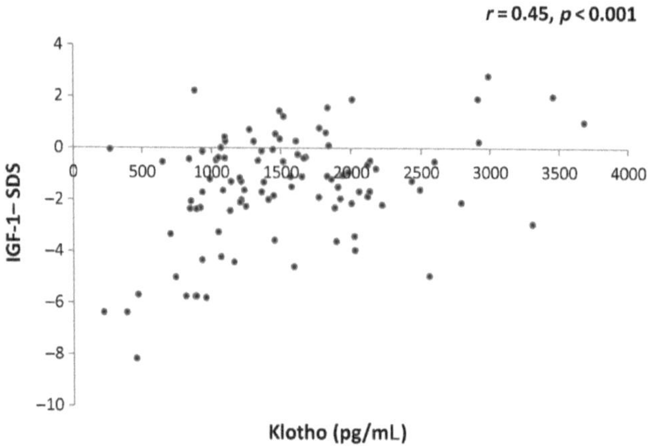

Fig. 3 The correlation between serum klotho levels and IGF-1 in children with short stature. A significant positive correlation was noted ($r=0.45$, $p<0.001$). *Wolf, I., Shahmoon, S., Ben Ami, M., Levy-Shraga, Y., Mazor-Aronovitch, K., Pinhas-Hamiel, O.,…, Modan-Moses, D. (2014). Association between decreased klotho blood levels and organic growth hormone deficiency in children with growth impairment.* PLoS One 9(9), *e107174. http://dx.doi.org/10.1371/journal.pone.0107174, copyright (2014).* (See the color plate.)

significance. As there is no "gold-standard" for the diagnosis of GHD (Hokken-Koelega, 2011) and there is a considerable overlap in the biochemical findings between children with normal growth and those with GHD, it is likely that the idiopathic GHD group included children who were not truly GH deficient. In order to address this problem we conducted an analysis in which children with idiopathic GHD were divided retrospectively according to clinical criteria and follow-up data, into likely "true" GHD ($n=6$) vs likely GH sufficient ($n=16$). More patients classified clinically as having true GHD had subnormal klotho levels (<1000 pg/mL) compared to those judged to be GH sufficient (4/6, 67% vs 2/16, 12.5%; $p=0.025$).

Locher et al. (2015) measured klotho levels in ten adults with severe GHD. While klotho levels of GHD patients were similar to GH sufficient controls, GH replacement treatment resulted in a significant increase in klotho levels.

As discussed earlier, there is an immense need for novel biomarkers to differentiate GH sufficient and deficient children (Bright et al., 1999; GH Research Society, 2000; Hokken-Koelega, 2011; Murray, Dattani, et al., 2015; Quigley et al., 2013; Tauber et al., 1997; Thomas et al., 2003; Toogood et al., 1994). Our results, if validated by additional studies, may indicate klotho as a potential novel biomarker of GHD and, possibly, as a tool for the monitoring of treatment with GH.

6.2.2 Klotho in Anorexia Nervosa

Anorexia nervosa (AN), which affects about 1% of adolescent females, is a chronic disorder with an onset usually during adolescence, characterized by self-imposed malnutrition, weight loss, and an intense fear of gaining weight, despite being underweight (American Psychiatric Association & American Psychiatric Association. Task Force on DSM-IV, 1994; Misra & Klibanski, 2014). The disease provides an attractive model for studying the effect of caloric restriction per se, since in AN patients under nutrition occurs in the absence of other physical illness, and weight rehabilitation is planned and controlled. AN is characterized by reduced body fat mass and skeletal muscle mass and strength (Rigaud et al., 1997), and is associated with severe medical complications, including growth retardation (Modan-Moses et al., 2012), electrolyte disorders, hematologic disorders, cardiac dysfunction, gastrointestinal disturbances, and severe bone loss (Fazeli & Klibanski, 2014a; Miller et al., 2005).

Multiple endocrine abnormalities have been observed in AN, including hypothalamic amenorrhea, hypercortisolism, and GH resistance leading to low concentrations of IGF-1 despite increased GH levels (Misra & Klibanski, 2014). The mechanisms mediating GH resistance in AN may include downregulation of GHRs in the liver, reduced levels of binding proteins and increased levels of FGF21 (Fazeli & Klibanski, 2014b).

Several key characteristics of malnutrition and AN were noted in the *kl/kl* model, including decreased fat mass, growth retardation, reduced skeletal muscle mass, osteopenia, abnormalities in glucose metabolism, disturbances in gastric motility, and impaired fertility. We investigated klotho levels in 19 female adolescents (aged 16.1 ± 1.8 years) hospitalized because of AN (Table 2). Blood samples were obtained on admission and after weight restoration (interval 4.0 ± 2.3 months). Klotho levels on admission were considerably lower than expected for age and increased significantly during hospitalization (from 1061 ± 421 to 1519 ± 781 pg/mL, $p = 0.008$). Consistent with previous studies (Fazeli & Klibanski, 2014b; Hotta et al., 2000), IGF-1 was low at the time of admission, and increased significantly during hospitalization (from 44 ± 17 to 53 ± 11 nmol/L). We hypothesize that low klotho levels observed on admission were secondary to low IGF-1 levels associated with malnutrition, and that the increase in IGF-1 during hospitalization contributed to the increase in klotho levels. In accordance with this hypothesis, the increase in klotho levels following weight gain paralleled the increase in IGF-1 levels. Similar findings were reported by Amitani et al. who noted low klotho levels in a group of 12 AN compared to controls. In six of

Table 2 Clinical, Anthropometric, and Laboratory Characteristics of Study Participants ($n = 19$)

	Admission	Discharge	*p* Value
Age	16.1 ± 1.8	16.5 ± 1.8	<0.001
Height-SDS	−0.3 ± 0.9	−0.3 ± 0.8	0.44
Weight (kg)	42.5 ± 5.0	50.5 ± 5.5	<0.001
Weight-SDS	−1.8 ± 1.1	−0.5 ± 0.8	<0.001
BMI (kg/m^2)	16.7 ± 1.7	19.7 ± 1.6	<0.001
BMI-SDS	−2.0 ± 1.4	−0.4 ± 0.7	<0.001
Calcium (mg/dL)	10.3 ± 0.5	9.9 ± 0.6	0.005
PO_4 (mg/dL)	4.4 ± 0.4	4.2 ± 0.5	0.14
ALP (IU/L)	85 ± 32	105 ± 58	0.08
IGF-1 (nmol/L)	44 ± 17	53 ± 11	0.008
IGF-I SDS	0.30 ± 1.3	1.43 ± 0.85	<0.001
Klotho (pg/mL)	1061 ± 421	1519 ± 781	0.008
L1–L4 *z*-score	1.2 ± 0.8	NA	–

BMI, body mass index; *SDS*, standard deviation score; *PO_4*, phosphorus; *ALP*, alkaline phosphatase; *IGF-1*, insulin growth factor-1; *L1–L4 z-score*, bone mineral density standard deviation score of the lumbar spine vertebrae 1–4.

these patients klotho levels were also measured 3 years after the initial evaluation and were found to have increased significantly. IGF-1 levels were not evaluated in this study (Amitani et al., 2013).

Klotho-deficient mice (*kl/kl*) have barely detectable amounts of adipose tissue, and several studies suggest that klotho may be involved in and promote adipocyte differentiation, as klotho mRNA and protein expression increase during the initial stages of adipocyte differentiation (Chihara et al., 2006). In addition, mRNA expression of adipocyte differentiation markers, including such as CCAAT/enhancer-binding protein (C/EBP)α and C/EBPβ, were decreased by klotho suppression, and increased by klotho overexpression (Chihara et al., 2006). Altogether, these findings point to the possibility that reduced klotho levels may directly contribute to the reduction in fat mass seen in patients with AN. By the same token, the increase in klotho levels following dietary rehabilitation may enhance adipocyte differentiation and mediate the rapid increase in adipose tissue seen in AN patients who gain weight (Misra et al., 2003).

6.2.3 Klotho in Obesity

Despite considerable body of evidence pointing to an association between the state of adipose tissue depots and regulation of the somatotropic axis, the relationship between obesity and GH status is still incompletely understood. Patients with GHD commonly present with a metabolic syndrome characterized by increased fat mass, reduced muscle mass, unfavorable plasma lipid profile, and increased cardiovascular morbidity. Conversely, endogenous GH secretion is markedly reduced in obese individuals, with a reduction in both frequency and amplitude of GH secretory bursts and an increase in GH metabolic clearance rate. Additionally, the GH response to pharmacological challenges such as the insulin tolerance test is blunted. This phenomenon has been attributed to dysregulation of GHRH, somatostatin and ghrelin pathways, hyperinsulinemia and excess of circulating FFA. Whether the reduced GH secretion in obesity actually result in reduced serum IGF-1 levels remains controversial. Several studies demonstrated normal or even increased levels in obese individuals, while others reported IGF-1 to be low-normal or low (Di Somma et al., 2008; Eden Engstrom et al., 2006; Juul, 2003; Rasmussen, 2010; Savastano, Di Somma, Barrea, & Colao, 2014). Experimental studies have also shown that obese people have blunted IGF-1 generation in response to GH stimulation, suggesting bidirectionality in the association between IGF-1 and fat mass (Berryman, Glad, List, & Johannsson, 2013). Hence, it appears that there is at least a subset of obese individuals with combined functional GHD and IGF-1 deficiency.

To the best of our knowledge, the association between klotho and obesity has been investigated in a single study assessing 11 obese women (aged 21.27 ± 0.95 years, BMI $= 35.72 \pm 3.17$ kg/m^2). In this study, obese participants had significantly lower klotho levels compared with control participants. GH and IGF-1 levels were not assessed (Amitani et al., 2013).

6.2.4 Klotho and Aging

Klotho serum levels decrease with aging (Koyama et al., 2015; Pedersen, Pedersen, Brasen, & Rasmussen, 2013; Yamazaki et al., 2010), and it has recently been suggested that klotho may serve as a biomarker of aging (Koyama et al., 2015). Furthermore, lower klotho levels were associated with poor skeletal muscle strength (Semba et al., 2012), disability in activities of daily living (Crasto et al., 2012), and increased mortality (Semba et al., 2011b) in the elderly. Aging may also be viewed as an IGF-1 deficiency state, as GH and IGF-1 concentrations are lower in old age (Bann et al., 2015;

Bidlingmaier et al., 2014; O'Connor et al., 1998; Vestergaard et al., 2014). Furthermore, diminished GH/IGF-1 axis activity has been implicated in age-related changes, including decreased bone and lean body mass, and increased fat mass (O'Connor et al., 1998). Possibly, some of these effects are mediated by klotho, rather than directly through IGF-1. This remains to be tested in future studies.

6.3 Klotho and GH/IGF-1 Excess: Acromegaly

Acromegaly is a rare disorder with an annual incidence of approximately five cases/million, characterized by excess secretion of GH, which induces hypersecretion of IGF-1. Complications and comorbidities associated with acromegaly include altered glucose metabolism, cardiovascular disease, arthropathy, lipid abnormalities, sleep apnea, and increased mortality. Over 95% of acromegaly cases are caused by GH-secreting pituitary adenomas; rare causes include GHRH secreting hypothalamic tumors and ectopic GH or GHRH hypersecretion from extrapituitary tumors. Treatment options include surgery, somatostatin analogs, GH receptor antagonists, dopamine agonists, and radiotherapy (Ben-Shlomo, Sheppard, Stephens, Pulgar, & Melmed, 2011; Melmed, Kleinberg, Bonert, & Fleseriu, 2014).

Sze et al. reported excessively high klotho levels in 24 patients with acromegaly ($n=15$, aged 28–76 years), which normalized following transsphenoidal surgery and normalization of GH and IGF-1 levels. Baseline klotho levels correlated significantly with GH ($r=0.64$), IGF-1 ($r=0.57$), and tumor size ($r=0.5$) (Sze et al., 2012). In three subsequent studies, the same group demonstrated that increased serum klotho levels were specific to GH-producing adenomas (Neidert et al., 2013); that for a given GH excess females had lower IGF-1 but higher klotho levels compared to males (Sze et al., 2014); and that klotho levels increased in patients with disease recurrence (Kohler et al., 2013). Taken together, these studies support the role of the GH/IGF-1 axis in regulating klotho circulating levels. From the clinical standpoint, they suggest that klotho may aid in establishing the diagnosis of acromegaly and may be used as a potential marker of disease activity, response to treatment, and recurrence (Table 3).

7. CONCLUDING REMARKS

Taken together, currently available studies support a tight interaction between klotho and the GH/IGF-1 axis, with a complex reciprocal regulation between them. On one hand, increased activity of the GH/IGF-1 axis

Table 3 Serum Klotho Levels in Different States of IGF-1-Related Conditions

Other	Klotho/IGF-1 Correlation	IGF-1 Levels	Klotho pg/mL	*M*	Age, Years	*N*	Subjects	References
No association with age Higher levels in pubertal vs prepubertal children and in girls vs boys Klotho was positively correlated with 1,25OHD	$r=0.323$ $p<0.001$	Normal	1945 (372–5866)	82	8.78 ± 3.4 7 (2–18)	159	Healthy children	Gkentzi et al. (2014)
No association with age, gender, pubertal status	$r=0.45$ $p<0.001$	−1.4 ± 1.5 SD −2.0 ± 2.8 SD −3.7 ± 2.5 SD −1.7 ± 2.2 SD	1645 ± 778 1497 ± 754 727 ± 273 1500 ± 754	29 15 5 49	8.9 ± 3.3 8.9 ± 4.3 9.7 ± 4.0 9 (1.2–17.4)	59 29 11 99	Short GHS IGHD OGHD Toal	Wolf, Shahmoon, et al. (2014)
Klotho was negatively correlated with lumbar spine BMD z-score and positively correlated with ALP levels	Klotho levels increased in Parallel to increase in IGF-1	Low[a] Normal[b]	1061 ± 421[a] 1519 ± 781[b]	-	16.1 ± 1.8	19	Female with AN	Wolf et al. (2015)
GH treatment resulted in a significant increase in klotho levels	NA	497 (419–658) 558 (462–631)	497 (419–581) 558 (462–631)	6 6	42.6 ± 12.5 matched	10 10	GHD GHS	Locher et al. (2015)

Continued

Table 3 Serum Klotho Levels in Different States of IGF-1-Related Conditions—cont'd

Other	Klotho/ IGF-1 Correlation	IGF-1 Levels	Klotho pg/mL	*M*	Age, Years	*N*	Subjects	References
Klotho levels increased significantly after the recovery of BMI in AN patients	NA	NA	764.6 ± 65.4 847.1 ± 111.3 1391.6 ± 145	– – –	21 ± 1.65 21.3 ± 0.95 21 ± 1.29	12 11 11	AN Obese Controls	Amitani et al. (2013)
Klotho levels correlated with GH and tumor size	$r=0.57$, $p=0.003$ NA NA	588 ± 35[c] 193 ± 12	4200 ± 700[c] 700 ± 100[d] 600 ± 200	14 15	28–76 39 ± 8	24 26	Acromegaly Controls	Sze et al. (2012)
For a given GH status, klotho concentrations were higher and IGF-1 lower in women than in men	NA	604 ± 33 490 ± 33	3439 ± 431 5171 ± 590	31 M 31 F	45.6 ± 2.8 47.6 ± 2.4	62	Acromegaly	Sze et al. (2014)

[a]On admission.
[b]After weight rehabilitation.
[c]Before surgery.
[d]After surgery.

and elevated IGF-1 levels upregulate serum klotho levels, which in turn inhibits the peripheral activity of IGF-1, thus forming a negative feedback loop. On the other hand, klotho abrogates the inhibitory effect of IGF-1 in the pituitary, leading to enhanced GH secretion and further increase of IGF-1 production, thus forming a positive feedback loop (Fig. 4).

Currently available data cannot discern whether klotho levels are affected directly by GH or by IGF-1. However, as klotho levels are reduced in AN patients, a state of GH resistance, and treatment with IGF-1 is associated

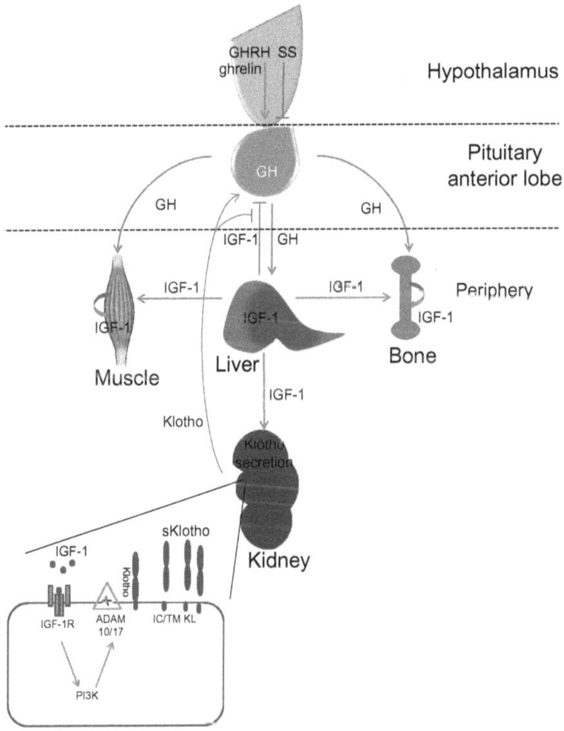

Fig. 4 Regulation of GH/IGF and klotho in the circulation. GH secretion is regulated by hypothalamic factors (GHRH, somatostatin, and ghrelin). GH activates GH receptors (GHR) in the liver, which stimulate synthesis and secretion of IGF-1. GH may also activate GHR in muscles and bones. IGF-1 binds to IGF-1Rs in target tissues like muscles and bones and induces growth responses, and also serves as a negative regulator for GH secretion. Klotho is highly expressed in the kidney and insulin, and IGF-1 trigger its shedding, through the action of the metalloproteases ADAM 10 and ADAM 17 (see inset). Klotho enhances GH secretion from the pituitary directly and by abrogating the inhibitory effect of IGF-1. (See the color plate.)

with increased klotho levels, it is likely that klotho is affected directly by IGF-1. Since cleavage and shedding of klotho can be mediated through activation of ADAM10 and ADAM17 (Chen et al., 2007) by activation of the insulin pathway, it is possible that similar activity is also mediated by IGF-1. Indeed, preliminary experiments in our lab showed that treatment of klotho-transfected COS7 kidney cells with IGF-1 induces secretion of klotho into the media (unpublished results).

Interestingly, ADAM17 not only participates in klotho shedding but also participates in the shedding of GHR, thus generating GHBPs which enhance bioavailability of GH, which in turn may increase IGF-1 levels and further enhance the shedding of klotho, forming another positive feedback loop leading to increased GH secretion.

The physiological role of klotho as a downstream effector of the GH/IGF-1 axis remains to be elucidated (Sperling, 2015). The GH/IGF-1 axis is pivotal for proper development and growth of bones and skeletal muscles. Klotho activity is associated with building and maintenance of muscle mass. Thus, a decline in muscle strength and running endurance was observed in *kl/kl* mice, and low klotho serum levels in humans was associated with reduced knee and hand grip strength (Crasto et al., 2012; Phelps, Pettan-Brewer, Ladiges, & Yablonka-Reuveni, 2013; Semba et al., 2011b). Klotho is also involved in bone mineralization (Sasaki et al., 2013). *Kl/kl* mice show a low-turnover osteoporosis with a decreased number of osteoblasts and osteoclasts and klotho levels are reduced in patients with osteoporosis (John, Cheng, & Kuro-o, 2011). It is, therefore, possible that some of the anabolic effects of the GH/IGF-1 axis on bone and muscle are mediated through klotho.

Finally, we speculate that klotho may have a role in modifying the promalignant effects of GH and IGF-1. The GH/IGF-1 axis has been implicated in cancer progression, including enhanced tumorigenesis, metastasis, resistance to chemotherapy and radiotherapy, and transformation (Yang & Yee, 2012). Both GH and IGF-1 activate multiple downstream intermediates relevant to tumorigenesis, such as MAPK, ERK, PI3K, AKT, and mTOR (Gallagher & LeRoith, 2011; Rosenfeld & Hwa, 2009). Furthermore, the GH/IGF-1 system interacts with oncogenes and tumor suppressors, while some tumor suppressors function in part by inhibiting IGF action (Werner, 2012). Still, there is no convincing epidemiological evidence that GH excess states such as acromegaly or treatment with GH result in an increased risk of neoplasia (Raman et al., 2015). Thus, elevated klotho levels under these conditions may inhibit these promalignant actions. We and others have shown that treatment of breast, ovarian, and pancreatic cancer cells with klotho inhibits the IGF-1 signaling pathway in these cells as well as their growth (Abramovitz et al., 2011; Chen, Wang, Zhao, & Wu, 2010; Kuro-o, 2008; Lu, Katsaros, et al., 2008; Wolf et al., 2008; Xie, Chen, Liu, & Zhan, 2013). Hence, elevated klotho levels following activation of the GH/IGF-1 axis may serve as a means to restrain the malignant potential of the axis.

From the clinical standpoint, the striking effect of the GH/IHG-1 axis on klotho levels might be used as a biomarker for GH status. Direct GH measurements are useless due to the pulsatile nature of GH secretion, and indirect measurements, including IGF-1 levels are unsatisfactory as well as they are influenced by age, gender (estrogens), race, genetic makeup, liver function, nutritional status, portal insulin, thyroid hormones, and concomitant inflammatory disease (Neidert et al., 2013). This hinders the use of IGF-1 as a biomarker for GH levels. Therefore, the use of klotho as a biomarker for activity of GH and IGF-1 could have an immense clinical significance in various diseases, including GHD, acromegaly, AN, and cachexia. Further studies are required to answer many questions that remain open.

REFERENCES

Abramovitz, L., Rubinek, T., Ligumsky, H., Bose, S., Barshack, I., Avivi, C., … Wolf, I. (2011). KL1 internal repeat mediates klotho tumor suppressor activities and inhibits bFGF and IGF-I signaling in pancreatic cancer. *Clinical Cancer Research*, *17*(13), 4254–4266. http://dx.doi.org/10.1158/1078-0432.ccr-10-2749.

American Psychiatric Association & American Psychiatric Association. Task Force on DSM-IV (1994). *Diagnostic and statistical manual of mental disorders. DSM-IV* (4th ed.). Washington, DC: American Psychiatric Association.

Amitani, M., Asakawa, A., Amitani, H., Kaimoto, K., Sameshima, N., Koyama, K. I., … Inui, A. (2013). Plasma klotho levels decrease in both anorexia nervosa and obesity. *Nutrition*, *29*(9), 1106–1109.

Atkin, S. L., Landolt, A. M., Jeffreys, R. V., Diver, M., Radcliffe, J., & White, M. C. (1993). Basic fibroblastic growth factor stimulates prolactin secretion from human anterior pituitary adenomas without affecting adenoma cell proliferation. *The Journal of Clinical Endocrinology and Metabolism*, 77(3), 831–837.

Bann, D., Holly, J. M., Lashen, H., Hardy, R., Adams, J., Kuh, D., … Ben-Shlomo, Y. (2015). Changes in insulin-like growth factor-I and -II associated with fat but not lean mass in early old age. *Obesity (Silver Spring)*, *23*(3), 692–698. http://dx.doi.org/10.1002/oby.21002.

Bartke, A., Sun, L. Y., & Longo, V. (2013). Somatotropic signaling: Trade-offs between growth, reproductive development, and longevity. *Physiological Reviews*, *93*(2), 571–598. http://dx.doi.org/10.1152/physrev.00006.2012.

Ben-Shlomo, A., Sheppard, M. C., Stephens, J. M., Pulgar, S., & Melmed, S. (2011). Clinical, quality of life, and economic value of acromegaly disease control. *Pituitary*, *14*(3), 284–294. http://dx.doi.org/10.1007/s11102-011-0310-7.

Berelowitz, M., Szabo, M., Frohman, L. A., Firestone, S., Chu, L., & Hintz, R. L. (1981). Somatomedin-C mediates growth hormone negative feedback by effects on both the hypothalamus and the pituitary. *Science*, *212*(4500), 1279–1281.

Berryman, D. E., Glad, C. A., List, E. O., & Johannsson, G. (2013). The GH/IGF-1 axis in obesity: Pathophysiology and therapeutic considerations. *Nature Reviews. Endocrinology*, *9*(6), 346–356. http://dx.doi.org/10.1038/nrendo.2013.64.

Bidlingmaier, M., Friedrich, N., Emeny, R. T., Spranger, J., Wolthers, O. D., Roswall, J., … Arafat, A. M. (2014). Reference intervals for insulin-like growth factor-1 (igf-i) from birth to senescence: Results from a multicenter study using a new automated chemiluminescence IGF-I immunoassay conforming to recent international recommendations. *The Journal of Clinical Endocrinology and Metabolism*, *99*(5), 1712–1721. http://dx.doi.org/10.1210/jc.2013-3059.

Black, E. G., Logan, A., Davis, J. R., & Sheppard, M. C. (1990). Basic fibroblast growth factor affects DNA synthesis and cell function and activates multiple signalling pathways in rat thyroid FRTL-5 and pituitary GH3 cells. *The Journal of Endocrinology*, *127*(1), 39–46.

Bodner, M., & Karin, M. (1987). A pituitary-specific trans-acting factor can stimulate transcription from the growth hormone promoter in extracts of nonexpressing cells. *Cell*, *50*(2), 267–275.

Brazeau, P., Vale, W., Burgus, R., Ling, N., Butcher, M., Rivier, J., & Guillemin, R. (1973). Hypothalamic polypeptide that inhibits the secretion of immunoreactive pituitary growth hormone. *Science*, *179*(4068), 77–79.

Bright, G. M., Julius, J. R., Lima, J., & Blethen, S. L. (1999). Growth hormone stimulation test results as predictors of recombinant human growth hormone treatment outcomes: Preliminary analysis of the National Cooperative Growth Study Database. *Pediatrics*, *104*(Suppl. 5), 1028–1031.

Camilli, T. C., Xu, M., O'Connell, M. P., Chien, B., Frank, B. P., Subaran, S., … Weeraratna, A. T. (2011). Loss of Klotho during melanoma progression leads to increased filamin cleavage, increased Wnt5A expression, and enhanced melanoma cell motility. *Pigment Cell & Melanoma Research*, *24*(1), 175–186. http://dx.doi.org/10.1111/j.1755-148X.2010.00792.x.

Carreira, M. C., Crujeiras, A. B., Andrade, S., Monteiro, M. P., & Casanueva, F. F. (2013). Ghrelin as a GH-releasing factor. *Endocrine Development*, *25*, 49–58. http://dx.doi.org/10.1159/000346052.

Cha, S.-K., Ortega, B., Kurosu, H., Rosenblatt, K. P., Kuro-o, M., & Huang, C.-L. (2008). Removal of sialic acid involving Klotho causes cell-surface retention of TRPV5 channel via binding to galectin-1. *Proceedings of the National Academy of Sciences of the United States of America*, *105*, 9805–9810.

Chang, Q., Hoefs, S., van der Kemp, A. W., Topala, C. N., Bindels, R. J., & Hoenderop, J. G. (2005). The βετα-glucuronidase klotho hydrolyzes and activates the TRPV5 channel. *Science*, *310*(5747), 490–493.

Chen, T. H., Kuro, O. M., Chen, C. H., Sue, Y. M., Chen, Y. C., Wu, H. H., & Cheng, C. Y. (2013). The secreted Klotho protein restores phosphate retention and suppresses accelerated aging in Klotho mutant mice. *European Journal of Pharmacology*, *698*(1–3), 67–73.

Chen, C.-D., Podvin, S., Gillespie, E., Leeman, S. E., & Abraham, C. R. (2007). Insulin stimulates the cleavage and release of the extracellular domain of Klotho by ADAM10 and ADAM17. *Proceedings of the National Academy of Sciences of the United States of America*, *104*(50), 19796–19801.

Chen, C. D., Tung, T. Y., Liang, J., Zeldich, E., Tucker Zhou, T. B., Turk, B. E., & Abraham, C. R. (2014). Identification of cleavage sites leading to the shed form of the anti-aging protein klotho. *Biochemistry*, *53*(34), 5579–5587. http://dx.doi.org/10.1021/bi500409n.

Chen, B., Wang, X., Zhao, W., & Wu, J. (2010). Klotho inhibits growth and promotes apoptosis in human lung cancer cell line A549. *Journal of Experimental & Clinical Cancer Research*, *29*, 99. http://dx.doi.org/10.1186/1756-9966-29-99. 1756-9966-29-99 [pii].

Chen, J., Zhang, X., Zhang, H., Lin, J., Zhang, C., Wu, Q., & Ding, X. (2013). Elevated Klotho promoter methylation is associated with severity of chronic kidney disease. *PLoS One*, *8*(11), e79856. http://dx.doi.org/10.1371/journal.pone.0079856.

Chihara, Y., Rakugi, H., Ishikawa, K., Ikushima, M., Maekawa, Y., Ohta, J., … Ogihara, T. (2006). Klotho protein promotes adipocyte differentiation. *Endocrinology*, *147*(8), 3835–3842.

Crasto, C. L., Semba, R. D., Sun, K., Cappola, A. R., Bandinelli, S., & Ferrucci, L. (2012). Relationship of low-circulating "anti-aging" klotho hormone with disability in activities of daily living among older community-dwelling adults. *Rejuvenation Research*, *15*(3), 295–301.

Denef, C. (2008). Paracrinicity: The story of 30 years of cellular pituitary crosstalk. *Journal of Neuroendocrinology*, *20*(1), 1–70. http://dx.doi.org/10.1111/j.1365-2826.2007.01616.x.

Dermaku-Sopjani, M., Sopjani, M., Saxena, A., Shojaiefard, M., Bogatikov, E., Alesutan, I., … Lang, F. (2011). Downregulation of NaPi-IIa and NaPi-IIb Na-coupled phosphate transporters by coexpression of Klotho. *Cellular Physiology and Biochemistry*, *28*(2), 251–258. http://dx.doi.org/10.1159/000331737.

Di Somma, C., Angrisani, L., Rota, F., Savanelli, M. C., Cascella, T., Belfiore, A., … Savastano, S. (2008). GH and IGF-I deficiency are associated with reduced loss of fat mass after laparoscopic-adjustable silicone gastric banding. *Clinical Endocrinology*, *69*(3), 393–399. http://dx.doi.org/10.1111/j.1365-2265.2008.03183.x.

Doi, S., Zou, Y., Togao, O., Pastor, J. V., John, G. B., Wang, L., … Kuro, O. M. (2011). Klotho inhibits transforming growth factor-{beta}1 (TGF-{beta}1) signaling and suppresses renal fibrosis and cancer metastasis in mice. *The Journal of Biological Chemistry*, *286*(10), 8655–8665. http://dx.doi.org/10.1074/jbc.M110.174037 M110.174037 [pii].

Eden Engstrom, B., Burman, P., Holdstock, C., Ohrvall, M., Sundbom, M., & Karlsson, F. A. (2006). Effects of gastric bypass on the GH/IGF-I axis in severe obesity—and a comparison with GH deficiency. *European Journal of Endocrinology*, *154*(1), 53–59. http://dx.doi.org/10.1530/eje.1.02069.

Eigler, T., & Ben-Shlomo, A. (2014). Somatostatin system: Molecular mechanisms regulating anterior pituitary hormones. *Journal of Molecular Endocrinology*, *53*(1), R1–R19. http://dx.doi.org/10.1530/jme-14-0034.

Fazeli, P. K., & Klibanski, A. (2014a). Anorexia nervosa and bone metabolism. *Bone*, *2014*, 39–45.

Fazeli, P. K., & Klibanski, A. (2014b). Determinants of GH resistance in malnutrition. *The Journal of Endocrinology*, *220*(3), R57–R65. 13-0477.

Gallagher, E. J., & LeRoith, D. (2011). Minireview: IGF, Insulin, and Cancer. *Endocrinology*, *152*(7), 2546–2551. http://dx.doi.org/10.1210/en.2011-0231.

GH Research Society. (2000). Consensus guidelines for the diagnosis and treatment of growth hormone (GH) deficiency in childhood and adolescence: Summary statement of the GH research society. *The Journal of Clinical Endocrinology and Metabolism*, *85*(11), 3990–3993. http://dx.doi.org/10.1210/jc.85.11.3990.

Gkentzi, D., Efthymiadou, A., Kritikou, D., & Chrysis, D. (2014). Fibroblast growth factor 23 and Klotho serum levels in healthy children. *Bone*, *2014*(29), 8–14.

Heijboer, A. C., Blankenstein, M. A., Hoenderop, J., de Borst, M. H., Vervloet, M. G., & consortium, on behalf of the NIGRAM. (2013). Laboratory aspects of circulating alpha-Klotho. *Nephrology, Dialysis, Transplantation*, *28*, 2283–2287. http://dx.doi.org/10.1093/ndt/gft236.

Hokken-Koelega, A. C. S. (2011). Diagnostic workup of the short child. *Hormone Research in Paediatrics*, *76*(Suppl. 3), 6–8.

Hotta, M., Fukuda, I., Sato, K., Hizuka, N., Shibasaki, T., & Takano, K. (2000). The relationship between bone turnover and body weight, serum insulin-like growth factor (IGF) I, and serum IGF-binding protein levels in patients with anorexia nervosa. *The Journal of Clinical Endocrinology and Metabolism, 85*(1), 200–206.

Hu, M. C., Shi, M., Zhang, J., Pastor, J., Nakatani, T., Lanske, B., … Moe, O. W. (2010). Klotho: A novel phosphaturic substance acting as an autocrine enzyme in the renal proximal tubule. *Faseb Journal, 24*(9), 3438–3450. http://dx.doi.org/10.1096/fj.10-154765.

Imura, A., Tsuji, Y., Murata, M., Maeda, R., Kubota, K., Iwano, A., … Nabeshima, Y.-i. (2007). {alpha}-Klotho as a regulator of calcium homeostasis. *Science, 316*(5831), 1615–1618.

John, G. B., Cheng, C. Y., & Kuro-o, M. (2011). Role of Klotho in aging, phosphate metabolism, and CKD. *American Journal of Kidney Diseases, 58*(1), 127–134. http://dx.doi.org/10.1053/j.ajkd.2010.12.027.

Juul, A. (2003). Serum levels of insulin-like growth factor I and its binding proteins in health and disease. *Growth Hormone & IGF Research, 13*(4), 113–170.

Kaplan, S. A., & Cohen, P. (2007). The somatomedin hypothesis 2007: 50 years later. *The Journal of Clinical Endocrinology and Metabolism, 92*(12), 4529–4535. http://dx.doi.org/10.1210/jc.2007-0526.

Keiser, U., & Ho, K. K. Y. (2011). Pituitary physiology and diagnostic evaluation. In S. Melmed, P. Polonsky, R. Larsen, & H. M. Kronenberg (Eds.), *William's textbook of endocrinology* (12th ed., pp. 176–231). Philadelphia: Elsevier.

King, G. D., Rosene, D. L., & Abraham, C. R. (2012). Promoter methylation and age-related downregulation of Klotho in rhesus monkey. *Age (Dordrecht, Netherlands), 34*(6), 1405–1419. http://dx.doi.org/10.1007/s11357-011-9315-4.

Kitagawa, M., Sugiyama, H., Morinaga, H., Inoue, T., Takiue, K., Ogawa, A., … Makino, H. (2013). A decreased level of serum soluble Klotho is an independent biomarker associated with arterial stiffness in patients with chronic kidney disease. *PLoS One, 8*(2), 19.

Kohler, S., Tschopp, O., Sze, L., Neidert, M., Bernays, R. L., Spanaus, K. S., … Schmid, C. (2013). Monitoring for potential residual disease activity by serum insulin-like growth factor 1 and soluble Klotho in patients with acromegaly after pituitary surgery: Is there an impact of the genomic deletion of exon 3 in the growth hormone receptor (d3-GHR) gene on "safe" GH cut-off values? *General and Comparative Endocrinology, 188*, 282–287. http://dx.doi.org/10.1016/j.ygcen.2013.04.024.

Koyama, D., Sato, Y., Aizawa, M., Maki, T., Kurosawa, M., Kuro, O. M., & Furukawa, Y. (2015). Soluble alphaKlotho as a candidate for the biomarker of aging. *Biochemical and Biophysical Research Communications, 467*(4), 1019–1025. http://dx.doi.org/10.1016/j.bbrc.2015.10.018.

Kuro-o, M. (2008). Klotho as a regulator of oxidative stress and senescence. *Biological Chemistry, 389*(3), 233–241. http://dx.doi.org/10.1515/BC.2008.028.

Kuro-o, M., Matsumura, Y., Aizawa, H., Kawaguchi, H., Suga, T., Utsugi, T., … Nabeshima, Y.-i. (1997). Mutation of the mouse klotho gene leads to a syndrome resembling ageing. *Nature, 390*(6655), 45–51.

Kurosu, H., & Kuro-o, M. (2008). The Klotho gene family and the endocrine fibroblast growth factors. *Current Opinion in Nephrology and Hypertension, 17*(4), 368–372. http://dx.doi.org/10.1097/MNH.0b013e3282ffd994. 00041552-200807000-00007 [pii].

Kurosu, H., Ogawa, Y., Miyoshi, M., Yamamoto, M., Nandi, A., Rosenblatt, K. P., … Kuro-o, M. (2006). Regulation of fibroblast growth factor-23 signaling by klotho. *The Journal of Biological Chemistry, 281*(10), 6120–6123.

Kurosu, H., Yamamoto, M., Clark, J. D., Pastor, J. V., Nandi, A., Gurnani, P., … Kuro-o, M. (2005). Suppression of aging in mice by the hormone klotho. *Science, 309*(5742), 1829–1833.

Lee, J., Jeong, D.-J., Kim, J., Lee, S., Park, J.-H., Chang, B., … Lee, M.-S. (2010). The anti-aging gene KLOTHO is a novel target for epigenetic silencing in human cervical carcinoma. *Molecular Cancer, 9*(1), 109–119.

Lee, E. Y., Kim, S. S., Lee, J. S., Kim, I. J., Song, S. H., Cha, S. K., … Chung, C. H. (2014). Soluble alpha-klotho as a novel biomarker in the early stage of nephropathy in patients with type 2 diabetes. *PLoS One, 9*(8), e102984. http://dx.doi.org/10.1371/journal.pone.0102984.

LeRoith, D., & Yakar, S. (2007). Mechanisms of disease: Metabolic effects of growth hormone and insulin-like growth factor 1. *Nature Clinical Practice. Endocrinology & Metabolism, 3*(3), 302–310. http://dx.doi.org/10.1038/ncpendmet0427.

Lewis, U. J., Singh, R. N., Bonewald, L. F., Lewis, L. J., & Vanderlaan, W. P. (1979). Human growth hormone: Additional members of the complex. *Endocrinology, 104*(5), 1256–1265. http://dx.doi.org/10.1210/endo-104-5-1256.

Li, X. X., Huang, L. Y., Peng, J. J., Liang, L., Shi, D. B., Zheng, H. T., & Cai, S. J. (2014). Klotho suppresses growth and invasion of colon cancer cells through inhibition of IGF1R-mediated PI3K/AKT pathway. *International Journal of Cancer/International Journal of Oncology (in net), 45*(2), 611–618. http://dx.doi.org/10.3892/ijo.2014.2430. Epub 2014 May 3899.

Lichanska, A. M., & Waters, M. J. (2008). How growth hormone controls growth, obesity and sexual dimorphism. *Trends in Genetics, 24*(1), 41–47. http://dx.doi.org/10.1016/j.tig.2007.10.006.

Ligumsky, H., Rubinek, T., Merenbakh-Lamin, K., Yeheskel, A., Sertchook, R., Shahmoon, S., … Wolf, I. (2015). Tumor suppressor activity of klotho in breast cancer is revealed by structure-function analysis. *Molecular Cancer Research, 13*(10), 1398–1407. http://dx.doi.org/10.1158/1541-7786.mcr-15-0141.

Lin, Y., & Sun, Z. (2012). Antiaging gene Klotho enhances glucose-induced insulin secretion by up-regulating plasma membrane levels of TRPV2 in MIN6 beta-cells. *Endocrinology, 153*(7), 3029–3039. http://dx.doi.org/10.1210/en.2012-1091.

Lin, Y., & Sun, Z. (2015). Antiaging gene klotho attenuates pancreatic beta-cell apoptosis in type 1 diabetes. *Diabetes, 64*(12), 4298–4311. http://dx.doi.org/10.2337/db15-0066.

Lindberg, K., Amin, R., Moe, O. W., Hu, M. C., Erben, R. G., Ostman Wernerson, A., … Larsson, T. E. (2014). The kidney is the principal organ mediating klotho effects. *Journal of the American Society of Nephrology, 25*(10), 2169–2175. http://dx.doi.org/10.1681/asn.2013111209.

Liu, H., Fergusson, M. M., Castilho, R. M., Liu, J., Cao, L., Chen, J., … Finkel, T. (2007). Augmented Wnt signaling in a mammalian model of accelerated aging. *Science, 317*(5839), 803–806.

Liu, H. Z., Luo, P., Chen, S. H., & Shang, J. H. (2013). Basic fibroblast growth factor upregulates expression of growth hormone gene through extracellular signal-regulated kinase 1/2 in GH4 cells. *Bratislavské Lékarske Listy, 114*(7), 376–379.

Locher, R., Egger, A., Zwimpfer, C., Sze, L., Schmid, C., & Christ, E. (2015). Effect of growth hormone replacement therapy on soluble klotho in patients with growth hormone deficiency. *Clinical Endocrinology, 83*(4), 593–595. http://dx.doi.org/10.1111/cen.12756.

Lu, P., Boros, S., Chang, Q., Bindels, R. J., & Hoenderop, J. G. (2008). The β-glucuronidase klotho exclusively activates the epithelial Ca2+ channels TRPV5 and TRPV6. *Nephrology, Dialysis, Transplantation, 23*(11), 3397–3402. http://dx.doi.org/10.1093/ndt/gfn291.

Lu, L., Katsaros, D., Wiley, A., de la Longrais, I. A., Puopolo, M., & Yu, H. (2008). Klotho expression in epithelial ovarian cancer and its association with insulin-like growth factors and disease progression. *Cancer Investigation, 26*(2), 185–192. http://dx.doi.org/10.1080/07357900701638343. 788769994 [pii].

Martial, J. A., Hallewell, R. A., Baxter, J. D., & Goodman, H. M. (1979). Human growth hormone: Complementary DNA cloning and expression in bacteria. *Science, 205*(4406), 602–607.

Matsumura, Y., Aizawa, H., Shiraki-Iida, T., Nagai, R., Kuro-o, M., & Nabeshima, Y.-i. (1998). Identification of the human klotho gene and its two transcripts encoding membrane and secreted klotho protein. *Biochemical and Biophysical Research Communications, 242*(3), 626–630.

Mayo, K. E. (1992). Molecular cloning and expression of a pituitary-specific receptor for growth hormone-releasing hormone. *Molecular Endocrinology, 6*(10), 1734–1744. http://dx.doi.org/10.1210/mend.6.10.1333056.

Melmed, S., Kleinberg, D. L., Bonert, V., & Fleseriu, M. (2014). Acromegaly: Assessing the disorder and navigating therapeutic options for treatment. *Endocrine Practice, 20*(Suppl. 1), 7–17. http://dx.doi.org/10.4158/ep14430.ra. quiz 18–20.

Miller, G. M., Alexander, J. M., Bikkal, H. A., Katznelson, L., Zervas, N. T., & Klibanski, A. (1995). Somatostatin receptor subtype gene expression in pituitary adenomas. *The Journal of Clinical Endocrinology and Metabolism, 80*(4), 1386–1392. http://dx.doi.org/10.1210/jcem.80.4.7714115.

Miller, W. L., & Eberhardt, N. L. (1983). Structure and evolution of the growth hormone gene family. *Endocrine Reviews, 4*(2), 97–130. http://dx.doi.org/10.1210/edrv-4-2-97.

Miller, K. K., Grinspoon, S. K., Ciampa, J., Hier, J., Herzog, D., & Klibanski, A. (2005). Medical findings in outpatients with anorexia nervosa. *Archives of Internal Medicine, 165*(5), 561–566.

Misra, M., & Klibanski, A. (2014). Endocrine consequences of anorexia nervosa. *The Lancet. Diabetes & Endocrinology, 2*(7), 581–592.

Misra, M., Soyka, L. A., Miller, K. K., Grinspoon, S., Levitsky, L. L., & Klibanski, A. (2003). Regional body composition in adolescents with anorexia nervosa and changes with weight recovery. *The American Journal of Clinical Nutrition*, 77(6), 1361–1367.

Modan-Moses, D., Yaroslavsky, A., Kochavi, B., Toledano, A., Segev, S., Balawi, F., … Stein, D. (2012). Linear growth and final height characteristics in adolescent females with anorexia nervosa. *PLoS One*, 7(9), 18.

Murray, P. G., Dattani, M. T., & Clayton, P. E. (2015). Controversies in the diagnosis and management of growth hormone deficiency in childhood and adolescence. *Archives of Disease in Childhood.* http://dx.doi.org/10.1136/archdischild-2014-307228.

Murray, P. G., Higham, C. E., & Clayton, P. E. (2015). 60 Years Of Neuroendocrinology: The hypothalamo-GH axis: The past 60 years. *The Journal of Endocrinology, 226*(2), T123–T140. http://dx.doi.org/10.1530/joe-15-0120.

Neidert, M. C., Sze, L., Zwimpfer, C., Sarnthein, J., Seifert, B., Frei, K., … Bernays, R. L. (2013). Soluble alpha-klotho: A novel serum biomarker for the activity of GH-producing pituitary adenomas. *European Journal of Endocrinology, 168*(4), 575–583. http://dx.doi.org/10.1530/eje-12-1045.

O'Connor, K. G., Tobin, J. D., Harman, S. M., Plato, C. C., Roy, T. A., Sherman, S. S., & Blackman, M. R. (1998). Serum levels of insulin-like growth factor-I are related to age and not to body composition in healthy women and men. *The Journals of Gerontology. Series A, Biological Sciences and Medical Sciences, 53*(3), M176–M182.

Pedersen, L., Pedersen, S. M., Brasen, C. L., & Rasmussen, L. M. (2013). Soluble serum Klotho levels in healthy subjects. Comparison of two different immunoassays. *Clinical Biochemistry, 46*(12), 1079–1083. http://dx.doi.org/10.1016/j.clinbiochem.2013.05.046.

Phelps, M., Pettan-Brewer, C., Ladiges, W., & Yablonka-Reuveni, Z. (2013). Decline in muscle strength and running endurance in klotho deficient C57BL/6 mice. *Biogerontology, 14*(6), 729–739. http://dx.doi.org/10.1007/s10522-013-9447-2.

Quigley, C. A., Zagar, A. J., Liu, C. C., Brown, D. M., Huseman, C., Levitsky, L., ... Chipman, J. J. (2013). United States multicenter study of factors predicting the persistence of GH deficiency during the transition period between childhood and adulthood. *International Journal of Pediatric Endocrinology, 2013*(1), 6.

Raman, S., Grimberg, A., Waguespack, S. G., Miller, B. S., Sklar, C. A., Meacham, L. R., & Patterson, B. C. (2015). Risk of neoplasia in pediatric patients receiving growth hormone therapy—A report from the pediatric endocrine society drug and therapeutics committee. *The Journal of Clinical Endocrinology and Metabolism, 100*(6), 2192–2203. http://dx.doi.org/10.1210/jc.2015-1002.

Rasmussen, M. H. (2010). Obesity, growth hormone and weight loss. *Molecular and Cellular Endocrinology, 316*(2), 147–153. http://dx.doi.org/10.1016/j.mce.2009.08.017.

Reichenbach, A., Steyn, F. J., Sleeman, M. W., & Andrews, Z. B. (2012). Ghrelin receptor expression and colocalization with anterior pituitary hormones using a GHSR-GFP mouse line. *Endocrinology, 153*(11), 5452–5466. http://dx.doi.org/10.1210/en.2012-1622.

Rigaud, D., Moukaddem, M., Cohen, B., Malon, D., Reveillard, V., & Mignon, M. (1997). Refeeding improves muscle performance without normalization of muscle mass and oxygen consumption in anorexia nervosa patients. *The American Journal of Clinical Nutrition, 65*(6), 1845–1851.

Romero, C. J., Pine-Twaddell, E., Sima, D. I., Miller, R. S., He, L., Wondisford, F., & Radovick, S. (2012). Insulin-like growth factor 1 mediates negative feedback to somatotroph GH expression via POU1F1/CREB binding protein interactions. *Molecular and Cellular Biology, 32*(21), 4258–4269. http://dx.doi.org/10.1128/mcb.00171-12.

Rosenfeld, R. G., & Hwa, V. (2009). The growth hormone cascade and its role in mammalian growth. *Hormone Research, 71*(Suppl. 2), 36–40. http://dx.doi.org/10.1159/000192434. Epub 000192009 Apr 000192429.

Rotondi, S., Pasquali, M., Tartaglione, L., Muci, M. L., Mandanici, G., Leonangeli, C., ... Mazzaferro, S. (2015). Soluble alpha-klotho serum levels in chronic kidney disease *International Journal of Endocrinology, 2015*, 872193. http://dx.doi.org/10.1155/2015/872193.

Rubinek, T., Rubinfeld, H., Hadani, M., Barkai, G., & Shimon, I. (2005). Nitric oxide stimulates growth hormone secretion from human fetal pituitaries and cultured pituitary adenomas. *Endocrine, 28*(2), 209–216. http://dx.doi.org/10.1385/ENDO:28:2:209. ENDO:28:2:209 [pii].

Rubinek, T., Shulman, M., Israeli, S., Bose, S., Avraham, A., Zundelevich, A., ... Wolf, I. (2011). Epigenetic silencing of the tumor suppressor klotho in human breast cancer. *Breast Cancer Research and Treatment, 133*, 649–657. http://dx.doi.org/10.1007/s10549-011-1824-4.

Rubinek, T., Yu, R., Hadani, M., Barkai, G., Nass, D., Melmed, S., & Shimon, I. (2003). The cell adhesion molecules N-cadherin and neural cell adhesion molecule regulate human growth hormone: A novel mechanism for regulating pituitary hormone secretion. *The Journal of Clinical Endocrinology and Metabolism, 88*(8), 3724–3730.

Sasaki, M., Hasegawa, T., Yamada, T., Hongo, H., de Freitas, P. H., Suzuki, R., ... Amizuka, N. (2013). Altered distribution of bone matrix proteins and defective bone mineralization in klotho-deficient mice. *Bone, 57*(1), 206–219.

Savastano, S., Di Somma, C., Barrea, L., & Colao, A. (2014). The complex relationship between obesity and the somatropic axis: The long and winding road. *Growth Hormone & IGF Research, 24*(6), 221–226. http://dx.doi.org/10.1016/j.ghir.2014.09.002.

Scholze, A., Liu, Y., Pedersen, L., Xia, S., Roth, H. J., Hocher, B., ... Tepel, M. (2014). Soluble alpha-klotho and its relation to kidney function and fibroblast growth factor-23. *The Journal of Clinical Endocrinology and Metabolism, 99*(5), E855–E861. http://dx.doi.org/10.1210/jc.2013-4171.

Semba, R. D., Cappola, A. R., Sun, K., Bandinelli, S., Dalal, M., Crasto, C., ... Ferrucci, L. (2011a). Plasma klotho and cardiovascular disease in adults. *Journal of the American Geriatrics Society, 59*(9), 1596–1601.

Semba, R. D., Cappola, A. R., Sun, K., Bandinelli, S., Dalal, M., Crasto, C., ... Ferrucci, L. (2011b). Plasma klotho and mortality risk in older community-dwelling adults. *The Journals of Gerontology. Series A, Biological Sciences and Medical Sciences, 66*(7), 794–800.

Semba, R. D., Cappola, A. R., Sun, K., Bandinelli, S., Dalal, M., Crasto, C., ... Ferrucci, L. (2012). Relationship of low plasma klotho with poor grip strength in older community-dwelling adults: The InCHIANTI study. *European Journal of Applied Physiology, 112*(4), 1215–1220. http://dx.doi.org/10.1007/s00421-011-2072-3.

Shahmoon, S., Rubinfeld, H., Wolf, I., Cohen, Z. R., Hadani, M., Shimon, I., & Rubinek, T. (2014). The aging suppressor klotho: A potential regulator of growth hormone secretion. *American Journal of Physiology. Endocrinology and Metabolism, 307*(3), 17.

Shimada, T., Kakitani, M., Yamazaki, Y., Hasegawa, H., Takeuchi, Y., Fujita, T., ... Yamashita, T. (2004). Targeted ablation of Fgf23 demonstrates an essential physiological role of FGF23 in phosphate and vitamin D metabolism. *The Journal of Clinical Investigation, 113*(4), 561–568. http://dx.doi.org/10.1172/jci19081.

Shiraki-Iida, T., Aizawa, H., Matsumura, Y., Sekine, S., Iida, A., Anazawa, H., ... Nabeshima, Y.-i. (1998). Structure of the mouse klotho gene and its two transcripts encoding membrane and secreted protein. *FEBS Letters, 424*(1–2), 6–10.

Siler, T. M., VandenBerg, G., Yen, S. S., Brazeau, P., Vale, W., & Guillemin, R. (1973). Inhibition of growth hormone release in humans by somatostatin. *The Journal of Clinical Endocrinology and Metabolism, 37*(4), 632–634. http://dx.doi.org/10.1210/jcem-37-4-632.

Sperling, M. A. (2015). Traditional and novel aspects of the metabolic actions of growth hormone. *Growth Hormone & IGF Research*. pii: S1096-6374(15)30007-1. http://dx.doi.org/10.1016/j.ghir.2015.06.005.

Steyn, F. J., Huang, L., Ngo, S. T., Leong, J. W., Tan, H. Y., Xie, T. Y., ... Chen, C. (2011). Development of a method for the determination of pulsatile growth hormone secretion in mice. *Endocrinology, 152*(8), 3165–3171.

Sun, C. Y., Chang, S. C., & Wu, M. S. (2012). Suppression of Klotho expression by protein-bound uremic toxins is associated with increased DNA methyltransferase expression and DNA hypermethylation. *Kidney International, 81*(7), 640–650. http://dx.doi.org/10.1038/ki.2011.445.

Sze, L., Bernays, R. L., Zwimpfer, C., Wiesli, P., Brändle, M., & Schmid, C. (2012). Excessively high soluble Klotho in patients with acromegaly. *Journal of Internal Medicine, 272*(1), 93–97. http://dx.doi.org/10.1111/j.1365-2796.2012.02542.x.

Sze, L., Neidert, M. C., Bernays, R. L., Zwimpfer, C., Wiesli, P., Haile, S. R., ... Schmid, C. (2014). Gender dependence of serum soluble Klotho in acromegaly. *Clinical Endocrinology, 80*(6), 869–873. http://dx.doi.org/10.1111/cen.12385.

Tauber, M., Moulin, P., Pienkowski, C., Jouret, B., & Rochiccioli, P. (1997). Growth hormone (GH) retesting and auxological data in 131 GH-deficient patients after completion of treatment. *The Journal of Clinical Endocrinology and Metabolism, 82*(2), 352–356.

Theill, L. E., & Karin, M. (1993). Transcriptional control of GH expression and anterior pituitary development. *Endocrine Reviews, 14*(6), 670–689. http://dx.doi.org/10.1210/edrv-14-6-670.

Thomas, M., Massa, G., Maes, M., Beckers, D., Craen, M., Francois, I., ... Bourguignon, J. P. (2003). Growth hormone (GH) secretion in patients with childhood-onset GH deficiency: Retesting after one year of therapy and at final height. *Hormone Research, 59*(1), 7–15.

Tohyama, O., Imura, A., Iwano, A., Freund, J.-N., Henrissat, B., Fujimori, T., & Nabeshima, Y.-i. (2004). Klotho is a novel beta-glucuronidase capable of hydrolyzing steroid beta-glucuronides. *The Journal of Biological Chemistry*, *279*(11), 9777–9784. http://dx.doi.org/10.1074/jbc.M312392200.

Toogood, A. A., Beardwell, C. G., & Shalet, S. M. (1994). The severity of growth hormone deficiency in adults with pituitary disease is related to the degree of hypopituitarism. *Clinical Endocrinology*, *41*(4), 511–516.

Urakawa, I., Yamazaki, Y., Shimada, T., Iijima, K., Hasegawa, H., Okawa, K., … Yamashita, T. (2006). Klotho converts canonical FGF receptor into a specific receptor for FGF23. *Nature*, *444*(7120), 770–774.

Utsugi, T., Ohno, T., Ohyama, Y., Uchiyama, T., Saito, Y., Matsumura, Y., … Nagai, R. (2000). Decreased insulin production and increased insulin sensitivity in the klotho mutant mouse, a novel animal model for human aging. *Metabolism*, *49*(9), 1118–1123.

Vestergaard, P. F., Hansen, M., Frystyk, J., Espelund, U., Christiansen, J. S., Jorgensen, J. O., & Fisker, S. (2014). Serum levels of bioactive IGF1 and physiological markers of ageing in healthy adults. *European Journal of Endocrinology*, *170*(2), 229–236. http://dx.doi.org/10.1530/eje-13-0661.

Waters, M. J., & Brooks, A. J. (2011). Growth hormone receptor: Structure function relationships. *Hormone Research in Paediatrics*, *76*(Suppl. 1), 12–16. http://dx.doi.org/10.1159/000329138.

Werner, H. (2012). Tumor suppressors govern insulin-like growth factor signaling pathways: Implications in metabolism and cancer. *Oncogene*, *31*(22), 2703–2714. http://dx.doi.org/10.1038/onc.2011.447.

Wolf, M. T., An, S. W., Nie, M., Bal, M. S., & Huang, C. L. (2014). Klotho up-regulates renal calcium channel transient receptor potential vanilloid 5 (TRPV5) by intra- and extracellular N-glycosylation-dependent mechanisms. *The Journal of Biological Chemistry*, *289*(52), 35849–35857. http://dx.doi.org/10.1074/jbc.M114.616649.

Wolf, I., Levanon-Cohen, S., Bose, S., Ligumsky, H., Sredni, B., Kanety, H., … Rubinek, T. (2008). Klotho: A tumor suppressor and a modulator of the IGF-1 and FGF pathways in human breast cancer. *Oncogene*, *27*(56), 7094–7105.

Wolf, I., Shahmoon, S., Ben Ami, M., Levy-Shraga, Y., Mazor-Aronovitch, K., Pinhas-Hamiel, O., … Modan-Moses, D. (2014a). Association between decreased klotho blood levels and organic growth hormone deficiency in children with growth impairment. *PLoS One*, *9*(9), e107174. http://dx.doi.org/10.1371/journal.pone.0107174.

Wolf, I., Stein, D., Shahmoon, S., Ziv, S. I., Hemi, R., Kanety, H., … Modan-Moses, D. (2015). Alteration in serum klotho levels in anorexia nervosa patients. *Clinical Nutrition*. pii: S0261-5614(15)00192-2. http://dx.doi.org/10.1016/j.clnu.2015.07.013.

Wu, C., Wang, Q., Lv, C., Qin, N., Lei, S., Yuan, Q., & Wang, G. (2014). The changes of serum sKlotho and NGAL levels and their correlation in type 2 diabetes mellitus patients with different stages of urinary albumin. *Diabetes Research and Clinical Practice*, *106*(2), 343–350. http://dx.doi.org/10.1016/j.diabres.2014.08.026.

Xie, B., Chen, J., Liu, B., & Zhan, J. (2013). Klotho acts as a tumor suppressor in cancers. *Pathology and Oncology Research*, *19*(4), 611–617.

Xu, Y., & Sun, Z. (2015). Molecular basis of Klotho: From gene to function in aging. *Endocrine Reviews*, *36*(2), 174–193. http://dx.doi.org/10.1210/er.2013-1079.

Yamada, Y., Post, S. R., Wang, K., Tager, H. S., Bell, G. I., & Seino, S. (1992). Cloning and functional characterization of a family of human and mouse somatostatin receptors expressed in brain, gastrointestinal tract, and kidney. *Proceedings of the National Academy of Sciences of the United States of America*, *89*(1), 251–255.

Yamazaki, Y., Imura, A., Urakawa, I., Shimada, T., Murakami, J., Aono, Y., … Nabeshima, Y.-i. (2010). Establishment of sandwich ELISA for soluble alpha-Klotho measurement: Age-dependent change of soluble alpha-Klotho levels in healthy subjects. *Biochemical and Biophysical Research Communications*, *398*(3), 513–518.

Yang, Y., & Yee, D. (2012). Targeting insulin and insulin-like growth factor signaling in breast cancer. *Journal of Mammary Gland Biology and Neoplasia*, *17*(3–4), 251–261. http://dx.doi.org/10.1007/s10911-012-9268-y.

Zhang, Y., Jiang, J., Black, R. A., Baumann, G., & Frank, S. J. (2000). Tumor necrosis factor-alpha converting enzyme (TACE) is a growth hormone binding protein (GHBP) sheddase: The metalloprotease TACE/ADAM-17 is critical for (PMA-induced) GH receptor proteolysis and GHBP generation. *Endocrinology*, *141*(12), 4342–4348. http://dx.doi.org/10.1210/endo.141.12.7858.

CHAPTER FIVE

Klotho Prevents Translocation of NFκB

P. Buendía*, R. Ramírez†, P. Aljama*, J. Carracedo*[,1]

*Instituto Maimónides de Investigación Biomédica de Córdoba (IMIBIC), Universidad de Córdoba/Hospital Universitario Reina Sofía, Córdoba, Spain
†Alcalá de Henares University, Madrid, Spain
[1]Corresponding author: e-mail address: julia.carracedo.exts@juntadeandalucia.es

Contents

Abstract

Klotho protein is a β-glucuronidase capable of hydrolyzing steroid β-glucuronides. Two molecules are produced by the Klotho gene, a membrane bound form and a circulating form. This protein is recognized as an antiaging gene with pleiotropic functions. The

Vitamins and Hormones, Volume 101
ISSN 0083-6729
http://dx.doi.org/10.1016/bs.vh.2016.02.005

activation of cellular systems is associated with the pathogenesis of several chronic and degenerative diseases associated with an inflammatory state. Inflammation is characterized by an activation of NFκB. Klotho suppresses nuclear factor NFκB activation and the subsequent transcription of proinflammatory genes. This review focuses on the current understanding of Klotho protein function and its relationship with NFκB regulation, emphasizing its potential involvement in the pathophysiologic process.

1. INTRODUCTION

The discovery of the Klotho gene in 1997 (Kuro-o et al., 1997) has garnered great interest and has increased our understanding of the aging process. The Klotho gene takes its name from the Greek goddess of destiny—the "one that weaves the thread of life." Klotho was originally identified as a gene in mice, mutations in which are associated with a premature aging phenotype, characterized by a shortened life span and several disorders that resemble human premature aging syndromes, including vascular endothelial dysfunction, increased apoptosis, arterial calcification, and atherosclerosis (Kuro-o et al., 1997; Nabeshima, 2002; Shimada, Hasegawa, et al., 2004; Shimada, Takeshita, et al., 2004).

The Klotho gene is expressed primarily in the kidney, choroid plexus of the brain, reproductive organs, pituitary, parathyroid glands, and heart (Kuro-o et al., 1997; Nabeshima, 2002; Takeshita et al., 2004) and is found at low levels in other tissues.

The functions of Klotho are related to its binding to fibroblast growth factor receptor (FGFR) and its involvement in various intracellular signal transduction pathways. In this regard, the effects of Klotho are associated with antioxidant activity through increased transcription of FoxO1 (forkhead box) and, conversely, antiinflammatory effects due to the negative regulation of nuclear factor κB (NFκB) (Kuro-o, 2008; Kurosu & Kuro-O, 2009).

2. KLOTHO STRUCTURE

The Klotho gene encodes a single-pass transmembrane protein that belongs to the glycosylase 1 family and comprises 1012 amino acids in human and 1014 amino acids in mouse. Most residues of Klotho (980 residues) lie in the N-terminal extracellular domain, which is followed by a 21-amino-acid transmembrane domain and a small domain of 11 residues, corresponding to the intracellular C-terminus (Kuro-o et al., 1997; Matsumuraa et al., 1998).

The extracellular domain contains two repeat sequences of 440 amino acids, termed KL1 and KL2, respectively (Kuro-o et al., 1997). The junction between these repeats contains a sequence of four basic amino acids (Lys-Lys-Arg-Lys), a potential proteolytic cleavage site (Fig. 1).

The extracellular domain of Klotho can be separated from the cell membrane by the metalloproteinases ADAM-10 and ADAM-17 (a disintegrin metalloprotease), a cleavage that insulin can stimulate (Torres et al., 2009). The secreted form of Klotho, consisting of the entire extracellular domain, has been detected in the blood, urine, and cerebrospinal fluid (Imura et al., 2004; Kurosu et al., 2005).

In human and mouse, the Klotho gene comprises five exons and can generate two transcripts (Shiraki-Iida et al., 1998). In addition to full-length Klotho—1012 residues in human and 1014 in mouse—an alternative transcript that generates KL1-repeating Klotho can be produced (Shiraki-Iida et al., 1998). The KL1 and KL2 fragments have been identified in human and mouse urine, but unlike the entire extracellular domain, they are not found in cerebrospinal fluid (Imura et al., 2004). The KL1 and KL2

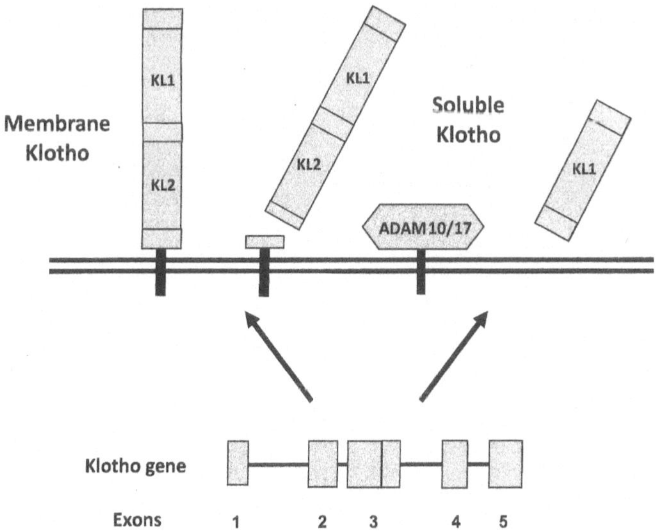

Fig. 1 Klotho structure. Domain structure of human Klotho. KL1 and KL2 indicate first and second Klotho repeat, respectively. The entire extracellular domain consisting of KL1 and KL2 repeats is cleaved by ADAM10 or ADAM17 and released into blood, urine, or cerebrospinal fluid. (See the color plate.)

fragments in urine result from the breakdown of the extracellular domain in the binding region between KL1 and KL2 by proteases, although it is possible that the alternative KL1 transcript produces a peptide that is secreted into urine (Huang, 2010).

3. THE FUNCTIONS OF KLOTHO

The various molecular forms of Klotho have been associated with disparate functions (Fig. 2).

3.1 Klotho Is a Specific Coreceptor for FGF-23

Fibroblast growth factors (FGFs) are proteins that increase mitotic activity and DNA synthesis, thereby facilitating the proliferation of various types of precursor cells. FGFs transduce signals via 4 receptor tyrosine kinases (FGFR1–4). FGFs mediate several types of responses, such as hematopoiesis, angiogenesis, and embryonic development. These factors are also associated with many oncogenic processes (Yamashita, 2005).

In addition to FGF-1 (FGFa) and FGF-2 (FGFb), other homologous proteins belong to the FGF family (4, 17, 18, 19), the most significant members of which include acid fibroblast growth factor (FGFa or FGF-1), basic fibroblast growth factor (FGFb or FGF-2), fibroblast growth factor 3 (FGF-3

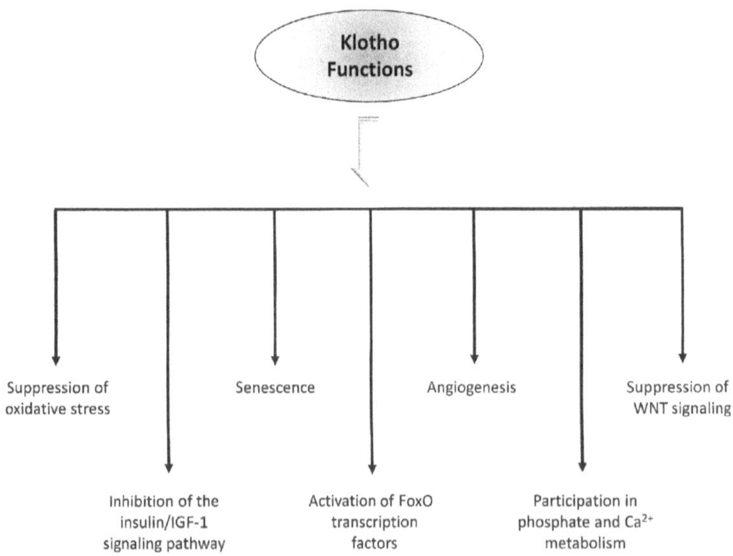

Fig. 2 Klotho functions. Partial list of the functions of klotho.

or INT-2), fibroblast growth factor 4 (FGF-4), fibroblast growth factor 5 (FGF-5), fibroblast growth factor 6 (FGF-6), and fibroblast growth factor 7 (FGF-7). The FGF family continues to expand, which 20 FGFs having been reported to date (Kirikoshi et al., 2000).

Cell surface receptors for FGFs are high affinity or low affinity. The high-affinity fibroblast growth factor receptors (FGFRs)—R1 to R4, encoded by four different genes—are transmembrane protein tyrosine kinases. FGF members and their receptors are redundant, based on their complexity, wherein the same factor can bind to several receptors and the same receptor can be activated by multiple FGFs (Coulier et al., 1991; Johnson & Williams, 1993). The low-affinity FGF receptors are heparan sulfate proteoglycans that lie on the cell surface and are expressed in the mitochondrial matrix (Rusnati, Urbinati, & Presta, 1993; Yayon, Klagsbrun, Esko, Leder, & Ornitz, 1991). Thus, the binding of FGF receptors is complicated and appears to involve more than a single receptor on the target cell.

The full-length Klotho is anchored to the surface of the cell membrane and interacts with the "canonical" FGFR, transforming it into a specific receptor for fibroblast growth factor 23 (FGF-23). FGF-23 has been studied extensively in recent years as a regulator of phosphorus metabolism, as have other factors, such as FGF-7 and secreted frizzled-related protein 4. Thus, Klotho is an essential coreceptor for the binding of FGF-23 to its receptor, FGFR (Kurosu et al., 2006; Urakawa et al., 2006)—FGF-23 requires Klotho to convert certain FGF receptors (FGFR1–4) with low affinity into their high-affinity versions (Komaba et al., 2010; Olauson et al., 2014) (Fig. 3).

In the distal tubule of the kidney, the activation of the FGFR1–Klotho complex by the binding of FGF-23 decreases serum phosphate levels through two mechanisms: downregulation of the phosphorus transporter NPT2a-2C on the luminal side of the proximal tubule, lowering the reabsorption of phosphate, and a decrease in the synthesis of calcitriol (CTR), effecting a decline in the absorption of phosphate from the intestine (Shimada, Hasegawa, et al., 2004; Shimada, Takeshita, et al., 2004; Shimada et al., 2005). The parathyroid glands respond to treatment with FGF-23 (Urakawa et al., 2006), indicating that they are a target organ for FGF-23 and thus should express the receptors that are necessary for such a response. Klotho has been reported to be expressed on the cell membrane of parathyroid cells, confirming that the parathyroid glands are targeted by FGF-23 (Ben-Dov et al., 2007).

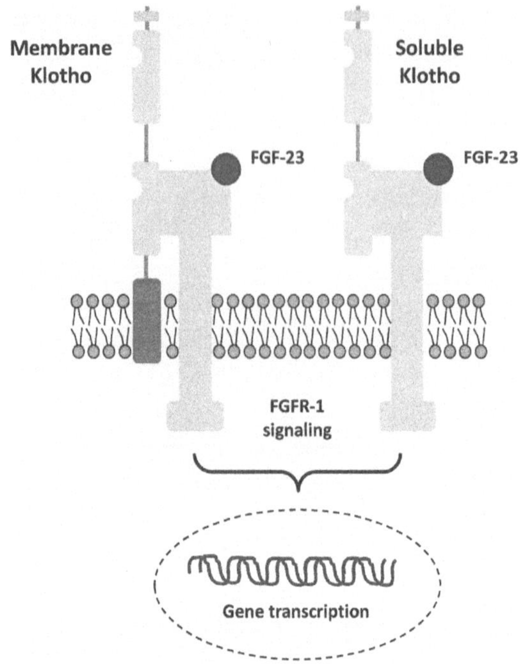

Fig. 3 Klotho-FGF-23 axis. FGF-23 can bind to the Klotho–FGFR1c complex on the surface of the parathyroid cell and thereby triggers downstream signaling. (See the color plate.)

3.2 Klotho Mediates Ion Exchange Processes

Klotho has several specific functions that are independent of FGF-23. In the distal convoluted tubule, the secreted form of Klotho removes the sialic acid terminus of the N-glycans from calcium channel TRPV5 (transient receptor potential cation channel subfamily V member 5) and potassium ROMK1 (renal outer medullary potassium channel 1), preventing their internalization. This activity raises the absorption of calcium and the secretion of potassium into the blood, respectively (Cha et al., 2009; Chang et al., 2005). In the proximal convoluted tubule, Klotho is upregulated in response to a decrease in calcium levels to mediate the rapid recruitment of the Na^+, K^+-ATPase pump to the cell membrane. This Na^+ gradient allows Ca cotransport by the Na^+/Ca^{2+} antiporter (Nabeshima & Imura, 2008).

3.3 Klotho Functions as a Hormone That Regulates Insulin

Klotho inhibits the intracellular insulin and insulin-like growth factor 1 (IGF-1) signaling pathway. Interruption of the insulin/IGF-1 route attenuates the aging phenotype in Klotho-deficient mice, suggesting that Klotho slows aging by impeding insulin/IGF-1 signals. Klotho may regulate IGF-1 activity through FoxO proteins and PI3K/Akt signaling pathways. Extended life span upon negative regulation of insulin and IGF1 signaling is an evolutionarily conserved mechanism to suppress aging. Examining the relationship between Klotho and glucose metabolism might increase our understanding of such pathologies as type II diabetes (Kurosu et al., 2005).

3.4 Klotho Regulates Intracellular Signaling

Klotho increases the production of nitric oxide and inhibits oxidative stress, protecting the cardiovascular system through these mechanisms (Wang & Sun, 2009). Klotho acts on other intracellular signaling pathways, such as p53/p21, cAMP, PKC, and Wnt (Wang & Sun, 2009). Further, Klotho diminishes the activity of the NFκB pathway, suggesting that it has antiinflammatory functions at the cellular level.

Angiotensin II negatively regulates the expression of Klotho (Mitani et al., 2002; Saito, Ishizaka, Mitani, Ohno, & Nagai, 2003); however, Klotho expression is stimulated by PPARγ agonists (Zhang et al., 2008) and CTR (Tsujikawa, Kurotaki, Fujimori, Fukuda, & Nabeshima, 2003). Proinflammatory cytokines, such as TWEAK (tumor necrosis factor-like weak inducer of apoptosis), downregulate Klotho through an NFκB-dependent mechanism (Moreno et al., 2011).

4. MECHANISMS OF KLOTHO

4.1 Klotho-FGF-23 Axis

The presence of the Klotho–FGFR1c complex in a tissue determines its specificity for FGF-23. Also, the binding of FGF-23 to this complex results in the phosphorylation of FGFR and activation of the ERK1/2-MAPK pathway (extracellular signal-regulated kinases, mitogen-activated protein kinase) (Kurosu et al., 2006; Urakawa et al., 2006). The Klotho–FGFR1c complex is expressed in the kidney, choroid plexus, sinoatrial node of the heart, pituitary gland, and parathyroid glands, rendering them the major target organs of FGF-23 (Takeshita et al., 2004).

FGF-23 is a phosphaturic factor and regulator of renal CTR production (Liu et al., 2006; Saito et al., 2003; Shimada, Hasegawa, et al., 2004; Shimada, Takeshita, et al., 2004; Shimada et al., 2005). FGF-23 exerts its effects on the kidney by binding to Klotho–FGFR1c complex in the distal tubule. However, the primary site of phosphate reabsorption in the kidney is the proximal tubule, which only expresses FGFR3. Thus, it is unknown how FGF-23–Klotho interacts with FGFR1c in the distal tubule, because it is in this part of the kidney where it activates the MAPK pathway, which has its effects on the proximal tubule. It has been proposed that a novel signaling mechanism exists in the distal tubule/proximal tubule that centers around an FGF-23-dependent paracrine factor (Liu, Vierthaler, Tang, Zhou, & Quarles, 2008).

The binding of FGF-23 to its receptor complex, Klotho–FGFR1c, appears to induce renal phosphate excretion by inhibiting the expression of the sodium/phosphate cotransporters NPT2a and 2C in the proximal tubule. Moreover, FGF-23 decreases the synthesis of CTR by suppressing renal 1α-hydroxylase (CYP27B1), an enzyme that converts 25-hydroxyvitamin D (25(OH)D) to 1,25-dihydroxyvitamin D (1,25(OH)2D or CTR). Further, FGF-23 stimulates the activity of 24-hydroxylase (CYP24), which modulates CTR to an inactive metabolite. According to these data, FGF-23 decreases serum phosphate levels by inhibiting NPT2a-2C, slowing the resorption of phosphorus, and thus mitigating their effects on CTR, which limits the decrease in phosphorus absorption from the intestine (Saito et al., 2003; Shimada, Hasegawa, et al., 2004; Shimada, Takeshita, et al., 2004; Shimada et al., 2005).

PTH (parathyroid hormone) also downregulates NPT2a-2c, thus reducing serum phosphate and FGF-23 levels. Studies in thyroparathyroidectomized rats reported that the activity of FGF-23 on sodium/phosphate cotransporter is independent of PTH (Shimada, Hasegawa, et al., 2004; Shimada, Takeshita, et al., 2004). Further, FGF-23, which decreases serum phosphorus and NPT2a expression, has renal effects in vitamin D receptor-deficient mice, demonstrating that its acts independently of vitamin D (Shimada et al., 2005).

Finally, it has been suggested that FGF-23 mediates PTH secretion, at least indirectly, by regulating serum phosphorus levels and CTR. However, several studies have shown that FGF-23 has a direct effect on the parathyroid glands in inhibiting the synthesis and secretion of PTH (Ben-Dov et al., 2007; Krajisnik et al., 2007).

4.2 Klotho Activates FoxO Transcription Factors

FoxO (forkhead box) activation is critical in the suppression of aging due to the inhibition of insulin-like signaling. In mammals, FoxO1, FoxO3a, and FoxO4 are negatively regulated by the insulin/IGF-1 pathway, the activation of which effects the phosphorylation and activation of Akt kinase (protein kinase B), which in turn phosphorylates and inactivates the FoxO transcription factors (Wang, Zhou, & Graves, 2014).

FoxO target genes include antioxidant enzymes, such as catalase and superoxide dismutase. The upregulation of these enzymes promotes the removal of reactive oxygen species (ROS) and lowers oxidative stress. The ability of FoxOs to protect cells against oxidative stress might be involved in the suppression of aging in mammals (Yamamoto et al., 2005).

Transgenic mice that overexpressed Klotho have higher SOD2 (superoxide dismutase 2) content, low levels of phosphorylated FoxOs, and reduced oxidative stress. Moreover, in HeLa cell cultures, the administration of the extracellular domain of Klotho induces the inhibition of the insulin/IGF-1 pathway, the activation of FoxOs, and the upregulation of SOD2. These mechanisms protect the cells against oxidative damage and apoptosis (Yamamoto et al., 2005).

4.3 Klotho and Nitric Oxide Production

Nitric oxide (NO) regulates vascular tone. In the aorta, endothelium-dependent vasodilation in response to acetylcholine is significantly attenuated in $kl^{-/-}$ mice, indicating that a deficiency in Klotho decreases NO synthesis in vascular endothelial cells. Klotho-deficient mice experience impairments in angiogenesis-dependent endothelial NO activity and synthesize less systemic NO (Ding & Ma, 2015; Martín-Núñez, Donate-Correa, Muros-de-Fuentes, Mora-Fernández, & Navarro-González, 2014).

4.4 Klotho in the Wnt Signaling Pathway

Soluble Klotho protein binds to multiple Wnt ligands, resulting in the suppression of Wnt activity. Wnt signaling is essential for the proliferation and survival of stem cells, but continuous exposure to Wnts can accelerate cellular depletion and senescence.

The increase in Wnt signaling in Klotho-deficient mice is associated with the activation of μ-calpain, a protease that mediates the degradation of β-catenin. This protein translocates to the nucleus and increases FoxO transcriptional activity, inducing cell cycle arrest and DNA repair. Thus, the rise in β-catenin degradation under Klotho deficiency can lower the activation of FoxOs and increase oxidative stress (Kuro-o, 2008; Sopjani, Rinnerthaler, Kruja, & Dermaku-Sopjani, 2015).

4.5 Klotho and TWEAK

TWEAK is a cytokine of the TNF superfamily. Data suggest that TWEAK mediates the pathogenesis of human diseases, such as atherosclerosis, rheumatoid arthritis, and autoimmune diseases that affect the kidneys. The functions of TWEAK vary by cell type and microenvironment and include cell proliferation, survival, migration, and apoptosis. TWEAK can promote or inhibit cell differentiation and upregulates proinflammatory molecules. In vivo, TWEAK stimulates angiogenesis and regulates neurovascular permeability.

TWEAK activates NFκB in renal tubular cells, thus increasing NFκB-DNA binding, NFκB-dependent transcription, IκBα phosphorylation, and nuclear translocation of RelA. Consequently, TWEAK upregulates the mRNA expression and secretion of MCP-1 (monocyte chemoattractant protein-1), IL-6 (interleukin-6), and RANTES (regulated on activation normal T cell expressed and secreted), in a RelA-dependent manner. The effects of TWEAK on these chemokines are notable with regard to the subsequent recruitment of inflammatory cells and rise in inflammation (Izquierdo et al., 2012).

The effects of TWEAK on Klotho expression are poorly understood. In animal models, exogenous TWEAK decreases renal expression of Klotho. In folic acid-induced acute renal lesions, blockade of TWEAK reverts Klotho levels. Moreover, inhibition of NFκB prevents the decrease in Klotho levels due to TWEAK. Thus, proinflammatory cytokines, such as TWEAK, negatively regulate the expression of Klotho through an NFκB-dependent mechanism, which in part explains the relationship between inflammation and diseases that are characterized by the accelerated aging of organs, including the chronic kidney disease (CKD) (Moreno et al., 2011).

4.6 Klotho and NFκB

Several studies have shown that Klotho expression is regulated by an NFκB-dependent mechanism. Inflammation is a source of oxidative stress, and inflammation and oxidative stress have been linked to many diseases that are associated with aging, such as cardiovascular disease (CVD) and cancer (Fig. 4).

NFκB is a nuclear factor that binds to the regulatory regions of promoters in immunoglobulin genes (Sen & Baltimore, 1986). It was initially believed to be a B-lymphocyte-specific transcription factor, having been found in

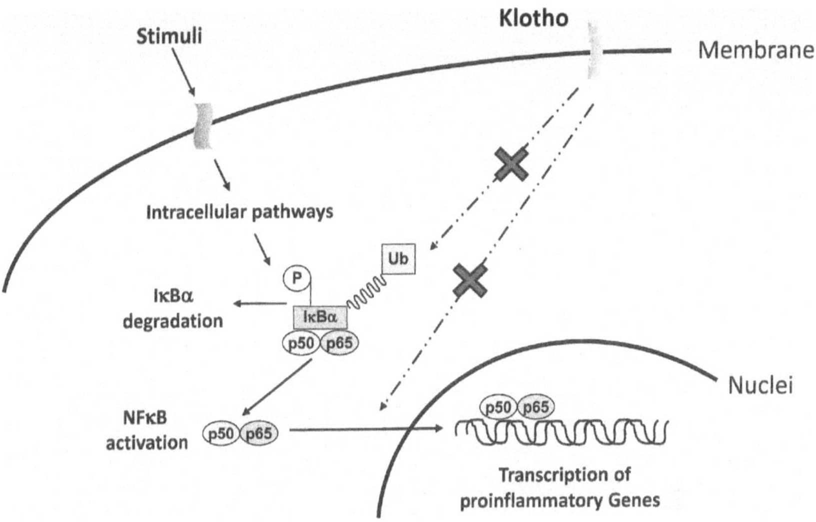

Fig. 4 Klotho and NFκB. Klotho suppresses nuclear NFκB activation and the subsequent transcription of proinflammatory genes. (See the color plate.)

B cell nuclei (Gugasyan et al., 2000). Subsequently, NFκB has been shown to be constitutively expressed in nearly all immune cells and other cell types. NFκB controls many cellular processes (Pahl, 1999), such as proliferation, oxidative stress, antiapoptotic responses, and immune and inflammatory responses, and regulates intracellular signaling during normal development in vertebrates (Lawrence, Gilroy, Colville-Nash, & Willoughby, 2001; Li et al., 1999; Li & Verma, 2002).

In humans, five proteins that belong to the superfamily of NFκB/Rel homology domain (Rel) transcription factors have been identified: RelA (p65), c-Rel, RelB, NFκB1 (p50/P105), and NFκB2 (p52/p100). All members are organized into homodimers or heterodimers, the most common forms of which are the p50/RelA and p52/RelA heterodimers. NFκB1 and NFκB2 are synthesized as 105- and 100-kDa precursors, respectively, and are subsequently processed into smaller active proteins (Sun & Andersson, 2002). Each NFκB family member has a conserved N-terminal region of 300 aa (Muller & Harrison, 1995), called the Rel homology domain, which contains domains for dimerization, interaction with inhibitory proteins or nuclear translocation, and DNA binding. The dimerization of NFκB subunits generates complexes with distinct DNA-binding specificities, with significant physiological implications (Li & Verma, 2002). The consensus DNA sequence to which all dimers binds is 5′-GGGRNNYYCC-3′ (where N denotes A, C, or T; R is a purine, and Y is a pyrimidine), for which various combinations have disparate affinities (Senftleben & Karin, 2002).

Under normal conditions, NFκB is found in the cytoplasm in an inactive form, linked to inhibitory proteins that are called IκBs, of which there are at least five isoforms: IκBα, IκBβ, IκBε, IκBγ, and Bcl-3 (Whiteside & Israel, 1997), and might have greater affinity for other NFκB dimmers (Tran, Merika, & Thanos, 1997). In general, IκBs regulate the subcellular localization and DNA binding of NFκB/Rel, masking certain areas of the RHD region that determines the nuclear destination of NFκB (Henkel et al., 1992).

In response to various stimuli, such as cytokines (TNF, IL-1), chemokines, viral or bacterial pathogens (LPS), ultraviolet radiation, and free radicals are activated by phosphorylation of two serine residues at positions 32 and 36 in the N-terminal region of IκB (Karin & Ben-Neriah, 2002; Karin & Delhase, 2000; Karin & Lin, 2002). This phosphorylation predisposes IκB to ubiquitination by the E3-IκB ubiquitin ligase complex, causing its degradation by the 26S proteasome (Karin & Ben-Neriah, 2002; Karin & Lin, 2002). This process activates NFKB, which then translocates to the nucleus and binds specific regions in the promoters of many genes to regulate their transcription (Abraham, 2000). Because the promoter of IκBα contains κB binding sites, the activation of NFκB increases the expression of IκBα, constituting a mechanism by which IκBα levels are regulated (Woronicz, Gao, Cao, Rothe, & Goeddel, 1997).

The transcriptional activity of NFκB is regulated by the degradation of IκB and based on composition of the NFκB complex. Over 150 genes have been reported to be governed by NFκB, including those that are involved in immunity and inflammation, antiapoptosis, cell proliferation, and downregulation of NFκB signaling. NFκB upregulates genes for cytokines (TNF, IL-1β, IL-6, IL-2, IL-12, IFNγ, CM-CSF), cell adhesion molecules (E-selectin, VCAM-1, ICAM-1), chemokines (IL-8, eolaxina), receptors (TF), and inducible enzymes (COX-2, iNOS). NFκB also increases the expression of important molecules that regulate cell proliferation, apoptosis, and cell cycle progression, such as c-IAP1 and 2, TRAF1 and 2, Bcl-2, Fas, c-myc, and cyclin D1 (May & Ghosh, 1998). The function of TNF-α, IL-1β, and cell adhesion molecules as "stimulators" and, simultaneously, as "final products" of the proinflammatory process that is mediated by NFκB suggests that there is a positive feedback loop that effects persistent inflammation, because the uncontrolled activation of NFκB upregulates key mediators of inflammation (Parry & Mackman, 1994).

NFκB activation is the final step in several signal transduction pathways that are initiated at the cell surface on the binding of effector molecules to their specific receptors (Mercurio & Manning, 1999). The key step in this cascade is the phosphorylation of IκB, mediated by a multimeric complex that is called IKK (IκB kinase) (Chen, Parent, & Maniatis, 1996). The IKK complex comprises two catalytic subunits—IKKα and IKKβ (also termed IKK-1 and IKK-2, respectively)—and a subunit, IKKγ, which does not have catalytic function but is essential for the activation of NFκB by multiple stimuli (Rothwarf, Zandi, Natoli, & Karin, 1998; Zandi, Rothwarf, Delhase, Hayakawa, & Karin, 1997). NIK (NFκB-inducible kinase) activates IKK and belongs to the MEKK (mitogen-activated protein kinase kinase) superfamily (Karin, 1999; Lee, Peters, Dang, & Maniatis, 1998).

The existence of various homodimers and heterodimers (each with a particular transactivator potential), isoforms of IκB that sequester specific dimers, and signal transduction pathways that induce nuclear translocation of NFκB suggests that each type of stimulus corresponds to a particular combination of NFκB isoforms (Baldwin, 1996). Moreover, NFκB activates genes that encode factors that mediate its activation, thus participating in its regulation by inducing Rel family genes, such as c-Rel and p105. These mechanisms ensure that the amplification signals are local and transient.

5. PATHOPHYSIOLOGICAL IMPLICATIONS OF KLOTHO/NFκB

5.1 Klotho in Inflammation

Inflammatory processes have been linked to lower Klotho expression, whereas the inhibition of proinflammatory cytokine production reverses this pattern and restores circulating Klotho levels. Thus, inflammation downregulates Klotho, generating oxidative stress (Xu & Sun, 2015; Zeng, Wang, Zhang, & Du, 2016).

Klotho is as an antiinflammatory modulator, negatively regulating NFκB and, consequently, decreasing proinflammatory gene transduction. We have noted high expression of adhesion molecules, such as ICAM-1 (intercellular adhesion molecule 1) and VCAM-1 (vascular cell adhesion protein 1), in endothelial cells that have been exposed to inflammatory stimuli, such as TNF, from animals that have experienced inflammation and human atherosclerotic plaques. Klotho suppresses TNF-induced ICAM-1 and VCAM-1 expression by attenuating NFκB activity, which has been associated with nitric oxide production. Thus, Klotho is essential in preventing the underlying endothelial dysfunction during inflammation. Further, the intracellular form of Klotho inhibits the expression of IL-6 and IL-8 in senescent cells (Wang & Sun, 2009).

Although greater NFκB activity increases the expression of proinflammatory mediators, such as cytokines and adhesion molecules, a normal immune response is required for optimal development of defensive functions. Normally, NFκB controls the expression of genes that mediate the generation of normal immune responses, during which rapid mobilization of NFκB is necessary to defend against external agents (eg, infection, stress). Excessive stimulation of NFκB, however, activates the immune response, which has been linked to the pathogenesis of several inflammatory diseases, such as rheumatoid arthritis, atherosclerosis, and asthma (Bacher & Schmitz, 2004; Baldwin, 1996; Dejardin, 2006; Mercurio & Manning, 1999; O'Sullivan, Thompson, & Thomas, 2007), and other pathologies, such as type 2 diabetes mellitus, cancer, AIDS, and degenerative diseases of the central nervous system (Aggarwal, Shishodia, Sandur, Pandey, & Sethi, 2006; Pande & Ramos, 2003; Pannu & Singh, 2004; Savoia & Schiffrin, 2007).

In cells that are associated with these diseases, NFκB is constitutively expressed in the nucleus, effecting greater transcriptional activity. It is believed that this chronic activation is caused by defects in the mechanisms that regulate it. For example, in atherosclerosis and its clinical manifestations (heart attack, stroke, or peripheral vascular disease), activated NFκB is present in human atherosclerotic plaques, as are molecules that stimulate it (Pateras, Giaginis, Tsigris, Patsouris, & Theocharis, 2014).

5.2 Klotho in Cellular Senescence

Initially, cultured cells have an infinite capacity to divide, but after several cycles, they enter replicative senescence. We analyzed the expression of Klotho in a model of endothelial cells (HUVECs) and found that gene and protein levels declined with age, preceding the typical manifestations of aging, such as telomere shortening and the expression of β-galactosidase. By flow cytometry, endothelial cells with short telomeres and high expression of β-galactosidase had lower levels of Klotho (Carracedo et al., 2012).

Moreover, cells that are subjected to stress can undergo stress-induced-premature senescence (SIPS). In this regard, experimental models have shown that soluble Klotho protects endothelial cells from SIPS (Carracedo et al., 2012).

The circulating form of Klotho is a humoral factor that inhibits intracellular signaling cascades, such as insulin/IGF-1, and other pathways, resulting in the formation of radicals that promote cellular aging. Thus, Klotho might increase the longevity of certain species due to greater resistance to oxidative stress. Inhibition of Klotho mRNA expression induces premature senescence in human cells. Several factors in certain diseases regulate Klotho expression through DNA methylation, accelerating the aging process. Also, NFκB is sensitive to changes in endothelial redox that are related to age, and its activation stimulates the transcription of proinflammatory cytokines that can further suppress endothelial function, creating a vicious cycle of inflammation and oxidative stress in the endothelium of elderly persons (Kuro-o, 2011; Maekawa et al., 2011).

Physiological aging and CKD are associated with low levels of Klotho, and it appears that Klotho influences cell death and survival pathways through its antisenescent and antioxidative effects (Hu, Kuro-o, & Moe, 2010; John, Cheng, & Kuro-o, 2011; Sugiura, Tsuchiya, & Nitta, 2011). Klotho also exerts its antiaging effects by suppressing the inflammatory activities of substances that are secreted by senescent cells. Klotho interacts with RIG-1 (retinoic acid-inducible gene 1), inhibiting the expression of IL-6 and IL-8 in vivo and in vitro, and Klotho levels are lower in plasma and urine in CKD models (Zeng et al., 2016). Low renal Klotho expression has also been observed in senescent compared with young mice, wherein plasma concentrations are twice as high in young animals than in adults (Rotondi et al., 2015).

Human fibroblasts in which Klotho is blocked undergo premature senescence and upregulate p21 (de Oliveira, 2006). Klotho regulates cellular senescence by suppressing the p53/p21 pathway and might attenuate apoptosis and senescence in human endothelial cells through the MAK and ERK pathways (Sopjani et al., 2015).

Based on these effects, we hypothesize that Klotho deficiency in uremia and aging accelerates cell senescence and apoptosis and decreases the levels of stem cells—all of which are hallmarks of both conditions.

In cultured HUVECs, TNF (tumor necrosis factor alpha) induces cell senescence in association with stress by downregulating Klotho, increasing β-galactosidase activity, and expanding the proportion of cells with short telomeres. When these cells are treated with extracellular Klotho (ie, exogenous Klotho), the percentage of cells with low levels of Klotho, high expression of β-galactosidase, and short telomeres is unaltered. However, exogenous Klotho prevents the rise in the percentage of cells that develop these characteristics. It also halts the increase in ROS activity and apoptosis in endothelial cells that are treated with TNFα. Exposure of HUVECs to proinflammatory stimuli, such as TNF, downregulates Klotho. In this case, exogenous Klotho does not slow cellular aging, but it prevents cellular changes that accelerate the development of cell senescence (Carracedo et al., 2012).

Conversely, oxidative stress increases progressively with age. Extracellular Klotho does not affect oxidative stress processes that are associated with age but protects endothelial cells from TNFα-induced oxidative damage. Thus, we hypothesize that exogenous klotho mitigates the proinflammatory effects of TNFα. It has been suggested that TNF and other proinflammatory cytokines regulate Klotho expression in cultured cells by an NFκB-dependent mechanism, implicating other regulatory factors that are not related to cell senescence.

Cellular changes during TNF-induced premature senescence are related to the induction of apoptosis by TNF—we have reported that telomere shortening precedes apoptosis. Other studies have shown that recombinant Klotho attenuates apoptosis and senescence in vascular cells through the ERK and MAPK pathways. In age-related senescence, telomerase activity decreases but not in SIPS. That treatment with exogenous Klotho does not affect this activity explains why it is unable to modulate the effects of TNFα with regard to senescence. But, it is possible that extracellular Klotho prevents the reduction in endogenous Klotho levels, protecting cells from SIPS (Buendía et al., 2015).

Experimental studies have demonstrated that various processes and intracellular pathways that are involved in the pathogenesis of endothelial dysfunction (eg, oxidative stress, impaired NO, vascular calcification, inflammation) are associated with changes in the homeostasis of Klotho. Based on findings that Klotho functions through the NFκB pathway, the relationship between Klotho and this transcription factor must be analyzed (Buendía et al., 2015).

6. KLOTHO AND PATHOLOGIES

6.1 Cardiovascular Disease

CVD is a common disease in the general population and the leading cause of mortality in developed countries. As a coreceptor for FGF-23, Klotho has been implicated as a key molecule in the development of CVD. In clinical studies, low levels of soluble Klotho correlate with the occurrence and severity of cardiovascular disease. In addition, several human Klotho gene polymorphisms have been associated with the incidence of cardiovascular events (Martín-Núñez et al., 2014).

Several studies suggest that Klotho maintains vascular homeostasis, improves endothelial dysfunction (by inducing NO production), and has antiinflammatory and antiaging effects, by suppressing the expression of adhesion molecules, tempering NFκB activation, and inhibiting Wnt signaling. Further, Klotho is linked to the attenuation of vascular calcification and prevention of cardiac hypertrophy. Understanding the mechanisms of this protein might help guide us in the treatment of vascular disorders (Maekawa et al., 2009; Maltese & Karalliedde, 2012).

6.2 CKD and Klotho

Inflammation is a frequent condition in CKD, especially in its terminal stages, and might contribute to the high risk of vascular calcification in CKD patients. Several factors, including genetic load, oxidative stress, and decreased clearance of proinflammatory cytokines (such as TNF-α, IL-6) or endothelial dysfunction, might promote the inflammatory state in CKD. Further, in CKD, uremic toxins increase the levels of proinflammatory cytokines and oxidative stress, downregulating Klotho at the protein and gene levels (Hu, Kuro-o, & Moe, 2013; Vervloet, Adema, Larsson, & Massy, 2014).

Experimentally, uremia induces premature senescence in endothelial cells, which explains the early development of cardiovascular events in patients with CKD. Uremic serum lowers renal expression of Klotho, decreases NFκB activation, and promotes cellular senescence that is associated with the activation of such proteins as p16, p21, and p53. The addition of exogenous Klotho to endothelial cell cultures that have been treated with uremic serum prevents oxidative stress and senescence. It appears that this antisenescent effect is related to the ability of Klotho to prevent nuclear translocation of NFκB (Hu et al., 2013; Koh et al., 2001; White, Yaqoob, & Harwood, 2015).

In a study by our group, senescence in uremia-induced endothelial cells correlated with decreased levels of Klotho. However, we are unable to conclude whether this decline is merely associated with senescence or a regulatory mechanism of this process. To determine whether the downregulation in intracellular Klotho is a cause or consequence of cellular aging, we blocked the expression of Klotho by siRNA-KL and analyzed its function in the development of senescence. Blockade of Klotho activity resulted in the shortening of telomeres, suggesting that partial inhibition of Klotho is linked to aging (Buendía et al., 2015).

Uremia-induced senescence in endothelial cells is caused primarily by the increase in oxidative stress and activation of the NFκB pathway. But, oxidative stress levels rise progressively with age and by certain stimuli, such as proinflammatory cytokines and uremic serum. Exogenous Klotho does not affect cellular aging and has no effect on the levels of oxidative stress that is associated with aging. However, it protects cells from damage due to specific stimuli, keeping endogenous levels of Klotho constant. Extracellular Klotho prevents the development of the classical signs of senescence by maintaining levels of soluble Klotho. By this mechanism, Klotho becomes protective through the regulation of FGF-R (Buendía et al., 2015).

Decreased klotho levels due to uremia are mediated by NFκB-dependent mechanisms. NFκB activity is associated with inflammation and oxidative stress. Proinflammatory cytokines, such as IL-6 and TNFa, downregulate Klotho through NFκB-dependent pathways, which explains in part the relationship between inflammation and diseases that are characterized by premature aging, including the CKD. Further, high NFκB activity correlates with endothelial dysfunction due to oxidative stress or inflammatory stimuli. Our results support that and extend current findings, identifying NFκB activity as a potential therapeutic target in accelerated senescence of uremia-induced endothelial cells (Buendía et al., 2015; Zeng et al., 2016).

Further, we found that exogenous Klotho protects endothelial cells from uremia-induced senescence—without this molecule, cells are more susceptible to damage by uremic toxins and other factors, such as proinflammatory cytokines, and modified proteins. Exogenous Klotho might prevent the reduction in soluble Klotho and its accompanying problems (Buendía et al., 2015).

Several uremic toxins, such as indoxyl sulfate (IS), decrease the expression of Klotho in the kidney through the production of ROS and activation of NFκB. Thus, uremic toxins might be involved in the low levels of Klotho in CKD. In in vitro experiments with endothelial cells that are cultured in the presence of IS, Klotho is protective, minimizing the effects of oxidative stress. IS induces ROS release, phosphorylation of p38 MAPK, and NFκB nuclear translocation in endothelial cells, whereas treatment with recombinant klotho decreases ROS production, inhibits the activation of p38 MAPK and NFκB, and mitigates the reduction in nitric oxide liberation (Adijiang, Shimizu, Higuchi, Nishijima, & Niwa, 2011; Niwa & Shimizu,

2012; Shimizu et al., 2011; Sun, Chang, & Wu, 2012; Sun, Hsu, & Wu, 2012: Yang et al., 2012). Epigenetic regulation plays an important role in uremic toxin-induced CKD progression. Uremic toxin accumulation can induce DNA methyltransferase (DNMT) protein expression, which is involved in the silencing of Klotho, through hypermethylation. Treatment with DNMT inhibitors may induce a hypermethylated status of Klotho and suppress mRNA and protein expression. Epigenetic targeting of specific genes may become an effective strategy to prevent progression of uremia-related CKD (Chen et al., 2013; Sun, Chang, et al., 2012; Sun, Hsu, et al., 2012; Young & Wu, 2012).

A proper understanding of the interaction between Klotho and NFκB could aid in the design of new therapeutic approaches for pathologies that associated with premature aging and inflammation, such as CKD, inflammation, and aging in general.

In summary, many studies on uremic serum or uremic toxins in culture have shown that Klotho improves the endothelial dysfunction that is induced by these compounds through signaling pathways that converge on NFκB.

6.2.1 Klotho and Cardiovascular Disease in CKD

CVD is the leading cause of death worldwide and is responsible for over 50% of mortality in patients with CKD. The risk of mortality in patients with end-stage renal disease is 20–30 times higher than in the general population (Foley, Parfrey, & Sarnak, 1998). Although kidney disease is often considered an independent risk factor for CVD, there is a patent relationship between diseases, which increases as kidney function deteriorates (Weiner, Tighiouart, Amin, et al., 2004; Weiner, Tighiouart, Stark, et al., 2004). CVD is caused by many factors, such as the presence of periodontal pathogens, oxidative stress, dyslipidemia, diabetes, hypertension, homocysteine, hyperfibrinogenemia, aging, inflammatory factors, vascular calcification, and the uremic state in kidney patients. These elements are acknowledged as new cardiovascular risk factors, because they can contribute to endothelial damage and vascular disease. With regard to CVD and CKD, several studies have reported a link between FGF-23 and increased mortality risk from cardiovascular events (Khan, Chirinos, Litt, Yang, & Rosas, 2012).

Endothelial injury is believed to be the initial step in the development of other lesions that culminate in CVD. A common link between all risk factors is the ability to impair the function of endothelial cells, which maintain homeostasis of the vascular wall (Vervloet et al., 2014).

Patients with CKD develop an early form of atherosclerotic CVD that is associated with accelerated induction of senescence in mature endothelial cells and endothelial progenitor cells. Endothelial pathophysiology is complex and results from several mechanisms that can culminate in the severe impairment of vascular homeostasis and endothelial cells death by apoptosis. However, the endothelium engages cell turnover mechanisms by which endothelial progenitor cells differentiate into mature cells that replace aged or apoptotic cells. When these mechanisms fail, endothelial dysfunction occurs—constituting the first step in the development of atherogenesis (Jaradat & Molitoris, 2002; Paparello, Kshirsagar, & Batlle, 2002; Tonelli & Pfeffer, 2007).

In this regard, the development of premature CVD is frequently associated with more rapid induction of cellular senescence. The presence of senescent cells is common in patients with degenerative diseases, including atherosclerosis. These cells experience changes to their genomes and alterations in their phenotype, such as telomere shortening, similar to those in cells from elderly subjects. Thus, this disease can be considered to be a form of accelerated aging. Telomere shortening is postulated as one of the mechanisms of endothelial cells death due to atherogenic factors, such as the presence of uremic toxins and proinflammatory cytokines (Carracedo et al., 2013; Elewa et al., 2012; Kooman, Kotanko, Schols, Shiels, & Stenvinkel, 2014; Small, Coombes, Bennett, Johnson, & Gobe, 2012; Tsirpanlis, 2008).

Klotho, which is expressed in the kidney and prolongs the life span by controlling insulin, prevents senescence and apoptosis due to oxidative stress in ECs. The study of Klotho has attracted significant interest because animals that have mutations in this gene develop diseases that are associated with aging, including CVD. There are little data on the function of Klotho in endothelial damage in patients who develop premature atherosclerotic CVD, as in the case of those with CKD, but we hypothesize that it protects against cellular senescence process (Martín-Núñez et al., 2014).

The predominant source of Klotho is the kidney. Kidney cells secrete Klotho to the surrounding medium, entering the circulation and traveling to various organs, where it exerts its antiaging functions. Circulating levels of Klotho decline in patients with kidney disease. Persons with CKD and who are on dialysis experience accelerated aging, have a lower survival rate than the general population, and more frequently suffer diseases that are associated with aging, such as cardiovascular and bone diseases, because dialysis does not replace all functions of the kidney (Rotondi et al., 2015).

The primary function of the kidney is to maintain acid–base and electrolyte balances and excrete toxins and waste products of metabolism. The kidney also secretes important hormones that prevent disease. Dialysis replaces certain activities of the kidney functions but not its production of hormones (Gutierrez & Ix, 2013). Evidence suggests that pharmacological supplementation of Klotho in murine models of CKD improves many of the accompanying disorders, including hyperphosphoremia, hypercalcemia, and vascular calcification, despite maintaining high levels of creatinine in plasma, indicating that the uremic syndrome phenotype is attributed more to a lower filtration rate and Klotho levels—which is novel concept in our understanding of CKD. It is thus necessary to provide kidney disease patients with hormones that are normally produced by the kidneys, such as erythropoietin, which combats anemia, and active vitamin D, which improves bone health and protects against many diseases. However, it remains impossible to compensate for a deficiency in Klotho (Hu, Kuro-o, & Moe, 2012; Hu et al., 2013).

6.2.2 Vascular Calcification

Hyperphosphatemia and elevated $Ca \times P$ product in patients with renal insufficiency are closely related to the calcification of coronary arteries, peripheral arteries, and heart valves (Block, 2000; London, Pannier, Marchais, & Guerin, 2000; Shigematsu et al., 2003) and to mortality in hemodialysis patients (Block, 2000). In uremic patients, vascular calcification (VC) is characterized because it is more prevalent and widespread, developing faster and appearing at an earlier age compared with the general population. Also, calcification of coronary arteries in the general population and in uremic patients correlates strongly with the presence of atherosclerotic plaque (Rumberger, Simons, Fitzpatrick, Sheedy, & Schwartz, 1995; Sangiorgi et al., 1998), plaque instability (Burke, Taylor, Farb, Malcom, & Virmani, 2000; Fitzgerald, Ports, & Yock, 1992), and an increased risk of stroke infarction (Beadenkopf, Daoud, & Love, 1964; Loecker, Schwartz, Cotta, & Hickman, 1992).

In patients with CKD and those on hemodialysis, the degree of arterial calcification is a predictive factor of CVD and mortality (Blacher, Guerin, Pannier, Marchais, & London, 2001; Lee, Huang, et al., 2014; Lee, Kim, et al., 2014; London et al., 2003). For many years, it was believed that vascular calcification was attributed to a passive process and that it was a degenerative disease without treatment options (Virchow, 1989). Evidence from the past two decades has shown that it is an active process that is highly regulated and similar to the physiological process of bone mineralization (Demer & Tintut, 2008; Moe & Chen, 2004; Shanahan et al., 1999). Further, during the development of the CV, vascular smooth muscle cells (VSMCs) undergo a phenotypic transformation toward osteogenic lineage cells (Steitz et al., 2001; Tintut et al., 2003). Consequently, there are significant similarities between the formation of the CV and osteogenesis, because ossification mechanisms in CV approximate those of bone formation (Towler, 2005; Vattikuti & Towler, 2004). Nevertheless, the mechanisms that affect the CV remain undefined.

In healthy subjects, the kidneys regulate the homeostasis of calcium and phosphorus through renal tubular reabsorption. In CKD patients, such mechanisms are severely compromised, leading to various adaptive changes in the levels of calcium (Ca), phosphorus (P), PTH, vitamin D, fibroblast growth factor (FGF-23), and Klotho. Of the alterations in CKD, those that are related to bone and mineral metabolism (which induces vascular calcification) have a significant impact on morbidity and mortality (de Oliveira et al., 2013).

In CKD, VSMCs may be transdifferentiated into osteoblast-like cells. Vascular calcification is also caused by an imbalance between molecules, such as calcium, phosphate, bone morphogenic proteins (BMP-2, BMP-4), RANKL (ligand for receptor activator NFκB), IL-6, and TNFα, and the activation of transcription factors, such as CbFα1/RunX2/Osterix and inflammatory mediators, like NFκB and the inhibitors of the calcification process, such as fetuin A, MGP (matrix GLA protein), osteopontin, osteoprotegerin, BMP-7, or pyrophosphate. Other factors that favor the calcification of VSMCs in uremic medium are oxidative stress, AGEs (advanced glycation end-products), hypertension, adynamic bone disease, and dyslipidemia (de Oliveira et al., 2013).

Phosphorus enters the cell through cotransporter Na-P (NTP). NTP-III (Pit-1 and Pit-2) is expressed in smooth muscle cells (Pit-1 > Pit-2). Phosphorus crosses the membrane through Pit-1 and stimulates the production of superoxide radicals that activate the NFκB pathway, inducing the formation of mineralizing vesicles and osteochondrogenic transformation. In the smooth muscle cells of Klotho-deficient mice, these cotransporters are upregulated, leading to a rise in the intracellular levels of phosphorus and causing the development of vascular calcification. Further, the addition of Klotho to VSMC cultures decreases the activity of Pit-1 and Pit-2, lowering phosphorus levels and thus preventing osteogenic differentiation in this cell type (Vervloet et al., 2014).

Uremic toxins appear to induce the expression of DNMT, a protein that mediates the silencing of Klotho through hypermethylation (Young & Wu, 2012). As a result of these mechanisms and due to the loss of renal mass, the availability of Klotho and FGF-23 declines without its coreceptor to act. Thus, during the progression of renal disease, FGF-23 loses its phosphaturic capacity, and plasma levels of phosphorus climb (Olauson & Larsson, 2013).

We believe that FGF-23 benefits survival, maintaining normophosphatemia and preventing vascular calcification. However, as kidney injury progresses, excess active FGF-23 that is produced in response to hyperphosphatemia might have adverse effects. Thus, FGF-23 contributes to the reduction of CTR-inducing vascular dysfunction and atherosclerosis. Klotho participates in aging-associated phenomena, through FGF-23; therefore, molecules may be both potential biomarkers and therapeutic targets that improve survival.

6.3 Klotho and Chronic Obstructive Pulmonary Disease

Chronic obstructive pulmonary disease (COPD) is associated with chronic inflammation and accelerated aging of the lung. Klotho levels are lower in epithelial cells of the respiratory tract in COPD patients in association with increased oxidative stress, inflammation, and apoptosis—all of which are mediated by NFκB and MAPK kinases (Gao et al., 2015).

Another study has demonstrated that the level of Klotho protein was decreased in alveolar macrophages in peripheral lung tissue of smokers and patients with COPD and in peripheral blood mononuclear cells. NFκB is likely to be involved in the underlying mechanism by which Klotho inhibits stimulated macrophage in COPD (Blake, Reese, Garcia, Dahlmann, & Dean, 2015; Li et al., 2015).

6.4 Klotho and Diabetes

Klotho is an aging hormone produced in the kidney and extends the life span, regulates renal function, and modulates cellular response to oxidative stress. Therefore, the Klotho levels in kidneys of diabetic mice and their relation with inflammation have been investigated. It has been observed that diabetic mice show low levels of Klotho mRNA and protein. A similar decrease was observed in primary cultures of epithelial cells treated with modified albumin. In these cultures, the addition of exogenous Klotho suppressed NFκB activation and the subsequent production and release of proinflammatory cytokines.

Klotho specifically inhibits RelA phosphorylation (Ser 536) and RelA binding to DNA without affecting IκB degradation. Therefore in diabetic mice, Klotho acts like an antiinflammatory modulator as it downregulates the production of inflammatory proteins induced by NFκB activation through a mechanism involving phosphorylation of RelA (Zhao et al., 2011).

Klotho deficiency promoted β-cell apoptosis in two models of T1DM. Overexpression of Klotho improved β-cell function and protected pancreatic β-cells in an animal model. The beneficial effects of Klotho in β-cells are likely mediated via attenuating β-cell apoptosis. Hence, β-cell-specific expression of Klotho may find applications in the treatment of type 1 diabetes mellitus. Moreover, Klotho improves β-cell function and attenuates the development of type 2 diabetes mellitus (Lin & Sun, 2015a, 2015b). Therefore, soluble klotho levels in plasma and urine may be novel and early markers of renal injury in diabetes (Lee, Huang, et al., 2014; Lee, Kim, et al., 2014).

7. CONCLUSIONS AND PERSPECTIVES

Klotho has been described as an antiaging gene. The decrease of Klotho is associated with inflammatory processes and cellular senescence. Klotho acts as a coreceptor involved in important intracellular signaling pathways, and the inhibition of translocation of NFκB transcription factor is included in these pathways. Numerous studies are currently carried on the Klotho implication in different pathophysiological processes. Therefore, there are interesting prospects in relation to the role of Klotho as a therapeutic target.

With the recent escalation of research of Klotho on pathophysiology, the understanding of the relationship of Klotho with kidney disease has been advanced considerably. Klotho is not merely an early biomarker for CKD but also a pathogenetic intermediate for CKD progression and complication development. Plasma or urinary Klotho should be explored as markers for the presence of renal disease, predictors for progression, and extrarenal complications, particularly soft-tissue calcification. Besides the important observational database, it is critical to know how to translate the preclinical therapeutic results of Klotho in animal experiments to human CKD. Perhaps the extrarenal effects of Klotho still warrant its replacement even if there are no renal effects or salvage. The most important concept is that we will likely individualize therapy in mineral disturbances in CKD.

ACKNOWLEDGMENTS

This work was supported by Plan Nacional de I+D+i Proyectos de Investigación en Salud of Instituto de Salud Carlos III (ISCIII)-Subdirección General de Evaluación, Fondos de desarrollo regional (FEDER) (PI12/01489 and PI14/00806); Junta de Andalucía Grants (P010-CTS-6337, P11-CTS-7352); and Fundación Nefrológica. J.C. was supported by a contract from Fundación de Investigaciones Biomédicas de Córdoba (Servicio Andaluz de Salud, Programa Nicolás Monardes). P.B. is fellow from Consejería de Innovacion, Ciencia y Empresa, Junta de Andalucía.

REFERENCES

Abraham, E. (2000). NF-κB activation. *Critical Care Medicine, 28*, 100–104.

Adijiang, A., Shimizu, H., Higuchi, Y., Nishijima, F., & Niwa, T. (2011). Indoxyl sulfate reduces klotho expression and promotes senescence in the kidneys of hypertensive rats. *Journal of Renal Nutrition, 21*, 105–109.

Aggarwal, B. B., Shishodia, S., Sandur, S. K., Pandey, M. K., & Sethi, G. (2006). Inflammation and cancer: How hot is the link? *Biochemical Pharmacology, 72*, 1605–1621.

Bacher, S., & Schmitz, M. L. (2004). The NF-kappaB pathway as a potential target for autoimmune disease therapy. *Current Pharmaceutical Design, 10*, 2827–2837.

Baldwin, A. S. (1996). The NF-κB and IκB proteins: New discoveries and insights. *Annual Review of Immunology, 14*, 649–683.

Beadenkopf, W. G., Daoud, A. S., & Love, B. M. (1964). Calcification in the coronary arteries and its relationship to arteriosclerosis and myocardial infarction. *American Journal of Roentgenology, Radium Therapy, and Nuclear Medicine, 92*, 865–871.

Ben-Dov, I. Z., Galitzer, H., Lavi-Moshayoff, V., Goetz, R., Kuro-o, M., Mohammadi, M., et al. (2007). The parathyroid is a target organ for FGF23 in rats. *Journal of Clinical Investigation, 117*, 4003–4008.

Blacher, J., Guerin, A. P., Pannier, B., Marchais, S. J., & London, G. M. (2001). Arterial calcifications, arterial stiffness, and cardiovascular risk in end-stage renal disease. *Hypertension, 38*, 938–942.

Blake, D. J., Reese, C. M., Garcia, M., Dahlmann, E. A., & Dean, A. (2015). Soluble extracellular Klotho decreases sensitivity to cigarette smoke induced cell death in human lung epithelial cells. *Toxicology In Vitro, 29*, 1647–1652.

Block, G. A. (2000). Prevalence and clinical consequences of elevated Ca x P product in hemodialysis patients. *Clinical Nephrology, 54*, 318–324.

Buendía, P., Carracedo, J., Soriano, S., Madueño, J. A., Ortiz, A., Martín-Malo, A., et al. (2015). Klotho prevents NFκB translocation and protects endothelial cell from senescence induced by uremia. *The Journals of Gerontology. Series A, Biological Sciences and Medical Sciences, 70*, 1198–1209.

Burke, A. P., Taylor, A., Farb, A., Malcom, G. T., & Virmani, R. (2000). Coronary calcification: Insights from sudden coronary death victims. *Zeitschrift für Kardiologie, 89*, 49–53.

Carracedo, J., Buendía, P., Merino, A., Madueño, J. A., Peralbo, E., Ortiz, A., et al. (2012). Klotho modulates the stress response in human senescent endothelial cells. *Mechanisms of Ageing and Development, 133*, 647–654.

Carracedo, J., Buendía, P., Merino, A., Soriano, S., Esquivias, E., Martín-Malo, A., et al. (2013). Cellular senescence determines endothelial cell damage induced by uremia. *Experimental Gerontology, 48*, 766–773.

Cha, S. K., Hu, M. C., Kurosu, H., Kuro-o, M., Moe, O., & Huang, C. L. (2009). Regulation of renal outer medullary potassium channel and renal K(+) excretion by Klotho. *Molecular Pharmacology, 76*, 38–46.

Chang, Q., Hoefs, S., van der Kemp, A. W., Topala, C. N., Bindels, R. J., & Hoenderop, J. G. (2005). The beta-glucuronidase klotho hydrolyzes and activates the TRPV5 channel. *Science, 310*, 490–493.

Chen, Z. J., Parent, L., & Maniatis, T. (1996). Site-specific phosphorylation of IkappaBalpha by a novel ubiquitination-dependent protein kinase activity. *Cell, 84*, 853–862.

Chen, J., Zhang, X., Zhang, H., Lin, J., Zhang, C., Wu, Q., et al. (2013). Elevated Klotho promoter methylation is associated with severity of chronic kidney disease. *PLoS One, 8*, e79856.

Coulier, F., Ollendorff, V., Marics, I., Rosnet, O., Batoz, M., Planche, J., et al. (1991). The FGF6 gene within the FGF multigene family. *Annals of the New York Academy of Sciences, 638*, 53–61.

Dejardin, E. (2006). The alternative NF-kappaB pathway from biochemistry to biology: Pitfalls and promises for future drug development. *Biochemical Pharmacology, 72*, 1161–1179.

Demer, L. L., & Tintut, Y. (2008). Vascular calcification: Pathobiology of a multifaceted disease. *Circulation, 117*, 2938–2948.

de Oliveira, R. M. (2006). Klotho RNAi induces premature senescence of human cells via a p53/p21 dependent pathway. *FEBS Letters, 580*, 5753–5758.

de Oliveira, R. B., Okazaki, H., Stinghen, A. E., Drüeke, T. B., Maíz, Z. A., & Jorgetti, V. (2013). Vascular calcification in chronic kidney disease: A review. *The Jornal Brasileiro de Nefrologia, 35*, 147–161.

Ding, H. Y., & Ma, H. X. (2015). Significant roles of anti-aging protein klotho and fibroblast growth factor23 in cardiovascular disease. *Journal of Geriatric Cardiology, 12*, 439–447.

Elewa, U., Sanchez-Niño, M. D., Martin-Cleary, C., Fernandez-Fernandez, B., Egido, J., & Ortiz, A. (2012). Cardiovascular risk biomarkers in CKD: The inflammation link and the road less traveled. *International Urology and Nephrology, 44*, 1731–1744.

Fitzgerald, P. J., Ports, T. A., & Yock, P. G. (1992). Contribution of localized calcium deposits to dissection after angioplasty. An observational study using intravascular ultrasound. *Circulation, 86*, 64–70.

Foley, R. N., Parfrey, P. S., & Sarnak, M. J. (1998). Epidemiology of cardiovascular disease in chronic renal disease. *Journal of the American Society of Nephrology, 9*(Suppl. 12), S16–S23.

Gao, W., Yuan, C., Zhang, J., Li, L., Yu, L., Wiegman, C. H., et al. (2015). Klotho expression is reduced in COPD airway epithelial cells: Effects on inflammation and oxidant injury. *Clinical Science (London), 129*, 1011–1023.

Gugasyan, R., Grumont, R., Grossmann, M., Nakamura, Y., Pohl, T., Nesic, D., et al. (2000). Rel/NF-kappaB transcription factors: Key mediators of B-cell activation. *Immunological Reviews, 176*, 134–140.

Gutierrez, O. M., & Ix, J. H. (2013). α-Klotho and kidney function decline: An important step forward in understanding the link between mineral metabolism and kidney disease progression. *American Journal of Kidney Diseases, 61*, 855–857.

Henkel, T., Zabel, U., van Zee, K., Muller, J. M., Fanning, E., & Baeuerle, P. A. (1992). Intramolecular masking of the nuclear location signal and dimerization domain in the precursor for the p50 NF-kappa B subunit. *Cell, 68*, 1121–1133.

Hu, M. C., Kuro-o, M., & Moe, O. W. (2010). Klotho and kidney disease. *Journal of Nephrology, 23*(Suppl. 16), S136–S144.

Hu, M. C., Kuro-o, M., & Moe, O. W. (2012). The emerging role of Klotho in clinical nephrology. *Nephrology, Dialysis, Transplantation, 27*, 2650–2657.

Hu, M. C., Kuro-o, M., & Moe, O. W. (2013). Klotho and chronic kidney disease. *Contributions to Nephrology, 180*, 47–63.

Huang, C. L. (2010). Regulation of ion channels by secreted Klotho: Mechanisms and implications. *Kidney International, 77*, 855–860.

Imura, A., Iwano, A., Tohyama, O., Tsuji, Y., Nozaki, K., Hashimoto, N., et al. (2004). Secreted Klotho protein in sera and CSF: Implication for post-translational cleavage in release of Klotho protein from cell membrane. *FEBS Letters, 565*, 143–147.

Izquierdo, M. C., Perez-Gomez, M. V., Sanchez-Niño, M. D., Sanz, A. B., Ruiz-Andres, O., Poveda, J., et al. (2012). Klotho, phosphate and inflammation/ageing in chronic kidney disease. *Nephrology, Dialysis, Transplantation, 27*(Suppl. 4), iv6–iv10.

Jaradat, M. I., & Molitoris, B. A. (2002). Cardiovascular disease in patients with chronic kidney disease. *Seminars in Nephrology, 22*, 459–473.

John, G. B., Cheng, C. Y., & Kuro-o, M. (2011). Role of Klotho in aging, phosphate metabolism, and CKD. *American Journal of Kidney Diseases, 58*, 127–134.

Johnson, D. E., & Williams, L. T. (1993). Structural and functional diversity in the FGF receptor multigene family. *Advances in Cancer Research, 60*, 1–41.

Karin, M. (1999). The beginning of the end: IkappaB kinase (IKK) and NF-kappaB activation. *The Journal of Biological Chemistry, 274*, 27339–27342.

Karin, M., & Ben-Neriah, Y. (2002). Regulatory functions of ubiquitination in the immune system. *Nature Immunology, 3*, 20–26.

Karin, M., & Delhase, M. (2000). The I kappa B kinase (IKK) and NF-kappa B: Key elements of proinflammatory signalling. *Seminars in Immunology, 12*, 85–98.

Karin, M., & Lin, A. (2002). NF-kappaB at the crossroads of life and death. *Nature Immunology, 3*, 221–227.

Khan, A. M., Chirinos, J. A., Litt, H., Yang, W., & Rosas, S. E. (2012). FGF-23 and the progression of coronary arterial calcification in patients new to dialysis. *Clinical Journal of the American Society of Nephrology, 7*, 2017–2022.

Kirikoshi, H., Sagara, N., Saitoh, T., Tanaka, K., Sekihara, H., Shiokawa, K., et al. (2000). Molecular cloning and characterization of human FGF-20 on chromosome 8p21.3-p22. *Biochemical and Biophysical Research Communications, 274*, 337–343.

Koh, N., Fujimori, T., Nishiguchi, S., Tamori, A., Shiomi, S., Nakatani, T., et al. (2001). Severely reduced production of klotho in human chronic renal failure kidney. *Biochemical and Biophysical Research Communications, 280*, 1015–1020.

Komaba, H., Goto, S., Fujii, H., Hamada, Y., Kobayashi, A., Shibuya, K., et al. (2010). Depressed expression of Klotho and FGF receptor 1 in hyperplastic parathyroid glands from uremic patients. *Kidney International, 77*, 232–238.

Kooman, J. P., Kotanko, P., Schols, A. M., Shiels, P. G., & Stenvinkel, P. (2014). Chronic kidney disease and premature ageing. *Nature Reviews. Nephrology, 10*, 732–742.

Krajisnik, T., Björklund, P., Marsell, R., Ljunggren, O., Akerström, G., Jonsson, K. B., et al. (2007). Fibroblast growth factor-23 regulates parathyroid hormone and 1alpha-hydroxylase expression in cultured bovine parathyroid cells. *Journal of Endocrinology, 195*, 125–131.

Kuro-o, M. (2008). Klotho as a regulator of oxidative stress and senescence. *Biological Chemistry, 389*, 233–241.

Kuro-o, M. (2011). Klotho and the aging process. *The Korean Journal of Internal Medicine, 26*, 113–122.

Kuro-o, M., Matsumura, Y., Aizawa, H., Kawaguchi, H., Suga, T., Utsugi, T., et al. (1997). Mutation of the mouse klotho gene leads to a syndrome resembling ageing. *Nature, 390*, 45–51.

Kurosu, H., & Kuro-O, M. (2009). The Klotho gene family as a regulator of endocrine fibroblast growth factors. *Molecular and Cellular Endocrinology, 299*, 72–78.

Kurosu, H., Ogawa, Y., Miyoshi, M., Yamamoto, M., Nandi, A., Rosenblatt, K. P., et al. (2006). Regulation of fibroblast growth factor-23 signaling by klotho. *Journal of Biological Chemistry, 281*, 6120–6123.

Kurosu, H., Yamamoto, M., Clark, J. D., Pastor, J. V., Nandi, A., Gurnani, P., et al. (2005). Suppression of aging in mice by the hormone Klotho. *Science, 309*, 1829–1833.

Lawrence, T., Gilroy, D. W., Colville-Nash, P. R., & Willoughby, D. A. (2001). Possible new role for NF-kappaB in the resolution of inflammation. *Nature Medicine*, *7*, 1291–1297.

Lee, C. T., Huang, C. C., Hsu, C. Y., Chiou, T. T., Ng, H. Y., Wu, C. H., et al. (2014). Calcification of the aortic arch predicts cardiovascular and all-cause mortality in chronic hemodialysis patients. *Cardiorenal Medicine*, *4*, 34–42.

Lee, E. Y., Kim, S. S., Lee, J. S., Kim, I. J., Song, S. H., Cha, S. K., et al. (2014). Soluble α-klotho as a novel biomarker in the early stage of nephropathy in patients with type 2 diabetes. *PLoS One*, *9*, e102984.

Lee, F. S., Peters, R. T., Dang, L. C., & Maniatis, T. (1998). MEKK1 activates both IkappaB kinase alpha and IkappaB kinase beta. *Proceedings of the National Academy of Sciences of the United States of America*, *95*, 9319–9324.

Li, Z. W., Chu, W., Hu, Y., Delhase, M., Deerinck, T., Ellisman, M., et al. (1999). The IKKbeta subunit of IkappaB kinase (IKK) is essential for nuclear factor kappaB activation and prevention of apoptosis. *Journal of Experimental Medicine*, *189*, 1839–1845.

Li, Q., & Verma, I. M. (2002). NF-kappaB regulation in the immune system. *Nature Reviews Immunology*, *2*, 725–734.

Li, L., Wang, Y., Gao, W., Yuan, C., Zhang, S., Zhou, H., et al. (2015). Klotho reduction in alveolar macrophages contributes to cigarette smoke extract-induced inflammation in chronic obstructive pulmonary disease. *The Journal of Biological Chemistry*, *290*(46), 27890–27900.

Lin, Y., & Sun, Z. (2015a). Antiaging gene klotho attenuates pancreatic β-cell apoptosis in type 1 diabetes. *Diabetes*, *64*(12), 4298–4311.

Lin, Y., & Sun, Z. (2015b). In vivo pancreatic β-cell-specific expression of antiaging gene Klotho: A novel approach for preserving β-cells in type 2 diabetes. *Diabetes*, *64*, 1444–1458.

Liu, S., Tang, W., Zhou, J., Stubbs, J. R., Luo, Q., Pi, M., et al. (2006). Fibroblast growth factor 23 is a counter-regulatory phosphaturic hormone for vitamin D. *Journal of the American Society of Nephrology*, *17*, 1305–1315.

Liu, S., Vierthaler, L., Tang, W., Zhou, J., & Quarles, L. D. (2008). FGFR3 and FGFR4 do not mediate renal effects of FGF23. *Journal of the American Society of Nephrology*, *19*, 2342–2350.

Loecker, T. H., Schwartz, R. S., Cotta, C. W., & Hickman, J. R., Jr. (1992). Fluoroscopic coronary artery calcification and associated coronary disease in asymptomatic young men. *Journal of the American College of Cardiology*, *19*, 1167–1172.

London, G. M., Guérin, A. P., Marchais, S. J., Métivier, F., Pannier, B., & Adda, H. (2003). Arterial media calcification in end-stage renal disease: Impact on all-cause and cardiovascular mortality. *Nephrology, Dialysis, Transplantation*, *18*, 1731–1740.

London, G. M., Pannier, B., Marchais, S. J., & Guerin, A. P. (2000). Calcification of the aortic valve in the dialyzed patient. *Journal of the American Society of Nephrology*, *11*, 778–783.

Maekawa, Y., Ishikawa, K., Yasuda, O., Oguro, R., Hanasaki, H., Kida, I., et al. (2009). Klotho suppresses TNF-alpha-induced expression of adhesion molecules in the endothelium and attenuates NF-kappaB activation. *Endocrine*, *35*, 341–346.

Maekawa, Y., Ohishi, M., Ikushima, M., Yamamoto, K., Yasuda, O., Oguro, R., et al. (2011). Klotho protein diminishes endothelial apoptosis and senescence via a mitogen-activated kinase pathway. *Geriatrics & Gerontology International*, *11*, 510–516.

Maltese, G., & Karalliedde, J. (2012). The putative role of the antiageing protein klotho in cardiovascular and renal disease. *International Journal of Hypertension*, *2012*, 757469.

Martín-Núñez, E., Donate-Correa, J., Muros-de-Fuentes, M., Mora-Fernández, C., & Navarro-González, J. F. (2014). Implications of Klotho in vascular health and disease. *World Journal of Cardiology*, *6*, 1262–1269.

Matsumuraa, Y., Aizawaa, H., Shiraki-Iidaa, T., Nagaib, R., Kuro-o, M., & Nabeshima, Y. (1998). Identification of the human klotho gene and its two transcripts encoding membrane and secreted klotho protein. *Biochemical and Biophysical Research Communications, 242*, 626–630.

May, M. J., & Ghosh, S. (1998). Signal transduction through NF-κB. *Immunology Today, 19*, 80–88.

Mercurio, F., & Manning, A. M. (1999). Multiple signals converging on NF-κB. *Current Opinion in Cell Biology, 11*, 226–232.

Mitani, H., Ishizaka, N., Aizawa, T., Ohno, M., Usui, S., Suzuki, T., et al. (2002). In vivo klotho gene transfer ameliorates angiotensin II-induced renal damage. *Hypertension, 39*, 838–843.

Moe, S. M., & Chen, N. X. (2004). Pathophysiology of vascular calcification in chronic kidney disease. *Circulation Research, 95*, 560–567.

Moreno, J. A., Izquierdo, M. C., Sanchez-Niño, M. D., Suárez-Alvarez, B., Lopez-Larrea, C., Jakubowski, A., et al. (2011). The inflammatory cytokines TWEAK and TNFα reduce renal klotho expression through NFκB. *Journal of the American Society of Nephrology, 22*, 1315–1325.

Muller, C. W., & Harrison, S. C. (1995). Structure of the NF-kappa B p50 homodimer bound to DNA. *Nature, 373*, 311–317.

Nabeshima, Y. (2002). Klotho: A fundamental regulator of aging. *Ageing Research Reviews, 1*, 627–638.

Nabeshima, Y., & Imura, H. (2008). Alpha-Klotho: A regulator that integrates calcium homeostasis. *American Journal of Nephrology, 28*, 455–464.

Niwa, T., & Shimizu, H. (2012). Indoxyl sulfate induces nephrovascular senescence. *Journal of Renal Nutrition, 22*, 102–106.

Olauson, H., & Larsson, T. E. (2013). FGF23 and Klotho in chronic kidney disease. *Current Opinion in Nephrology and Hypertension, 22*, 397–404.

Olauson, H., Vervloet, M. G., Cozzolino, M., Massy, Z. A., Ureña Torres, P., & Larsson, T. E. (2014). New insights into the FGF23-Klotho axis. *Seminars in Nephrology, 34*, 578–585.

O'Sullivan, B., Thompson, A., & Thomas, R. (2007). NF-kappa B as a therapeutic target in autoimmune disease. *Expert Opinion on Therapeutic Targets, 11*, 111–122.

Pahl, H. L. (1999). Activators and target genes of Rel/NF-κB transcription factors. *Oncogene, 18*, 6853–6866.

Pande, V., & Ramos, M. J. (2003). Nuclear factor kappa B: A potential target for anti-HIV chemotherapy. *Current Medicinal Chemistry, 10*, 1603–1615.

Pannu, R., & Singh, I. (2004). Pharmacological strategies for the regulation of inducible nitric oxide synthase: Neurodegenerative versus neuroprotective mechanisms. *Neurochemistry International, 49*, 170–182.

Paparello, J., Kshirsagar, A., & Batlle, D. (2002). Comorbidity and cardiovascular risk factors in patients with chronic kidney disease. *Seminars in Nephrology, 22*, 494–506.

Parry, G. C., & Mackman, N. (1994). A set of inducible genes expressed by activated human monocytic and endothelial cells contain κB-like sites that specifically bind c-Rel-p65 heterodimers. *The Journal of Biological Chemistry, 44*, 2870–2878.

Pateras, I., Giaginis, C., Tsigris, C., Patsouris, E., & Theocharis, S. (2014). NF-κB signaling at the crossroads of inflammation and atherogenesis: Searching for new therapeutic links. *Expert Opinion on Therapeutic Targets, 18*, 1089–1101.

Rothwarf, D. M., Zandi, E., Natoli, G., & Karin, M. (1998). IKK-gamma is an essential regulatory subunit of the IkappaB kinase complex. *Nature, 395*, 297–300.

Rotondi, S., Pasquali, M., Tartaglione, L., Muci, M. L., Mandanici, G., Leonangeli, C., et al. (2015). Soluble α-Klotho serum levels in chronic kidney disease. *International Journal of Endocrinology, 2015*, 872193.

Rumberger, J. A., Simons, D. B., Fitzpatrick, L. A., Sheedy, P. F., & Schwartz, R. S. (1995). Coronary artery calcium area by electron-beam computed tomography and coronary atherosclerotic plaque area. A histopathologic correlative study. *Circulation, 92*, 2157–2162.

Rusnati, M., Urbinati, C., & Presta, M. (1993). Internalization of basic fibroblast growth factor (bFGF) in cultured endothelial cells: Role of the low affinity heparin-like bFGF receptors. *Journal of Cellular Physiology, 154*, 152–161.

Saito, K., Ishizaka, N., Mitani, H., Ohno, M., & Nagai, R. (2003). Iron chelation and a free radical scavenger suppress angiotensin II-induced downregulation of klotho, an anti-aging gene, in rat. *FEBS Letters, 551*, 58–62.

Sangiorgi, G., Rumberger, J. A., Severson, A., Edwards, W. D., Gregoire, J., Fitzpatrick, L. A., et al. (1998). Arterial calcification and not lumen stenosis is highly correlated with atherosclerotic plaque burden in humans: A histologic study of 723 coronary artery segments using nondecalcifying methodology. *Journal of the American College of Cardiology, 31*, 126–133.

Savoia, C., & Schiffrin, E. L. (2007). Vascular inflammation in hypertension and diabetes: Molecular mechanisms and therapeutic interventions. *Clinical Science (London), 112*, 375–384.

Sen, R., & Baltimore, D. (1986). Multiple nuclear factors interact with the immunoglobulin enhancer sequences. *Cell, 46*, 705–716.

Senftleben, U., & Karin, M. (2002). The IKK/NF-kappaB pathway. *Critical Care Medicine, 30*, S18–S26.

Shanahan, C. M., Cary, N. R., Salisbury, J. R., Proudfoot, D., Weissberg, P. L., & Edmonds, M. E. (1999). Medial localization of mineralization-regulating proteins in association with Mönckeberg's sclerosis: Evidence for smooth muscle cell-mediated vascular calcification. *Circulation, 100*, 2168–2176.

Shigematsu, T., Kono, T., Satoh, K., Yokoyama, K., Yoshida, T., Hosoya, T., et al. (2003). Phosphate overload accelerates vascular calcium deposition in end-stage renal disease patients. *Nephrology, Dialysis, Transplantation, 18*, iii86–iii89.

Shimada, T., Hasegawa, H., Yamazaki, Y., Muto, T., Hino, R., Takeuchi, Y., et al. (2004). FGF-23 is a potent regulator of vitamin D metabolism and phosphate homeostasis. *Journal of Bone and Mineral Research, 19*, 429–435.

Shimada, T., Takeshita, Y., Murohara, T., Sasaki, K., Egami, K., Shintani, S., et al. (2004). Angiogenesis and vasculogenesis are impaired in the precocious-aging klotho mouse. *Circulation, 110*, 1148–1155.

Shimada, T., Yamazaki, Y., Takahashi, M., Hasegawa, H., Urakawa, I., Oshima, T., et al. (2005). Vitamin D receptor-independent FGF23 actions in regulating phosphate and vitamin D metabolism. *American Journal of Physiology. Renal Physiology, 289*, 1088–1095.

Shimizu, H., Bolati, D., Adijiang, A., Adelibieke, Y., Muteliefu, G., Enomoto, A., et al. (2011). Indoxyl sulfate downregulates renal expression of Klotho through production of ROS and activation of nuclear factor-κB. *American Journal of Nephrology, 33*, 319–324.

Shiraki-Iida, T., Aizawa, H., Matsumura, Y., Sekine, S., Iida, A., Anazawa, H., et al. (1998). Structure of the mouse klotho gene and its two transcripts encoding membrane and secreted protein. *FEBS Letters, 424*, 6–10.

Small, D. M., Coombes, J. S., Bennett, N., Johnson, D. W., & Gobe, G. C. (2012). Oxidative stress, anti-oxidant therapies and chronic kidney disease. *Nephrology (Carlton, Vic.), 17*, 311–321.

Sopjani, M., Rinnerthaler, M., Kruja, J., & Dermaku-Sopjani, M. (2015). Intracellular signaling of the aging suppressor protein Klotho. *Current Molecular Medicine, 15*, 27–37.

Steitz, S. A., Speer, M. Y., Curinga, G., Yang, H. Y., Haynes, P., Aebersold, R., et al. (2001). Smooth muscle cell phenotypic transition associated with calcification: Upregulation of Cbfa1 and downregulation of smooth muscle lineage markers. *Circulation Research, 89*, 1147–1154.

Sugiura, H., Tsuchiya, K., & Nitta, K. (2011). Circulating levels of soluble α-Klotho in patients with chronic kidney disease. *Clinical and Experimental Nephrology, 15*, 795–796.

Sun, Z., & Andersson, R. (2002). NF-κB activation and inhibition: A review. *Shock, 18*, 99–106.

Sun, C. Y., Chang, S. C., & Wu, M. S. (2012). Suppression of Klotho expression by protein-bound uremic toxins is associated with increased DNA methyltransferase expression and DNA hypermethylation. *Kidney International, 81*, 640–650.

Sun, C. Y., Hsu, H. H., & Wu, M. S. (2012). p-Cresol sulfate and indoxyl sulfate induce similar cellular inflammatory gene expressions in cultured proximal renal tubular cells. *Nephrology, Dialysis, Transplantation, 28*, 70–78.

Takeshita, K., Fujimori, T., Kurotaki, Y., Honjo, H., Tsujikawa, H., Yasui, K., et al. (2004). Sinoatrial node dysfunction and early unexpected death of mice with a defect of klotho gene expression. *Circulation, 109*, 1776–1782.

Tintut, Y., Alfonso, Z., Saini, T., Radcliff, K., Watson, K., Boström, K., et al. (2003). Multilineage potential of cells from the artery wall. *Circulation, 108*, 2505–2510.

Tonelli, M., & Pfeffer, M. A. (2007). Kidney disease and cardiovascular risk. *Annual Review of Medicine, 58*, 123–139.

Torres, P. U., Prié, D., Beck, L., De Brauwere, D., Leroy, C., & Friedlander, G. (2009). Klotho gene, phosphocalcic metabolism, and survival in dialysis. *Journal of Renal Nutrition, 19*, 50–56.

Towler, D. A. (2005). Inorganic pyrophosphate: A paracrine regulator of vascular calcification and smooth muscle phenotype. *Arteriosclerosis, Thrombosis, and Vascular Biology, 25*, 651–654.

Tran, K., Merika, M., & Thanos, D. (1997). Distinct functional properties of IκBα and IκBβ. *Molecular and Cell Biology, 17*, 5386–5399.

Tsirpanlis, G. (2008). Cellular senescence, cardiovascular risk, and CKD: A review of established and hypothetical interconnections. *American Journal of Kidney Diseases, 51*, 131–144.

Tsujikawa, H., Kurotaki, Y., Fujimori, T., Fukuda, K., & Nabeshima, Y. (2003). Klotho, a gene related to a syndrome resembling human premature aging, functions in a negative regulatory circuit of vitamin D endocrine system. *Molecular Endocrinology, 17*, 2393–2403.

Urakawa, I., Yamazaki, Y., Shimada, T., Iijima, K., Hasegawa, H., Okawa, K., et al. (2006). Klotho converts canonical FGF receptor into a specific receptor for FGF23. *Nature, 444*, 770–774.

Vattikuti, R., & Towler, D. A. (2004). Osteogenic regulation of vascular calcification: An early perspective. *American Journal of Physiology. Endocrinology and Metabolism, 286*, E686–E696.

Vervloet, M. G., Adema, A. Y., Larsson, T. E., & Massy, Z. A. (2014). The role of klotho on vascular calcification and endothelial function in chronic kidney disease. *Seminars in Nephrology, 34*, 578–585.

Virchow, R. (1989). Cellular pathology. As based upon physiological and pathological histology. Lecture XVI—Atheromatous affection of arteries. *Nutrition Reviews, 47*, 23–25.

Wang, Y., & Sun, Z. (2009). Current understanding of klotho. *Ageing Research Reviews, 8*, 43–51.

Wang, Y., Zhou, Y., & Graves, D. T. (2014). FOXO transcription factors: Their clinical significance and regulation. *BioMed Research International, 2014*, 925350.

Weiner, D. E., Tighiouart, H., Amin, M. G., Stark, P. C., MacLeod, B., Griffith, J. L., et al. (2004). Chronic kidney disease as a risk factor for cardiovascular disease and all-cause mortality: A pooled analysis of community-based studies. *Journal of the American Society of Nephrology*, *15*, 1307–1315.

Weiner, D. E., Tighiouart, H., Stark, P. C., Amin, M. G., MacLeod, B., Griffith, J. L., et al. (2004). Kidney disease as a risk factor for recurrent cardiovascular disease and mortality. *American Journal of Kidney Diseases*, *44*, 198–206.

White, W. E., Yaqoob, M. M., & Harwood, S. M. (2015). Aging and uremia: Is there cellular and molecular crossover? *World Journal of Nephrology*, *4*, 19–30.

Whiteside, S. T., & Israel, A. (1997). IκB proteins: Structure, function and regulation. *Seminars in Cancer Biology*, *8*, 75–82.

Woronicz, J. D., Gao, X., Cao, Z., Rothe, M., & Goeddel, D. V. (1997). IkappaB kinase-beta: NF-kappaB activation and complex formation with IkappaB kinase-alpha and NIK. *Science*, *278*, 866–869.

Xu, Y., & Sun, Z. (2015). Molecular basis of Klotho: From gene to function in aging. *Endocrine Reviews*, *36*, 174–193.

Yamamoto, M., Clark, J. D., Pastor, J. V., Gurnani, P., Nandi, A., Kurosu, H., et al. (2005). Regulation of oxidative stress by the anti-aging hormone klotho. *The Journal of Biological Chemistry*, *280*, 38029–38034.

Yamashita, T. (2005). Structural and biochemical properties of fibroblast growth factor 23. *Therapeutic Apheresis and Dialysis*, *9*, 313–318.

Yang, K., Nie, L., Huang, Y., Zhang, J., Xiao, T., Guan, X., et al. (2012). Amelioration of uremic toxin indoxyl sulfate-induced endothelial cell dysfunction by Klotho protein. *Toxicology Letters*, *215*, 77–83.

Yayon, A., Klagsbrun, M., Esko, J. D., Leder, P., & Ornitz, D. M. (1991). Cell surface, heparin-like molecules are required for binding of basic fibroblast growth factor to its high affinity receptor. *Cell*, *64*, 841–848.

Young, G. H., & Wu, V. C. (2012). KLOTHO methylation is linked to uremic toxins and chronic kidney disease. *Kidney International*, *81*, 611–612.

Zandi, E., Rothwarf, D. M., Delhase, M., Hayakawa, M., & Karin, M. (1997). The IkappaB kinase complex (IKK) contains two kinase subunits, IKKalpha and IKKbeta, necessary for IkappaB phosphorylation and NF-kappaB activation. *Cell*, *91*, 243–252.

Zeng, Y., Wang, P. H., Zhang, M., & Du, J. R. (2016). Aging-related renal injury and inflammation are associated with downregulation of Klotho and induction of RIG-I/NF-κB signaling pathway in senescence-accelerated mice. *Aging Clinical and Experimental Research*, *28*(1), 69–76.

Zhang, H., Li, Y., Fan, Y., Wu, J., Zhao, B., Guan, Y., et al. (2008). Klotho is a target gene of PPAR-gamma. *Kidney International*, *74*, 732–739.

Zhao, Y., Banerjee, S., Dey, N., LeJeune, W. S., Sarkar, P. S., Brobey, R., et al. (2011). Klotho depletion contributes to increased inflammation in kidney of the db/db mouse model of diabetes via RelA (serine)536 phosphorylation. *Diabetes*, *60*, 1907–1916.

CHAPTER SIX

The FGF23/KLOTHO Regulatory Network and Its Roles in Human Disorders

S. Kinoshita, M. Kawai[1]
Osaka Medical Center and Research Institute for Maternal and Child Health, Izumi, Japan
[1]Corresponding author: e-mail address: kawaim@mch.pref.osaka.jp

Contents

Abstract

The functions of Klotho (KL) are multifaceted and include the regulation of aging and mineral metabolism. It was originally identified as the gene responsible for premature aging-like symptoms in mice and was subsequently shown to function as a coreceptor in the fibroblast growth factor (FGF) 23 signaling pathway. The discovery of KL as a partner for FGF23 led to significant advances in understanding of the molecular mechanisms underlying phosphate and vitamin D metabolism, and simultaneously clarified the pathogenic roles of the FGF23 signaling pathway in human diseases. These novel insights led to the development of new strategies to combat disorders associated with the dysregulated metabolism of phosphate and vitamin D, and clinical trials on the blockade of FGF23 signaling in X-linked hypophosphatemic rickets are ongoing.

Vitamins and Hormones, Volume 101
ISSN 0083-6729
http://dx.doi.org/10.1016/bs.vh.2016.02.001

Molecular and functional insights on KL and FGF23 have been discussed in this review and were extended to how dysregulation of the FGF23/KL axis causes human disorders associated with abnormal mineral metabolism.

1. INTRODUCTION

Klotho (KL), which was named after one of the Moirai who controlled aging in Greek mythology, was originally identified in 1997 as the gene responsible for premature aging-like symptoms in mice (Kuro-o et al., 1997). *klotho* mice, which have an insertion mutation in the promoter region of the *Klotho* gene, display multiple premature aging-like symptoms including a short life span, infertility, arteriosclerosis, skin atrophy, osteoporosis, and emphysema (Kuro-o et al., 1997). In addition, a previous study reported that the life span of *Klotho* transgenic mice was extended by 20–30% (Kurosu et al., 2005). These findings emphasized the pivotal role of KL in the regulation of aging. Although accumulating evidence has implicated KL in aging-related disorders (Kuro-o, 2012; Olauson et al., 2014; Xu & Sun, 2015), the mechanisms by which KL determines the life span of humans have not yet been elucidated in detail. After the discovery of KL, two proteins that shared homology with KL were identified and named β-Klotho (Ito et al., 2000) and γ-Klotho (Klotho/lactase-phlorizin hydrolase-related protein) (Ito, Fujimori, Hayashizaki, & Nabeshima, 2002), and KL was subsequently referred to as α-Klotho in order to avoid confusion.

The discovery of KL led a deeper understanding of the mechanisms responsible for the regulation of aging, and its identification as a coreceptor for fibroblast growth factor 23 (FGF23) signaling provided further insights into its critical functions in mineral metabolism (Urakawa et al., 2006). FGF23 was initially discovered as a novel member of the FGF family in 2000 (Yamashita, Yoshioka, & Itoh, 2000) and was identified as the gene responsible for hypophosphatemic rickets and osteomalacia (ADHR consortium, 2000; Shimada et al., 2001). Thus, the discovery of KL as a coreceptor for FGF23 initiated a new era of research on mineral metabolism and has accelerated understanding on the regulation of phosphate (inorganic phosphate, Pi) and vitamin D metabolism by this pathway. The critical functions of KL have been discussed in the present review, with a focus on how its interaction with FGF23 generates the regulatory networks of phosphate and vitamin D metabolism in addition to its role in human diseases associated with dysregulated mineral metabolism.

2. KLOTHO

2.1 Structure of KL

KL is a 135 kDa single-pass transmembrane protein with an extracellular domain that exhibits weak β-galactosidase activity; however, the premature form of 120 kDa KL has also been detected in the endoplasmic reticulum (Imura et al., 2007; Xu & Sun, 2015). KL is composed of an extracellular domain that contains KL1 and KL2 internal repeats, a transmembrane domain, and intracellular carboxyl domain (Ito et al., 2000). One of the features of KL is its soluble form (sKL), which is derived from alternative splicing or ectodomain shedding. Alternative splicing produces sKL, corresponding to N-terminus KL1 (65–70 kDa), which is secreted into the extracellular space; however, the presence of 65–70 kDa sKL has not yet been confirmed in the circulation in humans (Matsumura et al., 1998; Shiraki-Iida et al., 1998). Membrane-bound KL was previously shown to be cleaved by disintegrin and metalloproteinase (ADAM)-10, ADAM-17, and beta-site amyloid precursor protein-cleaving enzyme 1 (Bloch et al., 2009; Chen, Podvin, Gillespie, Leeman, & Abraham, 2007) in order to produce ~130 kDa sKL in the circulation, and this type of sKL is considered to be predominantly present in the urine, serum, and cerebrospinal fluid in humans (Imura et al., 2004). sKL is known to function as a humoral factor and plays multiple roles in antioxidation, antiapoptosis, and ion transport (Chang et al., 2005; Kurosu et al., 2005; Xu & Sun, 2015; Fig. 1).

2.2 Expression Profile and Function of KL

The expression profile of KL is very unique; it is mainly expressed in the distal tubular epithelial cells of the kidney, chief cells in the parathyroid gland, the pituitary, and the choroid plexus of the brain (Kuro-o et al., 1997). KL was recently reported to be expressed in the human placenta and has been suggested to play a role in vitamin D metabolism (Ohata et al., 2011, 2014). Although KL may be expressed in various cells, its physiological significance currently remains unknown. Urakawa et al. originally demonstrated that KL functioned as a coreceptor for FGF23/FGFR signaling, a finding that was subsequently confirmed in a series of

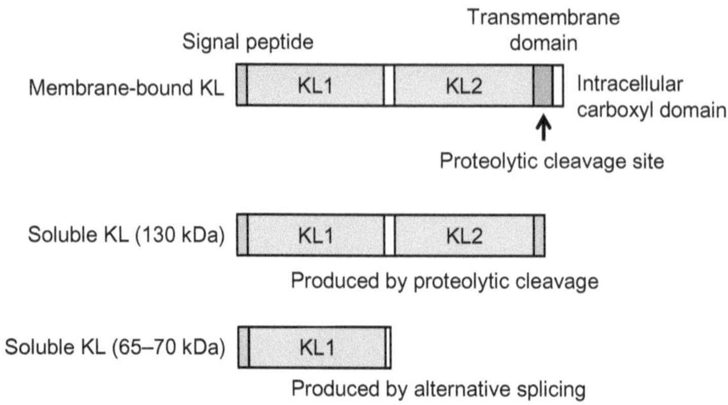

Fig. 1 Structure of Klotho (KL). KL is composed of two internal repeats (KL1 and KL2), a transmembrane domain and short intracellular carboxyl domain. KL is proteolytically cleaved at the transmembrane domain (*arrow*) to produce its soluble form. The shorter form of soluble KL is also produced by alternative splicing. (See the color plate.)

experiments (Kurosu et al., 2006). In accordance with these findings, *Fgf23*-deficient mice were shown to display very similar phenotypes to those of *Klotho*-deficient mice. The functions of KL as a mediator of the FGF23 signaling pathway have been discussed later in this review.

In addition to its role as an obligate coreceptor for FGF23, KL has been suggested to have multiple FGF23-independent functions. Accumulating evidence has revealed that KL inhibits the insulin/IGF-1 signaling pathway, which may be partly responsible for the short life span of *klotho* mice (Kurosu et al., 2005; Yamamoto et al., 2005). This may also be involved in the protective effects of KL against cardiac hypertrophy by downregulating TRPC6 channels (Xie et al., 2012); however, the underlying mechanisms by which KL suppresses the insulin/IGF-1 pathway remain largely unknown. KL was previously shown to form a protein complex with multiple Wnt molecules and suppresses the Wnt signaling pathway (Liu et al., 2007; Satoh et al., 2012; Zhou, Li, Zhou, Tan, & Liu, 2013). KL also plays a role in ion transport in the kidney. For example, sKL was found to increase Ca^{2+} transport by activating TRPV5 channels (Chang et al., 2005). Furthermore, KL may function as a quick regulator for Ca^{2+} homeostasis. Imura et al. demonstrated that KL formed a protein complex with Na^{+}, K^{+}-ATPase, which enhanced the localization of Na^{+}, K^{+}-ATPase at the plasma membrane and stimulated the secretion of PTH in response to low extracellular Ca^{2+} concentrations (Imura et al., 2007).

3. KL AS A CORECEPTOR FOR THE FGF23/FGFR PATHWAY

3.1 Structure of FGF23

In 2000, a novel member of the FGF family was identified and named FGF23 (Yamashita et al., 2000). FGF23 was simultaneously identified as the gene responsible for hypophosphatemic rickets and osteomalacia (ADHR consortium, 2000; Shimada et al., 2001). Human FGF23 is a 251 amino acid protein (32 kDa) with a signal peptide in its N-terminus region, with the rat and mouse FGF23 amino acid sequences being 72% and 71% homologous to human FGF23, respectively (Yamashita et al., 2000). FGF23 lacks the heparin-binding domain present in autocrine/paracrine FGFs (Goetz et al., 2007); therefore, FGF23 is not retained in the extracellular matrix space, which confers it with the ability to function as an endocrine molecule. Due to its extremely low affinity to FGFRs, FGF23 is known to require KL for binding to FGFRs (Hu, Shiizaki, Kuro-o, & Moe, 2013). FGF23 is proteolytically cleaved by proteins with endoprotease activity, such as furin, at the site of a subtilisin-like proprotein convertase (SPC) motif (^{176}RXXR^{179}) to produce N- and C-terminal fragments (Hu et al., 2013). The N-terminal region of FGF23 is a FGF core homology domain through which FGF23 interacts with FGFRs and the C-terminal domain is unique to FGF23; this region is also important for the FGF23/KL interaction (Yamazaki et al., 2008). Therefore, the cleavage of FGF23 at this site inactivates FGF23 activity. C-terminal fragments were previously found to be antagonistic to FGF23/KL binding and may play a role in another mechanism by which the cleavage of FGF23 blocks the FGF23 signaling pathway (Goetz et al., 2010).

FGF23 is subjected to posttranslational modifications such as glycosylation and phosphorylation, which affect its function and stability. GALNT3 (polypeptide *N*-acetylgalactosaminyltransferase 3) has been shown to glycosylate FGF23, and the glycosylation of Thr178 in the SPC motif may be critical for the production and secretion of full length, biologically active FGF23 (Frishberg et al., 2007; Kato et al., 2006). FGF23 has recently been reported to be phosphorylated at Ser180 within the SPC motif by FAM20C, which inhibits O-glycosylation by GALNT3, resulting in accelerated cleavage at this site (Lindberg et al., 2015; Tagliabracci et al., 2014; Fig. 2).

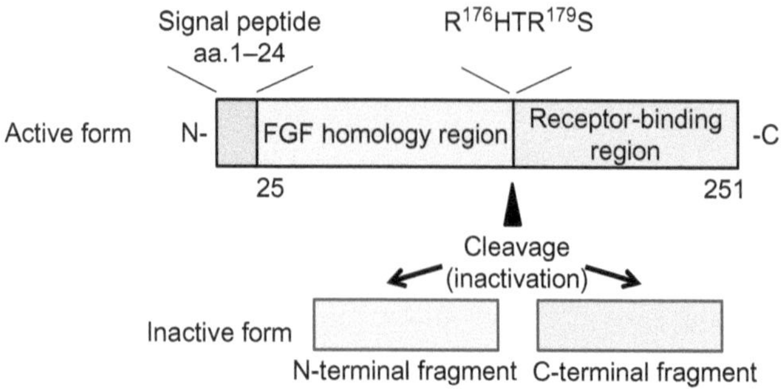

Fig. 2 Structure of FGF23. Intact FGF23 is proteolytically cleaved to generate inactive forms of FGF23. The site of cleavage is shown in the figure by an *arrowhead*. (See the color plate.)

3.2 Regulation of FGF23 Expression

FGF23 was originally considered to be expressed in multiple tissues including the brain, thymus, small intestine, and heart (ADHR consortium, 2000; Yamashita et al., 2000); however, the physiological relevance of FGF23 produced by these tissues has not yet been determined. Osteoblastic cells, especially osteocytes, were subsequently identified as a physiological source of FGF23 based on analyses of patients with fibrous dysplasia of bone (Collins et al., 2001; Riminucci et al., 2003).

Although the regulation of FGF23 in bone cells has been extensively examined, it has not yet been fully elucidated. The active form of vitamin D (1,25-dihydroxyvitamin D: 1,25-$(OH)_2D$) appears to increase FGF23 expression levels in vitro and in vivo by activating the vitamin D receptor (VDR) (Collins et al., 2005; Kolek et al., 2005; Liu et al., 2006). Although the VDR response element has been detected in the mouse *Fgf23* gene promoter (Liu et al., 2006), the mode of action of 1,25-$(OH)_2D$/VDR in the human *FGF23* gene has not been demonstrated in detail. A previous study reported that in vivo levels of FGF23 were markedly higher in *Klotho*-deficient mice, in which 1,25-$(OH)_2D$ concentrations were elevated, and the suppression of 1,25-$(OH)_2D$ by crossing *Klotho*-deficient mice with *Cyp27b1*-deficient mice restored FGF23 levels, suggesting that 1,25-$(OH)_2D$ signaling is critical for elevating the levels of FGF23 in *Klotho*-deficient mice; however, serum Pi levels were also restored in these double-mutant mice, indicating that correction of hyperphosphatemia may play a role in increasing in FGF23 levels in these mice (Ohnishi, Nakatani, Lanske, & Razzaque, 2009).

PTH has been suggested to regulate the expression of FGF23 in the skeleton. Elevations in circulating FGF23 levels have been reported under conditions in which the PTH signaling pathway is continuously active such as chronic kidney disease (CKD) and Jansen metaphyseal chondrodysplasia caused by a mutation in the *PTH1R* gene, suggesting that PTH is a positive regulator of the expression of FGF23 in bone (Brown et al., 2009; Lavi-Moshayoff, Wasserman, Meir, Silver, & Naveh-Many, 2010). In a CKD rat model, parathyroidectomy was found to suppress elevations in the concentration of FGF23 induced by kidney failure (Lavi-Moshayoff et al., 2010). In contrast to these models, another study showed that PTH failed to induce the expression of *Fgf23*, especially when PTH was intermittently administered (Samadfam, Richard, Nguyen-Yamamoto, Bolivar, & Goltzman, 2009). Furthermore, in vitro evidence for the induction of FGF23 by PTH is limited. FGF23 levels were previously reported to be similar in patients with primary hyperparathyroidism and healthy controls (Tebben, Singh, Clarke, & Kumar, 2004). Thus, the effects of PTH on the expression of *FGF23* are context specific and remain controversial (Brown et al., 2009; Lavi-Moshayoff et al., 2010; Saji et al., 2010; Samadfam et al., 2009).

In animal models, dietary Pi loads have been associated with increases in circulating FGF23 levels (Perwad et al., 2005; Saito et al., 2005), indicating that increases in the concentration of Pi trigger elevations in FGF23 levels. In human studies, Ferrari et al. showed that serum FGF23 levels in 29 healthy men were associated with the amount of dietary Pi consumed (Ferrari, Bonjour, & Rizzoli, 2005). Burnett also demonstrated that dietary phosphate was a key regulator of circulating FGF23 levels in humans (Burnett et al., 2006); however, there is also evidence reporting no association between dietary Pi and serum FGF23 levels (Larsson, Nisbeth, Ljunggren, Juppner, & Jonsson, 2003). Responses to an acute Pi load in humans were recently analyzed. In healthy participants, an intravenous infusion of Pi or duodenal Pi load increased FGF23 levels; however, this was preceded by an increase in PTH. 1,25-$(OH)_2D$ concentrations were found to decrease following elevations in FGF23 levels (Scanni, vonRotz, Jehle, Hulter, & Krapf, 2014). Although these findings indicated that the Pi load triggers increases in FGF23 levels, it currently remains unclear whether the induction of FGF23 is exerted by Pi itself or mediated through additional factors such as PTH.

One of the features of FGF23 is that it is regulated in a manner that involves circadian rhythms. Circulating Pi levels have been suggested to exhibit rhythmic patterns (Nelson, 1964); therefore, the metabolism of Pi may be under the control of circadian rhythms. In mice, the skeletal expression of *Fgf23* showed circadian rhythms, with peaks being observed in expression levels during the dark phase, which was associated with increases in the excretion of Pi in the urine (Kawai, Kinoshita, Shimba, Ozono, & Michigami, 2014). Consistent with these increases in skeletal *Fgf23* expression, circulating FGF23 levels were also found to be elevated during the dark phase. The rhythmic profile of circulating FGF23 has also been reported in humans (Carpenter et al., 2010; Vervloet et al., 2011). The rhythmic profile of FGF23 may be regulated by the time of food intake through food intake-driven sympathetic activation (Kawai et al., 2014). When mice were placed under *ad libitum* conditions, in which food intake was stimulated at the beginning of the dark phase, FGF23 peaked during the dark phase. However, when mice were placed under restricted feeding conditions, in which access to food was exclusively during the light phase, FGF23 levels peaked during the light phase, and this was associated with a phase shift in the peak activity of the sympathetic nervous system from the dark to the light phase. Although the pathogenic significance of sympathetic activity-regulated FGF23 expression remains to be determined, this system may be involved in the elevated levels of FGF23 observed in CKD and myocardial infarction.

The activation of FGFRs has been implicated in the regulation of FGF23 expression in bone, as evidenced in the human disease, osteoglophonic dysplasia, in which heterozygous missense mutations in *FGFR1* cause nonossifying bone lesions and abnormal mineral metabolism due to elevations in FGF23 levels (Farrow et al., 2006). An in vivo analysis of mouse models also supported the involvement of the signaling cascade downstream of FGFR1 in the regulation of FGF23 (Miyagawa et al., 2014; Xiao et al., 2014). The activation of FGFR by FGF2 was also reported to upregulate the expression of FGF23 (Han, Xiao, & Quarles, 2015; Xiao et al., 2010). The intranuclear activation of FGFR1 by high-molecular-weight FGF2 was also found to increase the expression of FGF23, possibly in a manner independent of the signaling cascades for the activation of membrane FGFR1 (Han et al., 2015). FGF23 levels have been positively associated with the biomarkers of inflammation, insulin resistance, myocardial infarction, and metabolic acidosis (Andrukhova, Slavic, Odorfer, & Erben, 2015; Hanks, Casazza, Judd, Jenny, & Gutierrez, 2015; Holecki et al., 2015; Krieger, Culbertson, Kyker-Snowman, & Bushinsky, 2012; Munoz Mendoza et al., 2012; Yamazaki et al., 2015); however, the molecular mechanisms underlying the induction of FGF23 under these conditions have yet to be determined (Fig. 3).

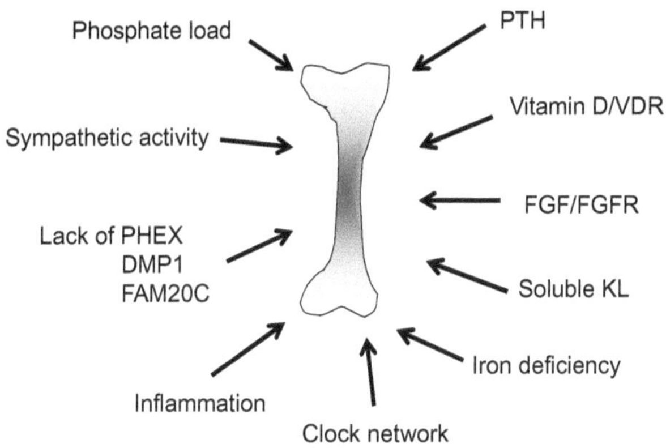

Fig. 3 FGF23 inducers. Accumulating evidence has identified the factors inducing FGF23 expression in the skeleton; however, further research is warranted and their roles in human physiology currently remain unclear.

3.3 FGF23 Signaling Pathway

3.3.1 KL-Dependent Pathway

Osteoblast lineage-specific cells, especially osteocytes, produce large amounts of FGF23, while other tissues including the small intestine, heart, ventrolateral thalamic nucleus, and thymus also produce FGF23; however, the physiological significance of FGF23 produced from these tissues currently remains unclear (Quarles, 2012; Yamashita et al., 2000). Since KL is needed by FGF23 for binding to FGFR, FGF23 target tissues are regarded as those expressing KL. In vitro experiments revealed that FGF23 formed a complex in the presence of KL with multiple FGFRs including FGFR1, 3c, and 4 (Kurosu et al., 2006), whereas the in vivo relevance of this complex may be restricted to FGFR1 (Gattineni et al., 2009; Liu, Vierthaler, Tang, Zhou, & Quarles, 2008). Upon forming a protein complex with FGFR and KL, FGF23 activates downstream signal cascades such as the ERK (extracellular signal-regulated kinases)/EGR-1 (early growth response-1) signal pathway (Kurosu et al., 2006; Urakawa et al., 2006; Fig. 4).

FGF23 may activate this signaling pathway in cooperation with sKL. Similar to membrane-bound KL, sKL has been shown to form a protein complex with FGFR and FGF23 (Kawai et al., 2013; Kurosu et al., 2006; Smith et al., 2012). As with membrane-bound KL, sKL also enhanced the binding of FGF23 to FGFRs (Kawai et al., 2013); however, the binding intensity between sKL and FGFR was weaker, and the resultant activation of the downstream pathway was not as strong as that of membrane-bound KL (Kurosu et al., 2006). The in vivo role of sKL in FGF23 signaling has been

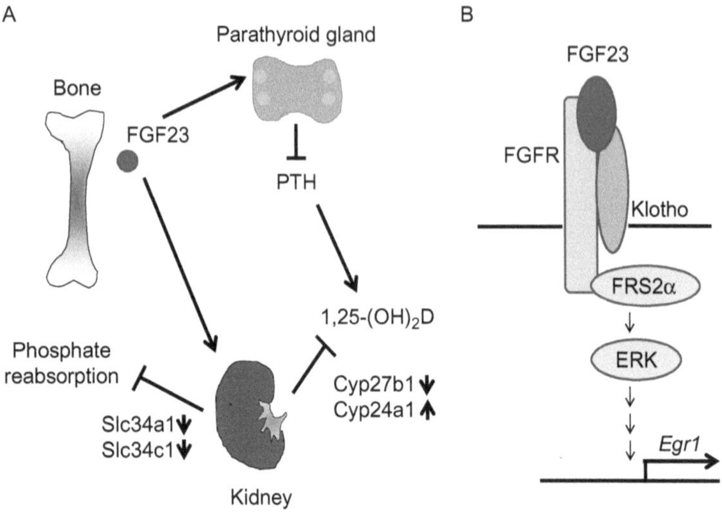

Fig. 4 Signaling cascades of the FGF23/Klotho pathway. (A) FGF23 produced from osteocytes mainly transduces its signals in the kidney and regulates phosphate reabsorption and the production of the active form of vitamin D. (B) FGF23 requires Klotho as a coreceptor to exert its effects and activate its downstream signaling molecules including FRS2α and ERK. (See the color plate.)

demonstrated in a mouse model highly overexpressing sKL (Smith et al., 2012). FGF23 levels are elevated in these mice and have been associated with the enhanced excretion of Pi. Mechanistically, the activation of FGFRs by FGF23 in cooperation with sKL has been suggested to induce the expression of FGF23 in bone cells (Smith et al., 2012). In chondrogenic cells, the FGF23/sKL/FGFR3 complex has been shown to suppress chondrocyte proliferation (Kawai et al., 2013). Thus, sKL may mediate the effects of FGF23 in cells in which membrane-bound KL is not expressed; however, because these models are based on the presence of supraphysiological levels of sKL, it currently remains unknown whether the FGF23/FGFR/sKL complex is present and operative under physiological conditions.

3.3.2 KL-Independent Pathway

An increasing amount of evidence has indicated that FGF23 exerts its effects in the absence of KL. Faul et al. demonstrated that FGF23-induced left ventricular hypertrophy through the activation of FGFRs and downstream cascades in the calcineurin–NFAT signaling pathway independent of KL (Faul et al., 2011). Furthermore, FGF23 has been suggested to suppress the secretion of PTH in KL-dependent and -independent manners, the latter of which also involves activation of the calcineurin–NFAT pathway (Olauson et al., 2013). The overexpression of FGF23 has been shown to suppress osteogenesis in osteoblastic cells, which do not express KL (Wang et al., 2008). Although these findings highlight the KL-independent signaling pathway of FGF23, the mechanisms by which FGF23 activates FGFR in the absence of KL have not yet been elucidated.

3.4 Functions of the FGF23/KL Axis

3.4.1 Regulation of Pi Metabolism

The involvement of FGF23 signaling in the metabolism of Pi has been demonstrated by in vivo analyses using *Fgf23*-deficient mice; the lack of *Fgf23* was previously shown to lead to the development of hyperphosphatemia (Shimada, Kakitani, et al., 2004) and *Fgf23*-transgenic mice developed hypophosphatemia (Larsson et al., 2004; Shimada, Urakawa, et al., 2004). In addition to these genetically engineered mouse models, an injection of recombinant FGF23 (rFGF23) into WT mice resulted in the development of hypophosphatemia (Shimada, Hasegawa, et al., 2004; Shimada et al., 2001). The effects of FGF23 on the handling of Pi are considered to be dependent on KL because the lack of KL in mice leads to hyperphosphatemia. Furthermore, the administration of rFGF23 to mice lacking the *Fgf23* and *Klotho* genes did not affect circulating Pi levels, whereas it decreased serum Pi levels in WT and *Fgf23*-deficient mice (Nakatani et al., 2009). The nodal point of the regulation of Pi by FGF23 appears to reside in its regulation of renal Npt2a and Npt2c, which are Na^+-dependent Pi cotransporters involved in the reabsorption of Pi in the kidney and are predominantly expressed at the brush border membrane of proximal tubular epithelial cells. FGF23 is known to suppress the expression of Npt2a and Npt2c and also reduce Na/Pi cotransporter activity, thereby suppressing the renal reabsorption of Pi and enhancing its excretion in the urine (Segawa et al., 2003; Shimada, Hasegawa, et al., 2004). Furthermore, the suppression of 1,25-$(OH)_2D$ generation by FGF23 also decreased intestinal Pi absorption and may be involved in the regulation of Pi by FGF23. The complexity of the role of FGF23 in the regulation of Pi reabsorption may reside in the predominant expression of KL in distal tubular epithelial cells. The significance of the expression of KL in distal tubular epithelial cells has also been confirmed in a mouse model in which the expression of *Klotho* was partially deleted in renal distal tubules under the Ksp-cadherin promoter (Olauson et al., 2012). Since FGF23 activates its downstream molecules such as ERK1/2 in KL-expressing cells in the distal tubular epithelial cells, which are distinct from, but adjacent to Npt2a-expressing proximal tubular epithelial cells (Farrow, Davis, Summers, & White, 2009; Farrow et al., 2010), it currently remains unclear how KL-dependent signal activation by FGF23 in the distal tubular epithelial cells influences the expression and function of Npt2a and Npt2c proteins in the proximal tubules; however, the existence of unknown secretory molecules mediating this connection has been proposed. KL is also weakly expressed in the proximal tubules, and this has been suggested to at least partly mediate the effects of FGF23 and regulate phosphate and vitamin D metabolism (Hu et al., 2010).

3.4.2 Regulation of Vitamin D Metabolism

FGF23 regulates the synthesis of 1,25-$(OH)_2$D in a KL-dependent manner. FGF23 was shown to decrease the circulating levels of 1,25-$(OH)_2$D by suppressing the expression of *Cyp27b1*, encoding for 1-α-hydroxylase, which converts 25-dihydroxyvitamin D to 1,25-$(OH)_2$D, and increasing the expression of *Cyp24a1*, coding for 24-hydroxylase, which is involved in the inactivation of 1,25-$(OH)_2$D (Shimada, Hasegawa, et al., 2004). Animal models in which the lack of FGF23/KL signaling, such as *Klotho*-deficient and *Fgf23*-deficient mice, led to increases in 1,25-$(OH)_2$D levels confirmed the in vivo relevance of this pathway. In addition to its direct effects on gene expression associated with vitamin D metabolism, FGF23 has been suggested to indirectly regulate 1,25-$(OH)_2$D synthesis by acting on chief cells in the parathyroid, as has been discussed later (Ben-Dov et al., 2007).

3.4.3 Regulation of PTH Synthesis and Secretion

Since KL is also expressed in the parathyroid glands, FGF23 may transduce its signals in the parathyroid and regulate the bioavailability of PTH. FGF23 has been shown to suppress the expression and secretion of PTH in the parathyroid glands, which was inhibited by blockade of the MAPK pathway (Ben-Dov et al., 2007). Since PTH is known to increase 1,25-$(OH)_2$D levels by enhancing the expression of *Cyp27b1*, FGF23 may decrease these levels, in part, by suppressing PTH levels. Despite accumulating evidence for the involvement of FGF23 in the regulation of PTH in animal studies, this has not yet been well supported in clinical studies. For example, secondary hyperparathyroidism is a common complication in patients with CKD in spite of elevated FGF23 levels. Thus, the physiological and/or pathogenic roles of FGF23 signaling in the parathyroid gland in the regulation of PTH expression and secretion in clinical settings have yet to be elucidated.

4. HUMAN DISEASES RELATED TO A DYSREGULATED FGF23/KL AXIS

4.1 Mutations in the *FGF23* Gene

Autosomal dominant hypophosphatemic rickets (ADHR) is a disorder that is characterized by rickets, osteomalacia, a short stature, bone pain, and dental abscesses due to the excessive excretion of Pi in the urine, and a gain-of-function mutation in the *FGF23* gene has been identified as the mutation responsible for this disorder (ADHR consortium, 2000). Responsible mutations include the substitution of Arg to Gln at position 176 (R176Q) in the SPC motif, which prevents the proteolytic cleavage of FGF23. Based on analyses using ADHR model mice in which the R176Q mutation was knocked in, an iron deficiency was found to play an important role in the development of ADHR phenotypes (Farrow et al., 2011). When mice were fed an iron-deficient diet, the expression of *Fgf23* mRNA increased in WT and ADHR mice. C-terminus FGF23 levels increased under these conditions, while serum levels of phosphate and intact FGF23 levels remained unchanged in WT mice, suggesting that proteolytic cleavage of FGF23 plays critical roles in maintaining the concentration of intact FGF23 levels in the circulation. On the other hand, ADHR mice developed hypophosphatemic rickets because the proteolytic cleavage of FGF23 is blocked due to R176Q mutation in these mice. This was also evident in a human ADHR patient who was able to successfully withdraw rickets medications following high-dose iron supplementation (Kapelari, Kohle, Kotzot, & Hogler, 2015). Mutations in the *FGF23* gene have also been implicated in hyperphosphatemic familial tumoral calcinosis (Araya et al., 2005; Benet-Pages, Orlik, Strom, & Lorenz-Depiereux, 2005). The mutated FGF23 protein is known to be susceptible to proteolytic cleavage, such that the concentrations of intact FGF23 are markedly reduced, and this has been associated with increases in C-terminal FGF23 levels.

4.2 Mutations that Elevate FGF23 Concentrations and/or Bioactivity

X-linked hypophosphatemic rickets (XLH) is caused by a loss-of-function mutation in the *PHEX* (a phosphate-regulating gene with homologies to endopeptidases on the X chromosome) gene, which results in elevated circulating FGF23 levels and causes hypophosphatemia (Jonsson et al., 2003). FGF23 was initially regarded as a substrate of the PHEX protein; however, subsequent studies proved that this concept was incorrect, and thus, the mechanism by which the lack of PHEX increases FGF23 production currently remains unclear. Since hypophosphatemia-associated clinical symptoms such as rickets and a short stature are predominantly caused by elevations in FGF23 levels, clinical studies to block FGF23 signaling have been attempted using neutralizing antibodies raised against FGF23 (KRN23). The monthly administration of KRN23 has been reported to successfully reverse hypophosphatemia and increase 1,25-$(OH)_2$D levels (Carpenter et al., 2014; Imel et al., 2015); however, it has not yet been established whether KRN23 effectively improves skeletal abnormalities in XLH patients, and future studies are needed to assess the effectiveness of this new therapy in XLH patients. Autosomal recessive hypophosphatemic rickets 1 and 2 are also characterized by increased FGF23 levels and caused by mutations in the *DMP1* (dentin matrix protein 1) and *ENPP1* (ectonucleotide pyrophosphatase/phosphodiesterase 1) genes, respectively; however, the mechanisms responsible for elevations in FGF23 levels in these diseases are currently unknown (Levy-Litan et al., 2010; Lorenz-Depiereux et al., 2006; Lorenz-Depiereux, Schnabel, Tiosano, Hausler, & Strom, 2010). The *FAM20C* gene, which is responsible for Raine syndrome, was recently shown to cause an autosomal recessive disorder characterized by hypophosphatemia, dental anomalies, and ectopic calcifications (Rafaelsen et al., 2013; Takeyari et al., 2014). FAM20C is a secretory protein with kinase activity that phosphorylates the small integrin-binding ligand N-linked glycoproteins such as DMP-1 and OPN (Tagliabracci et al., 2012). Recent studies reported that FAM20C phosphorylated FGF23 at Ser180, resulting in the accelerated cleavage of FGF23 by inhibiting O-glycosylation through GALNT3 (Lindberg et al., 2015; Tagliabracci et al., 2014); therefore, the loss-of-functional mutations in the *FAM20C* gene may increase FGF23 levels and cause hypophosphatemia.

Elevations in FGF23 levels have also been detected in some diseases including McCune–Albright syndrome, caused by an activating mutation in the *GNAS* gene (Riminucci et al., 2003), osteoglophonic dysplasia, caused by a heterozygous mutation in the *FGFR1* gene (Farrow et al., 2006), and Jansen metaphyseal chondrodysplasia (Brown et al., 2009). Tumor-induced osteomalacia is a rare condition in which increases in FGF23 levels are associated with hypophosphatemia, and inappropriately normal or low level of 1,25-$(OH)_2$D (Chong, Molinolo, Chen, & Collins, 2011). Abnormal mineral metabolism causes osteomalacia and rickets when mineral abnormalities appear before the closure of growth plates. Chromosomal translocations with a break point adjacent to the *KLOTHO* gene have been shown to increase the membrane expression of KL and cause hypophosphatemic rickets due to enhanced FGF23/KL signaling (Brownstein et al., 2008).

4.3 Mutations Reducing GF23 Concentrations and/or Bioactivity

Previous studies demonstrated that mutations in the *GALNT3* gene were responsible for familial tumoral calcinosis (Frishberg et al., 2007, 2005; Garringer et al., 2006; Ichikawa et al., 2010). Since GALNT3 is an enzyme that *O*-glycosylates FGF23 at Thr178, thereby protecting FGF23 from proteolytic cleavage, a mutation in the *GALNT3* gene enhances the cleavage of FGF23 and causes hyperphosphatemia. As expected, circulating levels of intact FGF23 were shown to be reduced in patients with familial tumoral calcinosis, whereas C-terminal fragments were increased. This was also recapitulated in a mouse model in which the *Galnt3* gene was deleted. These mice had hyperphosphatemia associated with decreased levels of intact FGF23 despite the upregulated expression of *Fgf23* mRNA (Ichikawa et al., 2009). Hyperphosphatemic phenotypes in *Galnt3*-deficient mice disappeared when an ADHR mutation (FGF23 R176Q) was knocked in (Ichikawa et al., 2014). The loss-of-functional mutations in the *KLOTHO* gene have also been reported to cause familial tumoral calcinosis; a H193R mutation in the *KLOTHO* gene in a patient with familial tumoral calcinosis destabilized the KL1 domain, which resulted in the attenuation of FGF23 signaling and hyperphosphatemia (Ichikawa et al., 2007).

4.4 CKD and FGF23

Accumulating evidence has shown that CKD is a state of excess FGF23 levels, increases in which have been shown to precede elevations in PTH and phosphate levels (Isakova, Wahl, et al., 2011; Larsson et al., 2003; Levin et al., 2007; Nakano et al., 2012; Portale et al., 2014). Based on the Chronic Renal Insufficiency Cohort, circulating FGF23 levels increase in the early stage of CKD (Isakova, Wahl, et al., 2011). Similar findings have also been reported in a rodent model of acute renal failure (Christov et al., 2013). Furthermore, increases in FGF23 levels have been associated with a higher risk of cardiovascular disease and mortality (Baia et al., 2013; Block et al., 2004; Gutierrez et al., 2008; Isakova, Xie, et al., 2011; Kendrick et al., 2011; Wolf et al., 2011). Thus, increases in FGF23 levels may be pathogenic to CKD-related complications, and FGF23 has been attracting increasing attention among researchers as a possible therapeutic target for CKD-related diseases. The efficacy of the blockade of FGF23 signaling has been examined in CKD rodent models; however, the use of FGF23-neutralizing antibodies to block FGF23 activity in a rat CKD model was found to increase mortality even though it ameliorated hyperparathyroidism (Shalhoub et al., 2012). Although the mechanisms by which FGF23 are elevated in the early stage of CKD remain largely unknown, increases in FGF23 mRNA levels in osteocytes and the prolonged half-life of FGF23 in the circulation have been reported (Christov et al., 2013). Furthermore, the expression of FGF23 in extraskeletal tissues such as the heart may also contribute to the increased levels of FGF23 in CKD (Olauson et al., 2014). As described earlier, FGF23 levels increase before elevations are detected in known FGF23 inducers, such as Pi and PTH, and the induction of FGF23 in a rodent model of acute renal failure was also evident in mice lacking PTH receptors on osteocytes and in *Pth*-deficient mice (Christov et al., 2013); hence, the early induction of FGF23 with renal failure may be independent of the induction by PTH or Pi. Thus, the physiological implications of the early increase in FGF23 levels in CKD patients have not yet been determined. This increase may be a compensatory pathway to reduce the retention of Pi; however, due to the concomitant decrease in *Klotho* expression in the kidney (Asai et al., 2012; Koh et al., 2001), this compensatory mechanism may be defective in CKD.

5. CONCLUSIONS AND FUTURE DIRECTIONS

Identification of the FGF23/KL system has yielded significant advances in understanding of the regulatory networks of phosphate metabolism and its roles in human disorders related to the disruption of this signaling pathway. The pathogenic implications of the overproduction of FGF23 in XLH have provided a therapeutic target for this disorder such that blockade of the FGF23/KL pathway may be beneficial. Clinical trials on the use of neutralizing antibodies against FGF23 are ongoing and have been successful for correcting abnormal mineral metabolism in these patients. In addition to XLH, a number of disorders have been attributed to a dysregulated FGF23/KL pathway, and therapeutic possibilities to target this pathway in these disorders are currently being investigated.

In addition to its FGF23-dependent functions, KL also has FGF23-independent roles, which include the regulation of oxidative stress, the IGF-1/insulin pathway, and calcium ion homeostasis. Although KL was originally identified as an aging-suppressor gene, the molecular mechanisms by which KL regulates the aging process remain largely unknown. Since increases in Pi levels in the circulation have been implicated in the pathogenesis of accelerated aging (Kuro-o, 2013), the FGF23-dependent functions of KL in the regulation of phosphate metabolism are, at least partly, responsible for the role of KL in aging processes. Nevertheless, further information on the critical roles of the FGF23-independent functions of KL in the regulation of aging processes will provide valuable insights into the mechanisms of aging.

ACKNOWLEDGMENT

This work is supported by grant from JSPS KAKENHI Grant Number 26461558.

REFERENCES

ADHR consortium (2000). Autosomal dominant hypophosphatemic rickets is associated with mutations in FGF23. *Nature Genetics*, *26*, 345–348.

Andrukhova, O., Slavic, S., Odorfer, K. I., & Erben, R. G. (2015). Experimental myocardial infarction upregulates circulating fibroblast growth factor-23. *Journal of Bone and Mineral Research*, *30*, 1831–1839.

Araya, K., Fukumoto, S., Backenroth, R., Takeuchi, Y., Nakayama, K., Ito, N., et al. (2005). A novel mutation in fibroblast growth factor 23 gene as a cause of tumoral calcinosis. *Journal of Clinical Endocrinology and Metabolism*, *90*, 5523–5527.

Asai, O., Nakatani, K., Tanaka, T., Sakan, H., Imura, A., Yoshimoto, S., et al. (2012). Decreased renal alpha-Klotho expression in early diabetic nephropathy in humans and mice and its possible role in urinary calcium excretion. *Kidney International*, *81*, 539–547.

Baia, L. C., Humalda, J. K., Vervloet, M. G., Navis, G., Bakker, S. J., & de Borst, M. H. (2013). Fibroblast growth factor 23 and cardiovascular mortality after kidney transplantation. *Clinical Journal of the American Society of Nephrology*, *8*, 1968–1978.

Ben-Dov, I. Z., Galitzer, H., Lavi-Moshayoff, V., Goetz, R., Kuro-o, M., Mohammadi, M., et al. (2007). The parathyroid is a target organ for FGF23 in rats. *Journal of Clinical Investigation*, *117*, 4003–4008.

Benet-Pages, A., Orlik, P., Strom, T. M., & Lorenz-Depiereux, B. (2005). An FGF23 missense mutation causes familial tumoral calcinosis with hyperphosphatemia. *Human Molecular Genetics*, *14*, 385–390.

Bloch, L., Sineshchekova, O., Reichenbach, D., Reiss, K., Saftig, P., Kuro-o, M., et al. (2009). Klotho is a substrate for alpha-, beta- and gamma-secretase. *FEBS Letters*, *583*, 3221–3224.

Block, G. A., Klassen, P. S., Lazarus, J. M., Ofsthun, N., Lowrie, E. G., & Chertow, G. M. (2004). Mineral metabolism, mortality, and morbidity in maintenance hemodialysis. *Journal of the American Society of Nephrology*, *15*, 2208–2218.

Brown, W. W., Juppner, H., Langman, C. B., Price, H., Farrow, E. G., White, K. E., et al. (2009). Hypophosphatemia with elevations in serum fibroblast growth factor 23 in a child with Jansen's metaphyseal chondrodysplasia. *Journal of Clinical Endocrinology and Metabolism*, *94*, 17–20.

Brownstein, C. A., Adler, F., Nelson-Williams, C., Iijima, J., Li, P., Imura, A., et al. (2008). A translocation causing increased alpha-klotho level results in hypophosphatemic rickets

and hyperparathyroidism. *Proceedings of the National Academy of Sciences of the United States of America, 105*, 3455–3460.
Burnett, S. M., Gunawardene, S. C., Bringhurst, F. R., Juppner, H., Lee, H., & Finkelstein, J. S. (2006). Regulation of C-terminal and intact FGF-23 by dietary phosphate in men and women. *Journal of Bone and Mineral Research, 21*, 1187–1196.
Carpenter, T. O., Imel, E. A., Ruppe, M. D., Weber, T. J., Klausner, M. A., Wooddell, M. M., et al. (2014). Randomized trial of the anti-FGF23 antibody KRN23 in X-linked hypophosphatemia. *Journal of Clinical Investigation, 124*, 1587–1597.
Carpenter, T. O., Insogna, K. L., Zhang, J. H., Ellis, B., Nieman, S., Simpson, C., et al. (2010). Circulating levels of soluble klotho and FGF23 in X-linked hypophosphatemia: Circadian variance, effects of treatment, and relationship to parathyroid status. *Journal of Clinical Endocrinology and Metabolism, 95*, E352–E357.
Chang, Q., Hoefs, S., van der Kemp, A. W., Topala, C. N., Bindels, R. J., & Hoenderop, J. G. (2005). The beta-glucuronidase klotho hydrolyzes and activates the TRPV5 channel. *Science, 310*, 490–493.
Chen, C. D., Podvin, S., Gillespie, E., Leeman, S. E., & Abraham, C. R. (2007). Insulin stimulates the cleavage and release of the extracellular domain of Klotho by ADAM10 and ADAM17. *Proceedings of the National Academy of Sciences of the United States of America, 104*, 19796–19801.
Chong, W. H., Molinolo, A. A., Chen, C. C., & Collins, M. T. (2011). Tumor-induced osteomalacia. *Endocrine-Related Cancer, 18*, R53–R77.
Christov, M., Waikar, S. S., Pereira, R. C., Havasi, A., Leaf, D. E., Goltzman, D., et al. (2013). Plasma FGF23 levels increase rapidly after acute kidney injury. *Kidney International, 84*, 776–785.
Collins, M. T., Chebli, C., Jones, J., Kushner, H., Consugar, M., Rinaldo, P., et al. (2001). Renal phosphate wasting in fibrous dysplasia of bone is part of a generalized renal tubular dysfunction similar to that seen in tumor-induced osteomalacia. *Journal of Bone and Mineral Research, 16*, 806–813.
Collins, M. T., Lindsay, J. R., Jain, A., Kelly, M. H., Cutler, C. M., Weinstein, L. S., et al. (2005). Fibroblast growth factor-23 is regulated by 1alpha,25-dihydroxyvitamin D. *Journal of Bone and Mineral Research, 20*, 1944–1950.
Farrow, E. G., Davis, S. I., Mooney, S. D., Beighton, P., Mascarenhas, L., Gutierrez, Y. R., et al. (2006). Extended mutational analyses of FGFR1 in osteoglophonic dysplasia. *American Journal of Medical Genetics. Part A, 140*, 537–539.
Farrow, E. G., Davis, S. I., Summers, L. J., & White, K. E. (2009). Initial FGF23-mediated signaling occurs in the distal convoluted tubule. *Journal of the American Society of Nephrology, 20*, 955–960.
Farrow, E. G., Summers, L. J., Schiavi, S. C., McCormick, J. A., Ellison, D. H., & White, K. E. (2010). Altered renal FGF23-mediated activity involving MAPK and Wnt: Effects of the Hyp mutation. *Journal of Endocrinology, 207*, 67–75.
Farrow, E. G., Yu, X., Summers, L. J., Davis, S. I., Fleet, J. C., Allen, M. R., et al. (2011). Iron deficiency drives an autosomal dominant hypophosphatemic rickets (ADHR) phenotype in fibroblast growth factor-23 (Fgf23) knock-in mice. *Proceedings of the National Academy of Sciences of the United States of America, 108*, E1146–E1155.
Faul, C., Amaral, A. P., Oskouei, B., Hu, M. C., Sloan, A., Isakova, T., et al. (2011). FGF23 induces left ventricular hypertrophy. *Journal of Clinical Investigation, 121*, 4393–4408.
Ferrari, S. L., Bonjour, J. P., & Rizzoli, R. (2005). Fibroblast growth factor-23 relationship to dietary phosphate and renal phosphate handling in healthy young men. *Journal of Clinical Endocrinology and Metabolism, 90*, 1519–1524.
Frishberg, Y., Ito, N., Rinat, C., Yamazaki, Y., Feinstein, S., Urakawa, I., et al. (2007). Hyperostosis-hyperphosphatemia syndrome: A congenital disorder of O-glycosylation

associated with augmented processing of fibroblast growth factor 23. *Journal of Bone and Mineral Research, 22*, 235–242.

Frishberg, Y., Topaz, O., Bergman, R., Behar, D., Fisher, D., Gordon, D., et al. (2005). Identification of a recurrent mutation in GALNT3 demonstrates that hyperostosis-hyperphosphatemia syndrome and familial tumoral calcinosis are allelic disorders. *Journal of Molecular Medicine (Berlin, Germany), 83*, 33–38.

Garringer, H. J., Fisher, C., Larsson, T. E., Davis, S. I., Koller, D. L., Cullen, M. J., et al. (2006). The role of mutant UDP-N-acetyl-alpha-D-galactosamine-polypeptide N-acetylgalactosaminyltransferase 3 in regulating serum intact fibroblast growth factor 23 and matrix extracellular phosphoglycoprotein in heritable tumoral calcinosis. *Journal of Clinical Endocrinology and Metabolism, 91*, 4037–4042.

Gattineni, J., Bates, C., Twombley, K., Dwarakanath, V., Robinson, M. L., Goetz, R., et al. (2009). FGF23 decreases renal NaPi-2a and NaPi-2c expression and induces hypophosphatemia in vivo predominantly via FGF receptor 1. *American Journal of Physiology. Renal Physiology, 297*, F282–F291.

Goetz, R., Beenken, A., Ibrahimi, O. A., Kalinina, J., Olsen, S. K., Eliseenkova, A. V., et al. (2007). Molecular insights into the klotho-dependent, endocrine mode of action of fibroblast growth factor 19 subfamily members. *Molecular and Cellular Biology, 27*, 3417–3428.

Goetz, R., Nakada, Y., Hu, M. C., Kurosu, H., Wang, L., Nakatani, T., et al. (2010). Isolated C-terminal tail of FGF23 alleviates hypophosphatemia by inhibiting FGF23-FGFR-Klotho complex formation. *Proceedings of the National Academy of Sciences of the United States of America, 107*, 407–412.

Gutierrez, O. M., Mannstadt, M., Isakova, T., Rauh-Hain, J. A., Tamez, H., Shah, A., et al. (2008). Fibroblast growth factor 23 and mortality among patients undergoing hemodialysis. *New England Journal of Medicine, 359*, 584–592.

Han, X., Xiao, Z., & Quarles, L. D. (2015). Membrane and integrative nuclear fibroblastic growth factor receptor (FGFR) regulation of FGF-23. *Journal of Biological Chemistry, 290*, 10447–10459.

Hanks, L. J., Casazza, K., Judd, S. E., Jenny, N. S., & Gutierrez, O. M. (2015). Associations of fibroblast growth factor-23 with markers of inflammation, insulin resistance and obesity in adults. *PLoS One, 10*, e0122885.

Holecki, M., Chudek, J., Owczarek, A., Olszanecka-Glinianowicz, M., Bozentowicz Wikarek, M., Dulawa, J., et al. (2015). Inflammation but not obesity or insulin resistance is associated with increased plasma fibroblast growth factor 23 concentration in the elderly. *Clinical Endocrinology, 82*, 900–909.

Hu, M. C., Shi, M., Zhang, J., Pastor, J., Nakatani, T., Lanske, B., et al. (2010). Klotho: A novel phosphaturic substance acting as an autocrine enzyme in the renal proximal tubule. *FASEB Journal, 24*, 3438–3450.

Hu, M. C., Shiizaki, K., Kuro-o, M., & Moe, O. W. (2013). Fibroblast growth factor 23 and Klotho: Physiology and pathophysiology of an endocrine network of mineral metabolism. *Annual Review of Physiology, 75*, 503–533.

Ichikawa, S., Baujat, G., Seyahi, A., Garoufali, A. G., Imel, E. A., Padgett, L. R., et al. (2010). Clinical variability of familial tumoral calcinosis caused by novel GALNT3 mutations. *American Journal of Medical Genetics. Part A, 152A*, 896–903.

Ichikawa, S., Gray, A. K., Padgett, L. R., Allen, M. R., Clinkenbeard, E. L., Sarpa, N. M., et al. (2014). Genetic rescue of glycosylation-deficient Fgf23 in the Galnt3 knockout mouse. *Endocrinology, 155*, 3891–3898.

Ichikawa, S., Imel, E. A., Kreiter, M. L., Yu, X., Mackenzie, D. S., Sorenson, A. H., et al. (2007). A homozygous missense mutation in human KLOTHO causes severe tumoral calcinosis. *Journal of Clinical Investigation, 117*, 2684–2691.

Ichikawa, S., Sorenson, A. H., Austin, A. M., Mackenzie, D. S., Fritz, T. A., Moh, A., et al. (2009). Ablation of the Galnt3 gene leads to low-circulating intact fibroblast growth factor 23 (Fgf23) concentrations and hyperphosphatemia despite increased Fgf23 expression. *Endocrinology, 150*, 2543–2550.

Imel, E. A., Zhang, X., Ruppe, M. D., Weber, T. J., Klausner, M. A., Ito, T., et al. (2015). Prolonged correction of serum phosphorus in adults with X-linked hypophosphatemia using monthly doses of KRN23. *Journal of Clinical Endocrinology and Metabolism, 100*, 2565–2573.

Imura, A., Iwano, A., Tohyama, O., Tsuji, Y., Nozaki, K., Hashimoto, N., et al. (2004). Secreted Klotho protein in sera and CSF: Implication for post-translational cleavage in release of Klotho protein from cell membrane. *FEBS Letters, 565*, 143–147.

Imura, A., Tsuji, Y., Murata, M., Maeda, R., Kubota, K., Iwano, A., et al. (2007). alpha-Klotho as a regulator of calcium homeostasis. *Science, 316*, 1615–1618.

Isakova, T., Wahl, P., Vargas, G. S., Gutierrez, O. M., Scialla, J., Xie, H., et al. (2011). Fibroblast growth factor 23 is elevated before parathyroid hormone and phosphate in chronic kidney disease. *Kidney International, 79*, 1370–1378.

Isakova, T., Xie, H., Yang, W., Xie, D., Anderson, A. H., Scialla, J., et al. (2011). Fibroblast growth factor 23 and risks of mortality and end-stage renal disease in patients with chronic kidney disease. *JAMA, 305*, 2432–2439.

Ito, S., Fujimori, T., Hayashizaki, Y., & Nabeshima, Y. (2002). Identification of a novel mouse membrane-bound family 1 glycosidase-like protein, which carries an atypical active site structure. *Biochimica et Biophysica Acta, 1576*, 341–345.

Ito, S., Kinoshita, S., Shiraishi, N., Nakagawa, S., Sekine, S., Fujimori, T., et al. (2000). Molecular cloning and expression analyses of mouse betaklotho, which encodes a novel Klotho family protein. *Mechanisms of Development, 98*, 115–119.

Jonsson, K. B., Zahradnik, R., Larsson, T., White, K. E., Sugimoto, T., Imanishi, Y., et al. (2003). Fibroblast growth factor 23 in oncogenic osteomalacia and X-linked hypophosphatemia. *New England Journal of Medicine, 348*, 1656–1663.

Kapelari, K., Kohle, J., Kotzot, D., & Hogler, W. (2015). Iron supplementation associated with loss of phenotype in autosomal dominant hypophosphatemic rickets. *Journal of Clinical Endocrinology and Metabolism, 100*(9), 3388–3392. jc20152391.

Kato, K., Jeanneau, C., Tarp, M. A., Benet-Pages, A., Lorenz-Depiereux, B., Bennett, E. P., et al. (2006). Polypeptide GalNAc-transferase T3 and familial tumoral calcinosis. Secretion of fibroblast growth factor 23 requires O-glycosylation. *Journal of Biological Chemistry, 281*, 18370–18377.

Kawai, M., Kinoshita, S., Kimoto, A., Hasegawa, Y., Miyagawa, K., Yamazaki, M., et al. (2013). FGF23 suppresses chondrocyte proliferation in the presence of soluble alpha-Klotho both in vitro and in vivo. *Journal of Biological Chemistry, 288*, 2414–2427.

Kawai, M., Kinoshita, S., Shimba, S., Ozono, K., & Michigami, T. (2014). Sympathetic activation induces skeletal Fgf23 expression in a circadian rhythm-dependent manner. *Journal of Biological Chemistry, 289*, 1457–1466.

Kendrick, J., Cheung, A. K., Kaufman, J. S., Greene, T., Roberts, W. L., Smits, G., et al. (2011). FGF-23 associates with death, cardiovascular events, and initiation of chronic dialysis. *Journal of the American Society of Nephrology, 22*, 1913–1922.

Koh, N., Fujimori, T., Nishiguchi, S., Tamori, A., Shiomi, S., Nakatani, T., et al. (2001). Severely reduced production of klotho in human chronic renal failure kidney. *Biochemical and Biophysical Research Communications, 280*, 1015–1020.

Kolek, O. I., Hines, E. R., Jones, M. D., LeSueur, L. K., Lipko, M. A., Kiela, P. R., et al. (2005). 1alpha,25-Dihydroxyvitamin D3 upregulates FGF23 gene expression in bone: The final link in a renal-gastrointestinal-skeletal axis that controls phosphate transport. *American Journal of Physiology Gastrointestinal and Liver Physiology, 289*, G1036–G1042.

Krieger, N. S., Culbertson, C. D., Kyker-Snowman, K., & Bushinsky, D. A. (2012). Metabolic acidosis increases fibroblast growth factor 23 in neonatal mouse bone. *American Journal of Physiology Renal Physiology, 303*, F431–F436.

Kuro-o, M. (2012). Klotho in health and disease. *Current Opinion in Nephrology and Hypertension, 21*, 362–368.

Kuro-o, M. (2013). Klotho, phosphate and FGF-23 in ageing and disturbed mineral metabolism. *Nature Reviews Nephrology, 9*, 650–660.

Kuro-o, M., Matsumura, Y., Aizawa, H., Kawaguchi, H., Suga, T., Utsugi, T., et al. (1997). Mutation of the mouse klotho gene leads to a syndrome resembling ageing. *Nature, 390*, 45–51.

Kurosu, H., Ogawa, Y., Miyoshi, M., Yamamoto, M., Nandi, A., Rosenblatt, K. P., et al. (2006). Regulation of fibroblast growth factor-23 signaling by klotho. *Journal of Biological Chemistry, 281*, 6120–6123.

Kurosu, H., Yamamoto, M., Clark, J. D., Pastor, J. V., Nandi, A., Gurnani, P., et al. (2005). Suppression of aging in mice by the hormone Klotho. *Science, 309*, 1829–1833.

Larsson, T., Marsell, R., Schipani, E., Ohlsson, C., Ljunggren, O., Tenenhouse, H. S., et al. (2004). Transgenic mice expressing fibroblast growth factor 23 under the control of the alpha1(I) collagen promoter exhibit growth retardation, osteomalacia, and disturbed phosphate homeostasis. *Endocrinology, 145*, 3087–3094.

Larsson, T., Nisbeth, U., Ljunggren, O., Juppner, H., & Jonsson, K. B. (2003). Circulating concentration of FGF-23 increases as renal function declines in patients with chronic kidney disease, but does not change in response to variation in phosphate intake in healthy volunteers. *Kidney International, 64*, 2272–2279.

Lavi-Moshayoff, V., Wasserman, G., Meir, T., Silver, J., & Naveh-Many, T. (2010). PTH increases FGF23 gene expression and mediates the high-FGF23 levels of experimental kidney failure: A bone parathyroid feedback loop. *American Journal of Physiology. Renal Physiology, 299*, F882–F889.

Levin, A., Bakris, G. L., Molitch, M., Smulders, M., Tian, J., Williams, L. A., et al. (2007). Prevalence of abnormal serum vitamin D, PTH, calcium, and phosphorus in patients with chronic kidney disease: Results of the study to evaluate early kidney disease. *Kidney International, 71*, 31–38.

Levy-Litan, V., Hershkovitz, E., Avizov, L., Leventhal, N., Bercovich, D., Chalifa-Caspi, V., et al. (2010). Autosomal-recessive hypophosphatemic rickets is associated with an inactivation mutation in the ENPP1 gene. *American Journal of Human Genetics, 86*, 273–278.

Lindberg, I., Pang, H. W., Stains, J. P., Clark, D., Yang, A. J., Bonewald, L., et al. (2015). FGF23 is endogenously phosphorylated in bone cells. *Journal of Bone and Mineral Research, 30*, 449–454.

Liu, H., Fergusson, M. M., Castilho, R. M., Liu, J., Cao, L., Chen, J., et al. (2007). Augmented Wnt signaling in a mammalian model of accelerated aging. *Science, 317*, 803–806.

Liu, S., Tang, W., Zhou, J., Stubbs, J. R., Luo, Q., Pi, M., et al. (2006). Fibroblast growth factor 23 is a counter-regulatory phosphaturic hormone for vitamin D. *Journal of the American Society of Nephrology, 17*, 1305–1315.

Liu, S., Vierthaler, L., Tang, W., Zhou, J., & Quarles, L. D. (2008). FGFR3 and FGFR4 do not mediate renal effects of FGF23. *Journal of the American Society of Nephrology, 19*, 2342–2350.

Lorenz-Depiereux, B., Bastepe, M., Benet-Pages, A., Amyere, M., Wagenstaller, J., Muller-Barth, U., et al. (2006). DMP1 mutations in autosomal recessive hypophosphatemia implicate a bone matrix protein in the regulation of phosphate homeostasis. *Nature Genetics, 38*, 1248–1250.

Lorenz-Depiereux, B., Schnabel, D., Tiosano, D., Hausler, G., & Strom, T. M. (2010). Loss-of-function ENPP1 mutations cause both generalized arterial calcification of infancy and autosomal-recessive hypophosphatemic rickets. *American Journal of Human Genetics, 86*, 267–272.

Matsumura, Y., Aizawa, H., Shiraki-Iida, T., Nagai, R., Kuro-o, M., & Nabeshima, Y. (1998). Identification of the human klotho gene and its two transcripts encoding membrane and secreted klotho protein. *Biochemical and Biophysical Research Communications, 242*, 626–630.

Miyagawa, K., Yamazaki, M., Kawai, M., Nishino, J., Koshimizu, T., Ohata, Y., et al. (2014). Dysregulated gene expression in the primary osteoblasts and osteocytes isolated from hypophosphatemic Hyp mice. *PLoS One, 9*, e93840.

Munoz Mendoza, J., Isakova, T., Ricardo, A. C., Xie, H., Navaneethan, S. D., Anderson, A. H., et al. (2012). Fibroblast growth factor 23 and inflammation in CKD. *Clinical Journal of the American Society of Nephrology*, 7, 1155–1162.

Nakano, C., Hamano, T., Fujii, N., Matsui, I., Tomida, K., Mikami, S., et al. (2012). Combined use of vitamin D status and FGF23 for risk stratification of renal outcome. *Clinical Journal of the American Society of Nephrology*, 7, 810–819.

Nakatani, T., Sarraj, B., Ohnishi, M., Densmore, M. J., Taguchi, T., Goetz, R., et al. (2009). In vivo genetic evidence for klotho-dependent, fibroblast growth factor 23 (Fgf23)-mediated regulation of systemic phosphate homeostasis. *FASEB Journal, 23*, 433–441.

Nelson, W. (1964). Aspects of circadian periodic changes in phosphorus metabolism in mice. *American Journal of Physiology, 206*, 589–598.

Ohata, Y., Arahori, H., Namba, N., Kitaoka, T., Hirai, H., Wada, K., et al. (2011). Circulating levels of soluble alpha-Klotho are markedly elevated in human umbilical cord blood. *Journal of Clinical Endocrinology and Metabolism, 96*, E943–E947.

Ohata, Y., Yamazaki, M., Kawai, M., Tsugawa, N., Tachikawa, K., Koinuma, T., et al. (2014). Elevated fibroblast growth factor 23 exerts its effects on placenta and regulates vitamin D metabolism in pregnancy of Hyp mice. *Journal of Bone and Mineral Research, 29*, 1627–1638.

Ohnishi, M., Nakatani, T., Lanske, B., & Razzaque, M. S. (2009). Reversal of mineral ion homeostasis and soft-tissue calcification of klotho knockout mice by deletion of vitamin D 1alpha-hydroxylase. *Kidney International, 75*, 1166–1172.

Olauson, H., Lindberg, K., Amin, R., Jia, T., Wernerson, A., Andersson, G., et al. (2012). Targeted deletion of Klotho in kidney distal tubule disrupts mineral metabolism. *Journal of the American Society of Nephrology, 23*, 1641–1651.

Olauson, H., Lindberg, K., Amin, R., Sato, T., Jia, T., Goetz, R., et al. (2013). Parathyroid-specific deletion of Klotho unravels a novel calcineurin-dependent FGF23 signaling pathway that regulates PTH secretion. *PLoS Genetics, 9*, e1003975.

Olauson, H., Vervloet, M. G., Cozzolino, M., Massy, Z. A., Urena Torres, P., & Larsson, T. E. (2014). New insights into the FGF23-Klotho axis. *Seminars in Nephrology, 34*, 586–597.

Perwad, F., Azam, N., Zhang, M. Y., Yamashita, T., Tenenhouse, H. S., & Portale, A. A. (2005). Dietary and serum phosphorus regulate fibroblast growth factor 23 expression and 1,25-dihydroxyvitamin D metabolism in mice. *Endocrinology, 146*, 5358–5364.

Portale, A. A., Wolf, M., Juppner, H., Messinger, S., Kumar, J., Wesseling-Perry, K., et al. (2014). Disordered FGF23 and mineral metabolism in children with CKD. *Clinical Journal of the American Society of Nephrology, 9*, 344–353.

Quarles, L. D. (2012). Skeletal secretion of FGF-23 regulates phosphate and vitamin D metabolism. *Nature Reviews. Endocrinology, 8*, 276–286.

Rafaelsen, S. H., Raeder, H., Fagerheim, A. K., Knappskog, P., Carpenter, T. O., Johansson, S., et al. (2013). Exome sequencing reveals FAM20c mutations associated with fibroblast growth factor 23-related hypophosphatemia, dental anomalies, and ectopic calcification. *Journal of Bone and Mineral Research, 28*, 1378–1385.

Riminucci, M., Collins, M. T., Fedarko, N. S., Cherman, N., Corsi, A., White, K. E., et al. (2003). FGF-23 in fibrous dysplasia of bone and its relationship to renal phosphate wasting. *Journal of Clinical Investigation, 112*, 683–692.

Saito, H., Maeda, A., Ohtomo, S., Hirata, M., Kusano, K., Kato, S., et al. (2005). Circulating FGF-23 is regulated by 1alpha,25-dihydroxyvitamin D3 and phosphorus in vivo. *Journal of Biological Chemistry, 280*, 2543–2549.

Saji, F., Shigematsu, T., Sakaguchi, T., Ohya, M., Orita, H., Maeda, Y., et al. (2010). Fibroblast growth factor 23 production in bone is directly regulated by 1{alpha},25-dihydroxyvitamin D, but not PTH. *American Journal of Physiology. Renal Physiology, 299*, F1212–F1217.

Samadfam, R., Richard, C., Nguyen-Yamamoto, L., Bolivar, I., & Goltzman, D. (2009). Bone formation regulates circulating concentrations of fibroblast growth factor 23. *Endocrinology, 150*, 4835–4845.

Satoh, M., Nagasu, H., Morita, Y., Yamaguchi, T. P., Kanwar, Y. S., & Kashihara, N. (2012). Klotho protects against mouse renal fibrosis by inhibiting Wnt signaling. *American Journal of Physiology. Renal Physiology, 303*, F1641–F1651.

Scanni, R., vonRotz, M., Jehle, S., Hulter, H. N., & Krapf, R. (2014). The human response to acute enteral and parenteral phosphate loads. *Journal of the American Society of Nephrology, 25*, 2730–2739.

Segawa, H., Kawakami, E., Kaneko, I., Kuwahata, M., Ito, M., Kusano, K., et al. (2003). Effect of hydrolysis-resistant FGF23-R179Q on dietary phosphate regulation of the renal type-II Na/Pi transporter. *Pflügers Archiv, 446*, 585–592.

Shalhoub, V., Shatzen, E. M., Ward, S. C., Davis, J., Stevens, J., Bi, V., et al. (2012). FGF23 neutralization improves chronic kidney disease-associated hyperparathyroidism yet increases mortality. *Journal of Clinical Investigation, 122*, 2543–2553.

Shimada, T., Hasegawa, H., Yamazaki, Y., Muto, T., Hino, R., Takeuchi, Y., et al. (2004). FGF-23 is a potent regulator of vitamin D metabolism and phosphate homeostasis. *Journal of Bone and Mineral Research, 19*, 429–435.

Shimada, T., Kakitani, M., Yamazaki, Y., Hasegawa, H., Takeuchi, Y., Fujita, T., et al. (2004). Targeted ablation of Fgf23 demonstrates an essential physiological role of FGF23 in phosphate and vitamin D metabolism. *Journal of Clinical Investigation, 113*, 561–568.

Shimada, T., Mizutani, S., Muto, T., Yoneya, T., Hino, R., Takeda, S., et al. (2001). Cloning and characterization of FGF23 as a causative factor of tumor-induced osteomalacia. *Proceedings of the National Academy of Sciences of the United States of America, 98*, 6500–6505.

Shimada, T., Urakawa, I., Yamazaki, Y., Hasegawa, H., Hino, R., Yoneya, T., et al. (2004). FGF-23 transgenic mice demonstrate hypophosphatemic rickets with reduced expression of sodium phosphate cotransporter type IIa. *Biochemical and Biophysical Research Communications, 314*, 409–414.

Shiraki-Iida, T., Aizawa, H., Matsumura, Y., Sekine, S., Iida, A., Anazawa, H., et al. (1998). Structure of the mouse klotho gene and its two transcripts encoding membrane and secreted protein. *FEBS Letters, 424*, 6–10.

Smith, R. C., O'Bryan, L. M., Farrow, E. G., Summers, L. J., Clinkenbeard, E. L., Roberts, J. L., et al. (2012). Circulating alphaKlotho influences phosphate handling by controlling FGF23 production. *Journal of Clinical Investigation, 122*, 4710–4715.

Tagliabracci, V. S., Engel, J. L., Wen, J., Wiley, S. E., Worby, C. A., Kinch, L. N., et al. (2012). Secreted kinase phosphorylates extracellular proteins that regulate biomineralization. *Science, 336*, 1150–1153.

Tagliabracci, V. S., Engel, J. L., Wiley, S. E., Xiao, J., Gonzalez, D. J., Nidumanda Appaiah, H., et al. (2014). Dynamic regulation of FGF23 by Fam20C phosphorylation, GalNAc-T3 glycosylation, and furin proteolysis. *Proceedings of the National Academy of Sciences of the United States of America, 111*, 5520–5525.

Takeyari, S., Yamamoto, T., Kinoshita, Y., Fukumoto, S., Glorieux, F. H., Michigami, T., et al. (2014). Hypophosphatemic osteomalacia and bone sclerosis caused by a novel homozygous mutation of the FAM20C gene in an elderly man with a mild variant of Raine syndrome. *Bone, 67*, 56–62.

Tebben, P. J., Singh, R. J., Clarke, B. L., & Kumar, R. (2004). Fibroblast growth factor 23, parathyroid hormone, and 1alpha,25-dihydroxyvitamin D in surgically treated primary hyperparathyroidism. *Mayo Clinic Proceedings, 79*, 1508–1513.

Urakawa, I., Yamazaki, Y., Shimada, T., Iijima, K., Hasegawa, H., Okawa, K., et al. (2006). Klotho converts canonical FGF receptor into a specific receptor for FGF23. *Nature, 444*, 770–774.

Vervloet, M. G., van Ittersum, F. J., Buttler, R. M., Heijboer, A. C., Blankenstein, M. A., & ter Wee, P. M. (2011). Effects of dietary phosphate and calcium intake on fibroblast growth factor-23. *Clinical Journal of the American Society of Nephrology, 6*, 383–389.

Wang, H., Yoshiko, Y., Yamamoto, R., Minamizaki, T., Kozai, K., Tanne, K., et al. (2008). Overexpression of fibroblast growth factor 23 suppresses osteoblast differentiation and matrix mineralization in vitro. *Journal of Bone and Mineral Research, 23*, 939–948.

Wolf, M., Molnar, M. Z., Amaral, A. P., Czira, M. E., Rudas, A., Ujszaszi, A., et al. (2011). Elevated fibroblast growth factor 23 is a risk factor for kidney transplant loss and mortality. *Journal of the American Society of Nephrology, 22*, 956–966.

Xiao, Z., Huang, J., Cao, L., Liang, Y., Han, X., & Quarles, L. D. (2014). Osteocyte-specific deletion of Fgfr1 suppresses FGF23. *PLoS One, 9*, e104154.

Xiao, L., Naganawa, T., Lorenzo, J., Carpenter, T. O., Coffin, J. D., & Hurley, M. M. (2010). Nuclear isoforms of fibroblast growth factor 2 are novel inducers of hypophosphatemia via modulation of FGF23 and KLOTHO. *Journal of Biological Chemistry, 285*, 2834–2846.

Xie, J., Cha, S. K., An, S. W., Kuro, O. M., Birnbaumer, L., & Huang, C. L. (2012). Cardioprotection by Klotho through downregulation of TRPC6 channels in the mouse heart. *Nature Communications, 3*, 1238.

Xu, Y., & Sun, Z. (2015). Molecular basis of Klotho: From gene to function in aging. *Endocrine Reviews, 36*, 174–193.

Yamamoto, M., Clark, J. D., Pastor, J. V., Gurnani, P., Nandi, A., Kurosu, H., et al. (2005). Regulation of oxidative stress by the anti-aging hormone klotho. *Journal of Biological Chemistry, 280*, 38029–38034.

Yamashita, T., Yoshioka, M., & Itoh, N. (2000). Identification of a novel fibroblast growth factor, FGF-23, preferentially expressed in the ventrolateral thalamic nucleus of the brain. *Biochemical and Biophysical Research Communications, 277*, 494–498.

Yamazaki, M., Kawai, M., Miyagawa, K., Ohata, Y., Tachikawa, K., Kinoshita, S., et al. (2015). Interleukin-1-induced acute bone resorption facilitates the secretion of fibroblast growth factor 23 into the circulation. *Journal of Bone and Mineral Metabolism, 33*, 342–354.

Yamazaki, Y., Tamada, T., Kasai, N., Urakawa, I., Aono, Y., Hasegawa, H., et al. (2008). Anti-FGF23 neutralizing antibodies show the physiological role and structural features of FGF23. *Journal of Bone and Mineral Research, 23*, 1509–1518.

Zhou, L., Li, Y., Zhou, D., Tan, R. J., & Liu, Y. (2013). Loss of Klotho contributes to kidney injury by derepression of Wnt/beta-catenin signaling. *Journal of the American Society of Nephrology, 24*, 771–785.

CHAPTER SEVEN

MicroRNA-34a and Impaired FGF19/21 Signaling in Obesity

T. Fu, J.K. Kemper[1]
University of Illinois at Urbana-Champaign, Urbana, IL, United States
[1]Corresponding author: e-mail address: jongsook@illinois.edu

Contents

Abstract

The obesity epidemic and the urgent need for effective and safe drugs to treat obesity-related diseases have greatly increased research interest in the metabolic hormones, fibroblast growth factor-19 (FGF19, FGF15 in mice), and FGF21. FGF19 and FGF21 function as endocrine hormones that play key roles in energy metabolism and counteract obesity. Importantly, in obese humans and lab animals, circulating FGF19 and FGF21 levels are elevated, and metabolic actions of these hormones are impaired but the underlying mechanisms remained unknown. Recent microRNA (miR) studies have revealed that aberrantly elevated miR-34a in obesity directly targets β-Klotho, the obligate coreceptor for both FGF19 and FGF21, and attenuates metabolic signaling of these hormones. In this review, we will discuss recent findings in the miR and FGF19/21 fields,

Vitamins and Hormones, Volume 101
ISSN 0083-6729
http://dx.doi.org/10.1016/bs.vh.2016.02.002

emphasizing the novel function of obesity-associated miR-34a in attenuation of FGF19/21 metabolic actions, and further discuss miRs, including miR-34a, as potential drug targets for obesity-related diseases.

1. INTRODUCTION

The prevalence of obesity has been rapidly increasing at an alarming rate which correlates with life-style changes in modern society (Grundy, 2008). Disruption of energy balance due to excessive food intake and/or insufficient physical activity plays a pivotal role in development of this global epidemic (Spiegelman & Flier, 2001). Obesity substantially increases risks for many human diseases, ranging from metabolic syndromes, including type 2 diabetes, fatty liver disease, and cardiovascular disease, to infertility, depression, neurological diseases, and even, certain types of cancer (Grundy, 2008; Kahn & Flier, 2000). Not surprisingly, the American Medical Association has officially identified obesity as a medical disease. The obesity epidemic and the urgent need for effective drugs to treat obesity-related metabolic diseases have greatly increased research interest in two metabolic hormones which play key roles in energy metabolism and counteract obesity, fibroblast growth factor-19 (FGF19, FGF15 in mice), and FGF21 (Canto & Auwerx, 2012; Potthoff, Kliewer, & Mangelsdorf, 2012).

The FGF19 subfamily members, FGF19, FGF21, and FGF23, are atypical FGFs that lack the conventional FGF heparin-binding domain and act as endocrine hormones (Beenken & Mohammadi, 2009; Kuro-o, 2008). FGF19 and FGF21 play key roles in regulation of lipid and carbohydrate metabolism in response to nutritional status. Binding of FGF19 and FGF21 to the FGF receptors (FGFRs) and the obligate coreceptor for these hormones, β-Klotho (βKL), triggers activation of intracellular signaling pathways to mediate metabolic actions of these hormones. Importantly, serum levels of FGF19 and FGF21 are highly elevated, and metabolic signaling of these hormones is impaired in obese patients and lab animals (Dushay et al., 2010; Mraz et al., 2011), but the underlying mechanisms remain unclear.

MicroRNAs (miRs) are small noncoding RNAs that act as regulatory molecules that control expression of their target genes (Hobert, 2008). Mature miRs bind to the 3′ untranslated region (UTR) of target transcripts and inhibit protein translation and/or mRNA stability (Lewis, Shih, Jones-Rhoades, Bartel, & Burge, 2003). Nearly 50% of all human genes are predicted to be regulated by miRs (Hobert, 2008). MiRs have been shown to play crucial roles in diverse biological pathways, including development, differentiation, cell proliferation, and metabolism. Importantly, miRs are aberrantly expressed in human diseases (Lee & Kemper, 2010; Rottiers & Naar, 2012; Rottiers et al., 2011; Zhang, Yang, Sun, Li, Xu, et al., 2013), revealing the therapeutic potential of targeting miRs.

In this review, we will describe the emerging role of miRs as powerful regulators of lipid and glucose metabolism. We will also discuss atypical FGFs, focusing on metabolic hormones regulating energy metabolism, FGF19 and FGF21, and their obligate coreceptor βKL, and further discuss the role of an obesity-associated miR-34a in down-regulating βKL expression, thus, attenuating metabolic action of FGF19 and FGF21. Finally, we will discuss the therapeutic potential of miR-34a as a novel drug target for treating obesity-related metabolic disorders, such as type 2 diabetes, nonalcoholic fatty liver disease (NAFLD), and cardiovascular disease.

2. MicroRNAs AS POWERFUL REGULATORS OF LIPID AND GLUCOSE METABOLISM

Accumulating evidence indicates that miRs play critical roles in regulation of lipid and glucose metabolism. In their original studies, Stoffel and his colleague demonstrated the functional importance of a pancreatic islet-specific miR, miR-375, in regulation of glucose-stimulated insulin secretion and exocytosis by directly targeting myotrophin (Poy et al., 2004). Esau et al. reported that miR-122, the most abundant miR in liver, plays important roles in liver cholesterol and fatty acid metabolism, and they further showed that miR-122 has a role in numerous hepatic functions, ranging from cholesterol metabolism, liver cancer, stress responses, and viral infection to circadian regulation (Esau et al., 2006). Also, miR-33 has been shown to contribute to the regulation of cholesterol homeostasis in concert with master transcriptional activators of cholesterol/lipid synthesis, SREBPs, by targeting the cholesterol transporter genes, ABCA1 and ABCG1, and interestingly, miR-33's is encoded within the introns of SREBP genes (Najafi-Shoushtari et al., 2010; Rayner et al., 2010). A recent study has shown that an SREBP-responsive miR operon expressing miR-182 and miR-96 constitutes a regulatory loop for intracellular lipid homeostasis (Jeon et al., 2013).

Aberrant expression of miRs has been detected in pathological conditions, such as obesity and diabetic conditions. MiR microarray studies in two obese mouse models, the leptin-deficient ob/ob mice and diet-induced obese mice, identified miR-34a as the miR with the highest relative increase (Lee & Kemper, 2010; Lee et al., 2010; Trajkovski et al., 2011). MiR-34a has been shown to act as a key regulator of hepatic lipid homeostasis (Choi et al., 2013; Fu et al., 2012) and also as a negative regulator of the formation of brown and beige fat in brown and white adipose tissue, respectively (Fu et al., 2014). In this review, we will focus on miR-34a, but other miRs have altered levels in obesity as briefly summarized later and as has been recently reviewed (Rottiers & Naar, 2012).

MiR-103 and 107 were upregulated in obese mice and silencing of miR-103/107 led to improved glucose homeostasis and insulin sensitivity (Trajkovski et al., 2011). Expression of both miR-378 and its host gene, PGC-1β, was elevated in obese mice. MiR-378 directly targets carnitine *O*-acetyltransferase, a key enzyme in mitochondrial metabolism of fatty acids (Carrer et al., 2012). Additional miRs that regulate glucose metabolism and insulin sensitivity and are increased in obesity include let-7, miR-143, miR-145, and miR-802 (Jordan et al., 2011; Kornfeld et al., 2013; Perez et al., 2013). MiR-155, miR-133, and miR-27 control adipocyte differentiation and that expression of miR-155 and miR-133 is increased, whereas that of miR-27 is decreased, in obesity (Chen et al., 2013; Sun & Trajkovski, 2014; Trajkovski, Ahmed, Esau, & Stoffel, 2012).

3. METABOLIC HORMONES FGF19/FGF21 AND THEIR OBLIGATE CORECEPTOR βKL

3.1 Atypical FGFs, FGF19, FGF21, and FGF23

FGFs are a family of growth factors that play critical roles in regulation of cell proliferation and differentiation, as well as normal development, tissue angiogenesis, and wound healing (Beenken & Mohammadi, 2009). Twenty-two FGF members have been identified in humans, and the FGF19 subfamily, the focus of this review, is comprised of the human FGF19 (FGF15 as mouse ortholog), FGF21, and FGF23 (Kir et al., 2011; Kuro-o, 2008). These FGF19 subfamily members act as endocrine hormones that control a variety of physiological processes, such as inhibiting bile acid synthesis in liver, promoting lipolysis in adipose tissue, and regulating vitamin D biosynthesis in kidney. The endocrine FGFs signal through cell surface receptor tyrosine kinases, FGFRs, including FGFR1, FGFR2, FGFR3, and FGFR4. The Klotho family members are membrane-spanning proteins that function as obligate coreceptors for the FGFRs and are required for the endocrine FGFs to bind to the receptor (Beenken & Mohammadi, 2009).

Intriguingly, expression of FGF19, FGF21, and FGF23 is transcriptionally regulated by nutrient-sensing nuclear receptors. Synthesis of FGF19 is induced by the bile acid-activated FXR in the small intestine in response to a meal, and circulating FGF21 is induced by fatty acid-activated peroxisome proliferator-activated receptor-α (PPARα) in the liver in response to prolonged fasting (Badman et al., 2007; Inagaki et al., 2007). FGF23 is induced by vitamin D nuclear receptor (Kuro-o, 2012). This review will focus on the physiological actions of FGF19 and FGF21 in regulation of lipid and carbohydrate metabolism in metabolic tissues and their dysregulation in obesity, in part due to elevated miR-34a.

3.2 βKL, the Obligate Coreceptor for FGF19 and FGF21

A unique feature of the endocrine FGFs is that they require the Klotho family proteins as coreceptors in order to exert their biological functions (Kuro-o, 2012; Ogawa et al., 2007). There are two Klotho proteins, Klotho and βKL, in the Klotho family. Klotho was initially discovered in 1997 in studies showing that mice with a mutation in this gene had shortened life spans and other phenotypes resembling human premature aging syndrome, "progeria" (Kuro-o, 2012; Kuro-o et al., 1997). The Klotho-null mice also exhibited high phosphate and calcium levels in the serum. Moreover, FGF23-knockout mice displayed strikingly similar phenotypes as the Klotho-null mice, suggesting that they might function in a same signal transduction pathway. Indeed, Klotho protein was identified as a coreceptor of FGF23, which was important for calcium and phosphate metabolism (Kurosu et al., 2006).

βKL has 41% amino acid sequence similarity with Klotho (Kuro-o, 2012). It is also a transmembrane protein and mainly distributed in metabolic organs, like pancreas, liver, and adipose tissue. However, βKL-null mice had a phenotype distinctly different from that of KL-null mice, most prominently, increased bile acid synthesis. In the βKL-null mice, expression of the Cyp7a1 gene, which encodes the rate-limiting enzyme of bile acid synthesis, was increased (Ito et al., 2005). Increased bile acid synthesis was also present in mice lacking FGF15 or its receptor, FGFR4, suggesting that the βKL-null phenotype was the result of impaired FGF15 signaling since βKL is the obligate coreceptor for FGF15 (FGF19 in human) in the liver. βKL also functions as the coreceptor of FGF21 in the liver and adipose tissue. While

FGFRs are widely expressed, βKL expression is restricted to liver, adipose tissue, pancreas, and brain and largely determines the tissue-specific metabolic actions of FGF15 and FGF21 (Kurosu et al., 2007).

4. MiR-34a AND IMPAIRED HEPATIC FGF19 SIGNALING IN OBESITY

4.1 Role of the miR-34a/βKL/FGF19 Axis in Liver

The metabolic response to FGF19 signaling in liver is impaired in obese patients with NAFLD (steatosis) and steatohepatitis, despite relatively normal FGF19 production and highly elevated serum levels of FGF19 (Schreuder et al., 2010). In genetic or diet-induced obese mice, serum FGF15 levels were also increased (Fisher et al., 2010; Fu et al., 2014), suggestive of FGF15 resistance in obesity, but the underlying mechanisms remained unclear. Recent studies from our group have suggested that elevated miR-34a in obesity attenuates FGF19 signaling by directly targeting and downregulating βKL (Fu et al., 2012). Remarkably, antisense inhibition of miR-34a in diet-induced obese mice partially restored βKL levels in the liver. Although FGFR4 is not a direct target of miR-34a, the stability of FGFR4 protein and localization of FGFR4 to cell membrane were also increased in a βKL-dependent manner (Fu, Kim, Byun, Kim, Seok, et al., 2016).

As illustrated in Fig. 1, treatment of dietary obese mice with antisense RNA for miR-34a (anti-miR-34a) resulted in improved metabolic outcomes, including decreased liver triglyceride and increased liver glycogen, which is consistent with improved regulation of expression of glucose and fatty acid metabolic genes by miR34a in the liver and increased insulin sensitivity. However, since miR-34a targets βKL and other potential components involved in FGF19 and FGF21 signaling in liver and other metabolic tissues, the effects of anti-miR34a in multiple tissues probably contribute to the overall beneficial outcome, including lipid-lowering effects and insulin sensitivity (Fu et al., 2012, 2014).

Recent studies in FGF19 transgenic mice have revealed a new function for FGF19 in energy homeostasis. FGF19 transgenic mice have lower body weight, reduced fat, enhanced hepatic lipid oxidation, increased brown adipose tissue (BAT), and elevated energy expenditure (Wu et al., 2010). These mice also had lower serum glucose and triglyceride levels, but normal insulin levels compared to control mice after feeding of a high-fat diet. In addition, a recent study has demonstrated that treatment with a synthetic FXR agonist, Fex, in mice activates the FXR/FGF15 pathway selectively in the intestine,

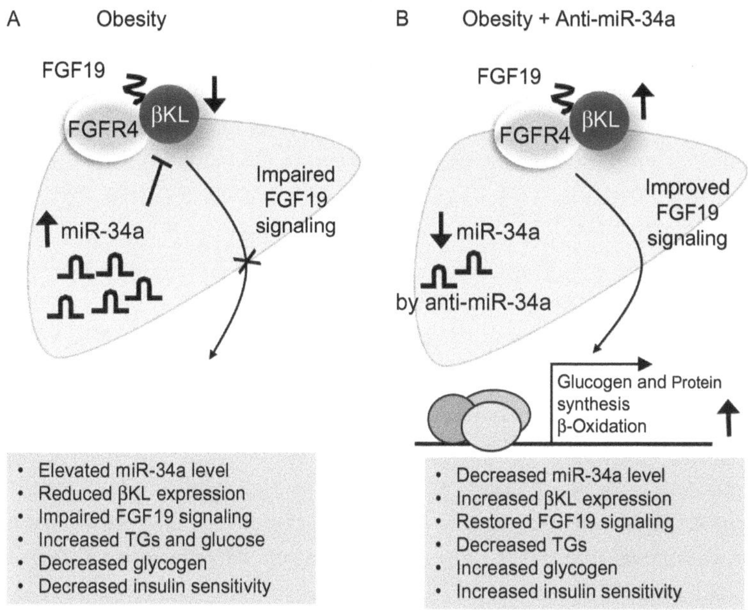

Fig. 1 Role of the miR-34a/βKL/FGF19 axis in liver. (A) In diet-induced obese mice, aberrantly elevated hepatic miR-34a represses the expression of the obligate coreceptor for FGF19, βKL, by directly binding to the 3′ UTR of the βKL transcript, resulting in attenuation of FGF19 signaling. The impaired FGF19 signaling in obesity leads to increased triglycerides/glucose levels, decreased glycogen levels in the liver, and decreased insulin sensitivity in fatty liver of obese mice. (B) Treatment with anti-miR-34a in dietary obese mice results in downregulation of the elevated miR-34a, increased expression of βKL, and subsequently, improves FGF19 signaling. The improved FGF19 signaling results in increased glycogen levels, decreased triglycerides levels in the liver, and increased insulin sensitivity.

resulting in increased serum FGF15 levels and counteracting obesity in part via increasing energy expenditure and promotion of fat browning in adipose tissue (Fang et al., 2015). It will be interesting to see whether miR-34a plays a role in these processes in the intestine and adipose tissue.

4.2 Regulation of Hepatic miR-34a Expression in Physiology and Obesity

It remains largely unknown how miR-34a expression is physiologically regulated and how miR-34a is abnormally upregulated in obesity, but recent studies from our group have provided evidence supporting the idea that the bile acid nuclear receptor FXR and SHP cascade pathway plays a role in inhibition of miR-34a expression in hepatocytes. In miR microarray studies, miR-34a exhibited the greatest fold-increase of liver miRs in

FXR-KO mice compared to wild-type mice, suggesting a potential role of FXR in inhibiting expression of miR-34a (Lee & Kemper, 2010; Lee et al., 2010). We further showed that activation of FXR induces expression of SHP, an orphan nuclear receptor and metabolic repressor, and SHP is recruited to the promoter of miR-34a and inhibits binding of p53, a key activator miR-34a expression (Lee & Kemper, 2010; Lee et al., 2010). Thus, activation of an FXR/SHP/p53 pathway appears to be involved in inhibition of miR-34a expression under physiological conditions.

In contrast, under pathophysiological conditions, such as NAFLD in obesity, this FXR/SHP/p53 cascade is defective, and transcription of miR-34a is not inhibited. The underlying mechanism is not clear, but decreased SIRT1 deacetylase and increased p300 acetylase function in fatty liver increase the acetylation status of p53 and FXR, which results in increased p53 and decreased FXR transcriptional activities (Kemper, Choi, & Kim, 2013). FXR, thus, appears to be a multifunctional regulator. On one hand, FXR ensures metabolic homeostasis by inducing the synthesis of a key metabolic hormone FGF19 in the intestine. On the other hand, FXR also inhibits expression of miR-34a in the liver, enhancing FGF19 signaling. However, a significant correlation between miR-34a and βKL levels was not observed in response to fasting/feeding in lean mice (T. Fu and K. Kemper, unpublished data), so the functional importance of this pathway in regulation of miR-34a expression in physiological conditions remains unclear and requires further study.

5. MiR-34a AND IMPAIRED FGF21 SIGNALING IN OBESITY

Synthesis of FGF21 is induced mainly in the liver during prolonged fasting but also induced in other metabolic tissues, such as adipose tissue, pancreas, and muscle, in response to physiological or pathological stressors, such as cold exposure, exercise, and autophagy deficiency (Badman et al., 2007; Bookout et al., 2013; Fisher et al., 2012; Inagaki et al., 2007; Kim et al., 2013). FGF21 binds mainly to FGFR1 in adipose tissue and to FGFR3 in liver, both of which require βKL as the obligate coreceptor. Binding of FGF21 to the receptor complex stimulates downstream signaling kinases, including ERK and GSK, which mediate the metabolic actions of this hormone.

5.1 Role of miR-34a in the βKL/FGF21 Axis in Liver

FGF21 has received tremendous attention in the obesity field because of its beneficial lipid-lowering and insulin-sensitizing effects. Administration of recombinant FGF21 to obese individuals has beneficial outcomes, such as weight loss and increased insulin sensitivity, although it has side effects, such as fragile bones and water retention (Wei et al., 2012). Physiologically, synthesis of FGF21 in the liver is induced during prolonged fasting through activation of a nuclear receptor, PPARα (Badman et al., 2007; Inagaki et al., 2007). In response to prolonged fasting, PPARα is activated by fatty acids, which are mostly released from adipocytes and taken up by hepatocytes, and binds directly to the promoter of the *Fgf21* gene to induce its transcription. Indeed, PPARα-KO mice are unable to catabolize fatty acids in the liver resulting in hepatic steatosis, but these effects are partially reversed by administration of FGF21. In addition, hepatic FGF21 induction is also significantly increased after challenged with a ketogenic diet or in suckling mouse neonates (Badman et al., 2007; Inagaki et al., 2007). In such conditions, fatty acids are elevated, which would active PPARα and induce synthesis of FGF21, resulting in increased fatty acid oxidation so that fatty acids are used as the primary metabolic fuel.

FGF21 signaling in the liver is mainly mediated by βKL and FGFR3 (Owen, Mangelsdorf, & Kliewer, 2015). Since βKL was shown to be a target of miR-34a, it is possible that elevated miR-34a levels in obesity may attenuate hepatic FGF21 signaling. In our studies, overexpression of miR-34a compromised FGF21 signaling in hepatocytes and dysregulated its downstream target genes, and conversely, downregulation of miR-34a by lentiviral-mediated expression of antisense-miR-34a in vivo resulted in improved hepatic FGF21 signaling (Fu et al., 2014). In addition, the expression of FGF21 target genes involved in hepatic fatty acid oxidation (*Cpt1*, *Mcad*, *Eci*, *CytC*, and *Pparα*) was increased, while that of lipogenic (*Fas* and *Srebp-1c*) and gluconeogenic (*G6pase* and *Pepck*) genes were decreased. These gene expression changes in the liver likely contribute to the systematic beneficial metabolic effects, such as weight loss, improved serum lipid and glucose profiles, and increased insulin sensitivity, which are observed after downregulation of miR-34a in obese mice.

MiR-34a likely targets multiple other genes in the liver in addition to βKL. We recently showed that hepatic miR-34a directly targets SIRT1, a NAD-dependent deacetylase, and NAMPT, a rate-limiting enzyme for NAD biosynthesis (Canto et al., 2012; Choi et al., 2013; Kemper et al.,

2013). Hepatic overexpression of miR-34a reduced NAMPT and subsequently, hepatic NAD levels and decreased SIRT1 deacetylase activity, which results in increased acetylation of SIRT1 target transcriptional regulators, such as PGC1-α, SRBEP-1c, FXR, and NF-κB (Choi et al., 2013). The overall effect of overexpression of miR-34a was to mimic metabolic conditions present in obesity. Conversely, antagonism of miR-34a in obese mice restored NAMPT/NAD levels and alleviated hepatic steatosis, inflammation, and glucose intolerance. It was also shown that miR-34a directly targets and downregulates hepatic PPARα in obese mice (Ding et al., 2015). Further, a recent study has shown that miR-34a directly inhibits expression of HNF-4, resulting in inhibition of VLDL secretion and promoting NAFLD (Xu et al., 2015). Therefore, a large network of genes regulated by the miR-34a/βKL axis may underlie the beneficial metabolic outcomes observed from the systematic anti-miR-34a experiments in diet-induced obese mice (Choi et al., 2013; Fu et al., 2012, 2014).

5.2 Role of miR-34a in βKL/FGF21 Signaling in Adipose Tissue

FGF21 is induced in the liver upon prolonged fasting (Inagaki et al., 2007), but it is also induced in response to feeding in white adipose tissue (WAT) which mediates insulin-sensitizing effects in WAT by activating PPARγ (Dutchak et al., 2012). Indeed, PPARγ agonists, such as a thiazolidinedione, reversed the diet-induced obese phenotype of FGF21-KO mice. Moreover, FGF21 increased the expression of adiponectin, which acts as a beneficial adipokine and maintains systematic glucose and lipid homeostasis (Yamauchi & Kadowaki, 2008). FGF21 expression is also induced upon cold exposure in both WAT and BAT, which results in increased expression of thermogenic transcriptional activators, such as PGC-1α, Prdm16, and mitochondrial uncoupling protein 1 (UCP1), and increased thermogenesis (Fisher et al., 2012).

FGF21 levels are highly elevated, and the expression of its receptor complex, both FGFR1 and βKL, is decreased in obesity, suggesting that FGF21 signaling may be impaired in obesity. In recent studies, our group has provided a mechanistic basis for FGF21 resistance by showing that elevated miR-34a contributes to attenuated FGF21 signaling in obese mice by downregulation of expression of the adipocyte FGF21 receptor complex components, both βKL and FGFR1, as illustrated in Fig. 2. Downregulation of miR-34a in obese mice had beneficial effects, including reduced adiposity, improved serum profiles, increased mitochondrial DNA copy number, and

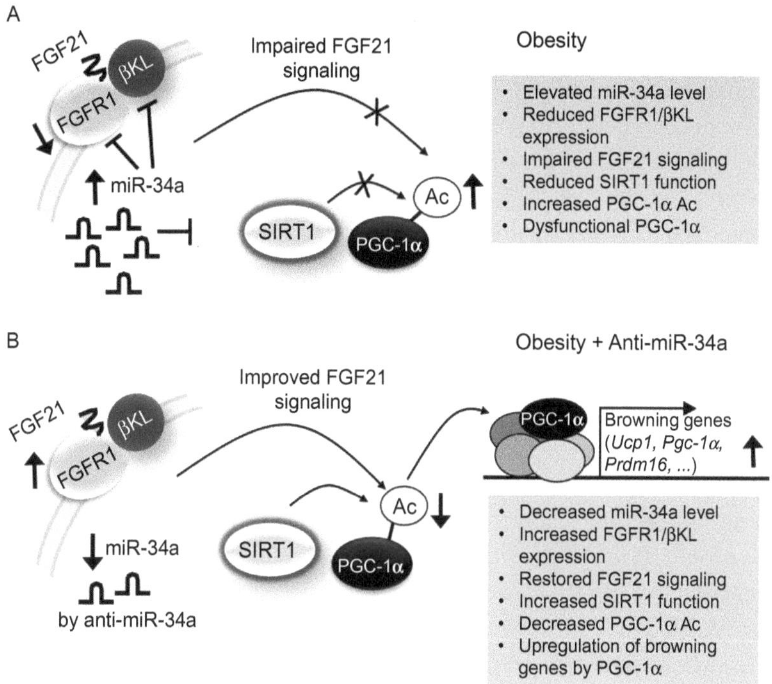

Fig. 2 Role of the miR-34a/βKL/FGF21 axis in adipose tissue. (A) In diet-induced obese mice, elevated miR-34a in adipocytes attenuates FGF21 signaling, at least in part, by directly targeting both components of the FGF21 receptor complex, both FGFR1 and βKL, and also by targeting SIRT1 deacetylase. Decreased SIRT1 function results in increased acetylation levels of PGC-1α and subsequently decreased transcriptional activity of PGC-1α in regulation of PGC-1α target browning-related genes. (B) Downregulation of elevated miR-34a in obese mice by treatment with anti-miR-34a improves FGF21 signaling, at least in part, by increasing expression of the FGF21 receptor complex, both FGFR1 and βKL, and also increases expression and activity of SIRT1, which contributes to deacetylation of PGC-1α and induction of PGC-1α target browning-related genes.

increased oxidative function in adipose tissue (Fu et al., 2014). Remarkably, downregulation of miR-34a increased coexpression of the beige fat-specific marker CD137 and the fat browning marker UCP1 in all types of white fat and also enhanced additional browning in BATs.

In addition to increased FGF21 signaling, downregulation of miR-34a also increased SIRT1-dependent deacetylation of PGC-1α, which contributed to the induction of the browning genes, *Ucp1*, *Pgc-1α*, and *Prdm16* (Fu et al., 2012; Kajimura, Seale, & Spiegelman, 2010). PGC-1α was previously shown to be critical for FGF21-mediated fat browning upon cold exposure, and FGF21 was shown to increase PGC1-α protein levels without affecting protein stability (Fu et al., 2012), but the mechanism for the upregulation of PGC-1α activity during brown adipocyte differentiation was not determined. Our studies suggest that deacetylation of PGC-1α plays a role in linking FGF21 signaling with the induction of PGC-1α target browning genes (Fu et al., 2014). Notably, PGC-1α acetylation levels were increased by downregulation of SIRT1 and also by downregulation of FGFR1/βKL, suggesting that both FGF21 signaling and SIRT1 function are important for deacetylation of PGC-1α and its browning gene activity (Fig. 2). Interestingly, each of these components, FGFR1, βKL, and SIRT1, is a direct target of miR-34a, suggesting that miR-34a is a key upstream regulator of multiple target genes that control energy metabolism in multiple metabolic tissues.

In addition to fat browning, FGF21 has also been reported to increase glucose uptake in adipocytes by enhancing the expression of glucose transporter-1 (GLUT-1). FGF21 induces the phosphorylation of the transcription factors, serum response factor and Ets-like protein 1 (Elk-1), which induce GLUT-1 expression (Ge et al., 2011). However, these effects of FGF21 are impaired in adipose tissue of diet-induced obese mice. Taken together, in vivo silencing of miR-34a appears to be an appealing option for treatment of obesity by dramatically increasing beige-like fat depots in all types of WATs, as well as increasing brown fat depots in BAT as shown in Fig. 2, together with its beneficial metabolic effects on insulin sensitivity and its effects in liver that result in amelioration of obesity-induced NAFLD as illustrated in Fig. 1.

5.3 Role of miR-34a in the βKL/FGF21 Axis in Other Tissues

FGF21, upon binding to its receptor, βKL and FGFR1, also exhibits beneficial metabolic actions in the pancreas. FGF21 improves pancreatic β-cell function and survival by activation of Erk and Akt signaling pathways (Wente et al., 2006). Studies from our group and others have shown that miR-34a levels are also elevated in pancreas of obese mice (Choi et al., 2013; Fu et al., 2014) and that elevated miR-34a leads to the dysfunction of pancreatic β-cell (Lovis et al., 2008). In pancreatic β-cell-derived MIN6B1 cells and pancreatic islets isolated from the leptin receptor-deficient db/db mice, chronic treatment with palmitic acid increased miR-34a levels in time- and dose-dependent manners, resulting in β-cell apoptosis and impaired insulin secretion (Lovis et al., 2008). Vesicle-associated membrane protein 2 (VAMP2), a direct target of miR-34a and a key player in β-cell exocytosis, was shown to mediate these effects. Further, FGF21 signaling in mouse pancreatic islets was impaired by high glucose levels which repressed expression of βKL, mediated by repression of PPARα, so that FGF21 signaling was attenuated (So et al., 2013). A role for miR-34a in regulation of βKL in the pancreas has not been reported, but since pancreatic miR-34a levels were abnormally increased in obesity (Fu et al., 2014), it is possible that miR-34a may underlie the decreased βKL expression in the pancreas that results in compromised FGF21 signaling in obesity.

The FGF21/βKL pathway also orchestrates metabolism through regulation in the central nervous system. FGF21 has been reported to alter the brain circadian clock to coordinate activity and reproduction as well as to increase systemic glucocorticoid levels and suppressing physical activity, as part of the adaption to fasting. In brain-specific βKL-null mice, these effects of FGF21 were not observed showing that the FGF21 effects were mediated by its FGFR/βKL receptor (Bookout et al., 2013). In the hypothalamus, FGF21 acts at the suprachiasmatic nucleus to suppress vasopressin-kisspeptin signaling which blocks the luteinizing hormone surge (Owen et al., 2013). Since miR-34a levels are increased in brain in obesity (Li, Khanna, Li, & Wang, 2011), especially in the aged brain, it will also be interesting to investigate the role of the brain FGF21/miR-34a axis in obesity and aging-related diseases.

It has been reported that FGF21 expression and secretion are induced upon prolonged starvation (Badman et al., 2007; Inagaki et al., 2007). It was also shown that autophagy deficiency and subsequent mitochondrial dysfunction promote Fgf21 gene expression, and thus, promote protection from diet-induced obesity and insulin resistance (Kim et al., 2013). FGF21 also increases glucose uptake in muscle through upregulation of GLUT4 (Cuevas-Ramos et al., 2012). Although miR-34a levels were not changed in skeletal muscle in obesity, it is possible that miR-34a expression is altered and thus, may affect FGF21 action in muscle in response to other biological cues or pathological conditions (Gan et al., 2013).

6. THERAPEUTIC AND DIAGNOSTIC POTENTIAL OF miRs

6.1 Therapeutic Potential of miRs

Since expression of miRs is abnormally downregulated or upregulated in several disease states, efforts have been made to develop therapeutic agents that can mimic or downregulate miRs (Olson, 2014). Treatment with antagomirs, antisense miR inhibitors, of several miRs has shown beneficial effects, demonstrating the therapeutic potential of targeting miRs (Rottiers et al., 2011). In general, there are three major antisense approaches, antagomiRs, which are conjugated to cholesterol to facilitate cellular uptake; locked nucleic acid (LNA) phosphorothioate chemistry (Rottiers et al., 2013); or chemical modification of the oligonucleotide at 2′-sugar and phosphate backbone moiety with MOE (2′-*O*-methoxyethylphosphorothioate) (Olson, 2014).

Antagonism of miR-122 in mice by systemically administered LNA-conjugated antisense miR resulted in upregulation of a large set of predicted target mRNAs in the liver (Esau et al., 2006). Antagonism of miR-33 promoted cholesterol transport and had beneficial effects on atherosclerosis in mice (Najafi-Shoushtari et al., 2010), and importantly, inhibition of miR-33a/b raised plasma HDL and lowered VLDL in nonhuman primates (Rayner et al., 2011). Antagonism of obesity-associated miR-103/miR-107 also resulted in improved insulin sensitivity (Trajkovski et al., 2011). Antagonism of miR-378 boosted activity of mitochondria, benefited obesity and cardiovascular disease (Trajkovski et al., 2011). Excitingly, recent in vivo antagonism or overexpression studies demonstrate that certain miRs are highly promising potential drug targets for treatment of metabolic diseases as summarized in Table 1.

As described earlier, downregulation of miR-34a in mice resulted in decreased adiposity, increased glucose tolerance and insulin sensitivity, and improved lipid profiles (Fu et al., 2014). Mechanistically, downregulation of miR-34a in obese mice resulted in improved FGF21 and FGF19 signaling by increasing expression of βKL and FGFR1. MiR-34a also has other targets, including SIRT1 and NAMPT, which would affect SIRT1 expression and activity (Choi et al., 2013). Therefore, the beneficial effects of downregulation of miR-34a in obesity may involve increased expression of multiple gene targets in multiple tissues, which may provide an effective therapeutic advantage compared to the classical approach that targets only one molecule. However, this promiscuous targeting by miR-34a, as well as other miRs, could have detrimental side effects. To produce anti-miRs with acceptable therapeutic indexes, a thorough understanding of miR-34a regulatory networks, as aided by bioinformatics and systems biology approaches, will be essential.

Furthermore, abnormal activation of FGF19 is associated with colon and liver cancer so the extent of increased FGF19 signaling by downregulation of miR-34a must be carefully considered (Desnoyers et al., 2008). Similarly, miR-34a is a well-known tumor suppressor, and deletion

Table 1 Metabolic microRNAs Currently in Development as Drug Targets

microRNA	Validated Target Genes	Function (Key References)	Anti-miR Drugs
miR-122	*Agpat1*, *Mogat1*, MTTP, Klf6	Increase hepatic TG level and HCC development (Hsu, Wang, Kota, Yu, Costinean, et al., 2012; Tsai, Hsu, Lai, Chau, Lin, et al., 2009)	RG-101 for Hepatitis C virus infection
miR-33	SREBP, HMGCR, LDL-R, ABCA1, ABCG1	Control cholesterol homeostasis (Esau et al., 2006; Najafi-Shoushtari et al., 2010; Rayner et al., 2011)	SPC-4955 for cholesterol, HDL, and cardiovascular diseases
miR-103/107	Caveolin-1	Regulate glucose homeostasis, insulin sensitivity (Trajkovski et al., 2011)	RG-125 (AZD4076) for NASH in T2D patients/prediabetes
miR-378	MED13	Mitochondrial metabolism (Carrer et al., 2012)	Obesity, cardiovascular disease
miR-208	Myh6, Myh7, Thrap1, Myostatin, MED13	Cardiac hypertrophy (Callis, Pandya, Seok, Tang, Tatsuguchi, et al., 2009; Grueter, van Rooij, Johnson, DeLeon, Sutherland, et al., 2012; Montgomery, Hullinger, Semus, Dickinson, Seto, et al., 2011)	Obesity, cardiovascular disease
miR-221	DDIT4, CDKN1, p27, p57	Cell cycle arrest, cell proliferation (Fornari, Gramantieri, Ferracin, Veronese, Sabbioni, et al., 2008; Pineau, Volinia, McJunkin, Marchio, Battiston, et al., 2010)	Hepatocelluar carcinoma
miR-21	PTEN	Cell proliferation, cell cycle arrest, Apoptosis (Meng, Henson, Wehbe-Janek, Ghoshal, Jacob, et al., 2007)	RG-012 for hepatocelluar carcinoma

of the *miR-34a* gene was associated with tumorigenesis (Yamakuchi, Ferlito, & Lowenstein, 2008). Thus, targeting the miR-34a/FGF19/21 axis will need a careful modulation of the extent of inhibition, because excess miR-34a is associated with metabolic diseases and extreme deficiency of miR-34a is associated with cancer.

6.2 Diagnostic Potential of miRs

Although miRs in general act inside cells, several miRs, such as miR-34a and miR-155, have been detected circulating in the blood (Pogribny et al., 2010). The circulating miRs have diagnostic potential since the levels correlated with tissue levels and they can, thus, serve as biomarkers for diseases. Although the mechanism of how these miRs travel through bloodstream is largely unknown, the discovery of circulating miRs has highlighted their potential as diagnostic markers for disease (Olson, 2014).

Human clinical studies show that miR-34a levels are increased more than sixfold in patients with NALFD and with chronic hepatitis C (Cermelli, Ruggieri, Marrero, Ioannou, & Beretta, 2011). In patients with nonalcoholic steatohepatitis (NASH), miR-34a levels were increased ~100-fold. Levels of miR-34a have been reported to increase ~10-fold in ob/ob mice and diet-induced obese mice and ~6-fold in STZ-induced diabetic mice (Owen et al., 2013). Moreover, clinical data from our group also demonstrated that miR-34a levels are highly correlated with the BMI index of obese patients (Choi et al., 2013). Strikingly, miR-34a levels in the serum of obese patients were significantly elevated compared to controls. In addition, it was shown that serum miR-34a levels correlate with progression of fatty liver disease, such as NAFLD, NASH, and cirrhosis (Cermelli et al., 2011; Cheung et al., 2008). This correlation suggests that serum miR-34a can serve as a potential diagnostic biomarker of liver diseases in place of liver biopsies.

7. CONCLUSION

Obesity has become a global pandemic, which has sparked research interest in energy metabolism. FGF19 and FGF21 are metabolic hormones that play key roles in energy metabolism and counteract obesity. Since βKL acts as the obligate coreceptor for both FGF19 and FGF21, βKL serves an effective potential therapeutic target for improving FGF19 and FGF21 signaling in obesity. Indeed, studies on agonist antibodies to FGFR1 or to both FGFR1 and βKL demonstrated that antibodies stably mimic the hormone action, resulting in beneficial metabolic effects in both mice and monkeys, although the antibody approach has the same side effects as recombinant FGF21, such as fragile bones and water retention (Foltz et al., 2012; Wu et al., 2011). As described earlier, targeting elevated miR-34a in obesity by anti-miR-34a holds promise as a potential antiobesity drug by targeting a network of genes, including βKL, in multiple tissues. The striking role of downregulation of miR-34a in promoting fat browning in all types of WATs and BAT (Fu et al., 2014), together with its beneficial effects on hepatic lipid metabolism, such as increased fatty acid oxidation, decreased lipogenesis, and improved steatosis (Fu et al., 2012), suggest that targeting a single microRNA, miR-34a, may provide an effective therapeutic option for combating obesity and treating obesity-related metabolic disease. Other small miRs play important roles in metabolic regulation and may also serve as attractive therapeutic drug targets for treatment of obesity-related disease and disease biomarkers.

ACKNOWLEDGMENTS

We thank Dr. Byron Kemper for critical reading of the manuscript. Research discussed from the authors' laboratory was supported by Grants DK062777 and NIH DK095842 from the National Institutes of Health.

REFERENCES

Badman, M. K., Pissios, P., Kennedy, A. R., Koukos, G., Flier, J. S., & Maratos-Flier, E. (2007). Hepatic fibroblast growth factor 21 is regulated by PPARalpha and is a key mediator of hepatic lipid metabolism in ketotic states. *Cell Metabolism, 6*, 426–437.

Beenken, A., & Mohammadi, M. (2009). The FGF family: Biology, pathophysiology and therapy. *Nature Reviews. Drug Discovery, 3*, 235–253.

Bookout, A. L., de Groot, M. H., Owen, B. M., Lee, S., Gautron, L., Lawrence, H. L., et al. (2013). FGF21 regulates metabolism and circadian behavior by acting on the nervous system. *Nature Medicine, 9*, 1147–1152.

Callis, T. E., Pandya, K., Seok, H. Y., Tang, R. H., Tatsuguchi, M., et al. (2009). MicroRNA-208a is a regulator of cardiac hypertrophy and conduction in mice. *The Journal of Clinical Investigation, 119*(9), 2772–2786.

Canto, C., & Auwerx, J. (2012). Cell biology. FGF21 takes a fat bite. *Science, 6082*, 675–676.

Canto, C., Houtkooper, R. H., Pirinen, E., Youn, D. Y., Oosterveer, M. H., Cen, Y., et al. (2012). The NAD(+) precursor nicotinamide riboside enhances oxidative metabolism and protects against high-fat diet-induced obesity. *Cell Metabolism, 6*, 838–847.

Carrer, M., Liu, N., Grueter, C. E., Williams, A. H., Frisard, M. I., Hulver, M. W., et al. (2012). Control of mitochondrial metabolism and systemic energy homeostasis by microRNAs 378 and 378*. *Proceedings of the National Academy of Sciences of the United States of America, 38*, 15330–15335.

Cermelli, S., Ruggieri, A., Marrero, J. A., Ioannou, G. N., & Beretta, L. (2011). Circulating microRNAs in patients with chronic hepatitis C and non-alcoholic fatty liver disease. *PLoS One, 8*, e23937.

Chen, Y., Siegel, F., Kipschull, S., Haas, B., Frohlich, H., Meister, G., et al. (2013). miR-155 regulates differentiation of brown and beige adipocytes via a bistable circuit. *Nature Communications, 4*, 1769.

Cheung, O., Puri, P., Eicken, C., Contos, M. J., Mirshahi, F., Maher, J. W., et al. (2008). Nonalcoholic steatohepatitis is associated with altered hepatic MicroRNA expression. *Hepatology, 6*, 1810–1820.

Choi, S. E., Fu, T., Seok, S., Kim, D. H., Yu, E., Lee, K. W., et al. (2013). Elevated microRNA-34a in obesity reduces NAD+ levels and SIRT1 activity by directly targeting NAMPT. *Aging Cell, 6*, 1062–1072.

Cuevas-Ramos, D., Almeda-Valdes, P., Meza-Arana, C. E., Brito-Cordova, G., Gomez-Perez, F. J., Mehta, R., et al. (2012). Exercise increases serum fibroblast growth factor 21 (FGF21) levels. *PLoS One, 5*, e38022.

Desnoyers, L. R., Pai, R., Ferrando, R. E., Hotzel, K., Le, T., Ross, J., et al. (2008). Targeting FGF19 inhibits tumor growth in colon cancer xenograft and FGF19 transgenic hepatocellular carcinoma models. *Oncogene, 1*, 85–97.

Ding, J., Li, M., Wan, X., Jin, X., Chen, S., Yu, C., et al. (2015). Effect of miR-34a in regulating steatosis by targeting PPARalpha expression in nonalcoholic fatty liver disease. *Scientific Reports, 5*, 13729.

Dushay, J., Chui, P. C., Gopalakrishnan, G. S., Varela-Rey, M., Crawley, M., Fisher, F. M., et al. (2010). Increased fibroblast growth factor 21 in obesity and nonalcoholic fatty liver disease. *Gastroenterology, 2*, 456–463.

Dutchak, P. A., Katafuchi, T., Bookout, A. L., Choi, J. H., Yu, R. T., Mangelsdorf, D. J., et al. (2012). Fibroblast growth factor-21 regulates PPARgamma activity and the antidiabetic actions of thiazolidinediones. *Cell, 3*, 556–567.

Esau, C., Davis, S., Murray, S. F., Yu, X. X., Pandey, S. K., Pear, M., et al. (2006). miR-122 regulation of lipid metabolism revealed by in vivo antisense targeting. *Cell Metabolism, 2*, 87–98.

Fang, S., Suh, J. M., Reilly, S. M., Yu, E., Osborn, O., Lackey, D., et al. (2015). Intestinal FXR agonism promotes adipose tissue browning and reduces obesity and insulin resistance. *Nature Medicine, 2*, 159–165.

Fisher, F. M., Chui, P. C., Antonellis, P. J., Bina, H. A., Kharitonenkov, A., Flier, J. S., et al. (2010). Obesity is a fibroblast growth factor 21 (FGF21)-resistant state. *Diabetes, 11*, 2781–2789.

Fisher, F. M., Kleiner, S., Douris, N., Fox, E. C., Mepani, R. J., Verdeguer, F., et al. (2012). FGF21 regulates PGC-1alpha and browning of white adipose tissues in adaptive thermogenesis. *Genes & Development, 3*, 271–281.

Foltz, I. N., Hu, S., King, C., Wu, X., Yang, C., Wang, W., et al. (2012). Treating diabetes and obesity with an FGF21-mimetic antibody activating the betaKlotho/FGFR1c receptor complex. *Science Translational Medicine, 162*, 162ra153.

Fornari, F., Gramantieri, L., Ferracin, M., Veronese, A., Sabbioni, S., et al. (2008). MiR-221 controls CDKN1C/p57 and CDKN1B/p27 expression in human hepatocellular carcinoma. *Oncogene, 27*, 5651–5661.

Fu, T., Choi, S. E., Kim, D. H., Seok, S., Suino-Powell, K. M., Xu, H. E., et al. (2012). Aberrantly elevated microRNA-34a in obesity attenuates hepatic responses to FGF19 by targeting a membrane coreceptor beta-Klotho. *Proceedings of the National Academy of Sciences of the United States of America, 40*, 16137–16142.

Fu, T., Kim, Y. C., Byun, S., Kim, D. H., Seok, S., et al. (2016). FXR primes the liver for intestinal FGF15 signaling by transient induction of β-klotho. *Molecular Endocrinology, 30*(1), 92–103.

Fu, T., Seok, S., Choi, S., Huang, Z., Suino-Powell, K., Xu, H. E., et al. (2014). MicroRNA 34a inhibits beige and brown fat formation in obesity in part by suppressing adipocyte fibroblast growth factor 21 signaling and SIRT1 function. *Molecular and Cellular Biology, 22*, 4130–4142.

Gan, Z., Rumsey, J., Hazen, B. C., Lai, L., Leone, T. C., Vega, R. B., et al. (2013). Nuclear receptor/microRNA circuitry links muscle fiber type to energy metabolism. *The Journal of Clinical Investigation, 6*, 2564–2575.

Ge, X., Chen, C., Hui, X., Wang, Y., Lam, K. S., & Xu, A. (2011). Fibroblast growth factor 21 induces glucose transporter-1 expression through activation of the serum response factor/Ets-like protein-1 in adipocytes. *The Journal of Biological Chemistry, 40*, 34533–34541.

Grueter, C. E., van Rooij, E., Johnson, B. A., DeLeon, S. M., Sutherland, L. B., et al. (2012). A cardiac microRNA governs systemic energy homeostasis by regulation of MED13. *Cell, 149*(3), 671–683.

Grundy, S. M. (2008). Metabolic syndrome pandemic. *Arteriosclerosis, Thrombosis, and Vascular Biology, 4*, 629–636.

Hobert, O. (2008). Gene regulation by transcription factors and microRNAs. *Science, 5871*, 1785–1786.

Hsu, S.-h., Wang, B., Kota, J., Yu, J., Costinean, S., et al. (2012). Essential metabolic, anti-inflammatory, and anti-tumorigenic functions of miR-122 in liver. *The Journal of Clinical Investigation, 122*(8), 2871–2883.

Inagaki, T., Dutchak, P., Zhao, G., Ding, X., Gautron, L., Parameswara, V., et al. (2007). Endocrine regulation of the fasting response by PPARalpha-mediated induction of fibroblast growth factor 21. *Cell Metabolism, 6*, 415–425.

Ito, S., Fujimori, T., Furuya, A., Satoh, J., Nabeshima, Y., & Nabeshima, Y. (2005). Impaired negative feedback suppression of bile acid synthesis in mice lacking betaKlotho. *The Journal of Clinical Investigation, 8*, 2202–2208.

Jeon, T. I., Esquejo, R. M., Roqueta-Rivera, M., Phelan, P. E., Moon, Y. A., Govindarajan, S. S., et al. (2013). An SREBP-responsive microRNA operon contributes to a regulatory loop for intracellular lipid homeostasis. *Cell Metabolism, 1*, 51–61.

Jordan, S. D., Kruger, M., Willmes, D. M., Redemann, N., Wunderlich, F. T., Bronneke, H. S., et al. (2011). Obesity-induced overexpression of miRNA-143 inhibits insulin-stimulated AKT activation and impairs glucose metabolism. *Nature Cell Biology, 4*, 434–446.

Kahn, B. B., & Flier, J. S. (2000). Obesity and insulin resistance. *The Journal of Clinical Investigation, 4*, 473–481.

Kajimura, S., Seale, P., & Spiegelman, B. M. (2010). Transcriptional control of brown fat development. *Cell Metabolism, 4*, 257–262.

Kemper, J. K., Choi, S. E., & Kim, D. H. (2013). Sirtuin 1 deacetylase: A key regulator of hepatic lipid metabolism. *Vitamins and Hormones, 91*, 385–404.

Kim, K. H., Jeong, Y. T., Oh, H., Kim, S. H., Cho, J. M., Kim, Y. N., et al. (2013). Autophagy deficiency leads to protection from obesity and insulin resistance by inducing Fgf21 as a mitokine. *Nature Medicine, 1*, 83–92.

Kir, S., Beddow, S. A., Samuel, V. T., Miller, P., Previs, S. F., Suino-Powell, K., et al. (2011). FGF19 as a postprandial, insulin-independent activator of hepatic protein and glycogen synthesis. *Science, 6024*, 1621–1624.

Kornfeld, J. W., Baitzel, C., Konner, A. C., Nicholls, H. T., Vogt, M. C., Herrmanns, K., et al. (2013). Obesity-induced overexpression of miR-802 impairs glucose metabolism through silencing of Hnf1b. *Nature, 7435*, 111–115.

Kuro-o, M. (2008). Endocrine FGFs and Klothos: Emerging concepts. *Trends in Endocrinology and Metabolism, 7*, 239–245.

Kuro-o, M. (2012). Klotho and betaKlotho. *Advances in Experimental Medicine and Biology, 728*, 25–40.

Kuro-o, M., Matsumura, Y., Aizawa, H., Kawaguchi, H., Suga, T., Utsugi, T., et al. (1997). Mutation of the mouse klotho gene leads to a syndrome resembling ageing. *Nature, 6655*, 45–51.

Kurosu, H., Choi, M., Ogawa, Y., Dickson, A. S., Goetz, R., Eliseenkova, A. V., et al. (2007). Tissue-specific expression of betaKlotho and fibroblast growth factor (FGF) receptor isoforms determines metabolic activity of FGF19 and FGF21. *The Journal of Biological Chemistry*, *37*, 26687–26695.

Kurosu, H., Ogawa, Y., Miyoshi, M., Yamamoto, M., Nandi, A., Rosenblatt, K. P., et al. (2006). Regulation of fibroblast growth factor-23 signaling by klotho. *The Journal of Biological Chemistry*, *10*, 6120–6123.

Lee, J., & Kemper, J. K. (2010). Controlling SIRT1 expression by microRNAs in health and metabolic disease. *Aging (Albany NY)*, *8*, 527–534.

Lee, J., Padhye, A., Sharma, A., Song, G., Miao, J., Mo, Y. Y., et al. (2010). A pathway involving farnesoid X receptor and small heterodimer partner positively regulates hepatic sirtuin 1 levels via microRNA-34a inhibition. *The Journal of Biological Chemistry*, *17*, 12604–12611.

Lewis, B. P., Shih, I. H., Jones-Rhoades, M. W., Bartel, D. P., & Burge, C. B. (2003). Prediction of mammalian microRNA targets. *Cell*, *115*(7), 787–798.

Li, X., Khanna, A., Li, N., & Wang, E. (2011). Circulatory miR34a as an RNA based, noninvasive biomarker for brain aging. *Aging (Albany NY)*, *10*, 985–1002.

Lovis, P., Roggli, E., Laybutt, D. R., Gattesco, S., Yang, J. Y., Widmann, C., et al. (2008). Alterations in microRNA expression contribute to fatty acid-induced pancreatic beta-cell dysfunction. *Diabetes*, *10*, 2728–2736.

Meng, F., Henson, R., Wehbe-Janek, H., Ghoshal, K., Jacob, S. T., et al. (2007). MicroRNA-21 regulates expression of the PTEN tumor suppressor gene in human hepatocellular cancer. *Gastroenterology*, *133*(2), 647–658.

Montgomery, R. L., Hullinger, T. G., Semus, H. M., Dickinson, B. A., Seto, A. G., et al. (2011). Therapeutic inhibition of miR-208a improves cardiac function and survival during heart failure. *Circulation*, *124*(14), 1537–1547.

Mraz, M., Lacinova, Z., Kavalkova, P., Haluzikova, D., Trachta, P., Drapalova, J., et al. (2011). Serum concentrations of fibroblast growth factor 19 in patients with obesity and type 2 diabetes mellitus: The influence of acute hyperinsulinemia, very-low calorie diet and PPAR-alpha agonist treatment. *Physiological Research*, *4*, 627–636.

Najafi-Shoushtari, S. H., Kristo, F., Li, Y., Shioda, T., Cohen, D. E., Gerszten, R. E., et al. (2010). MicroRNA-33 and the SREBP host genes cooperate to control cholesterol homeostasis. *Science*, *5985*, 1566–1569.

Ogawa, Y., Kurosu, H., Yamamoto, M., Nandi, A., Rosenblatt, K. P., Goetz, R., et al. (2007). BetaKlotho is required for metabolic activity of fibroblast growth factor 21. *Proceedings of the National Academy of Sciences of the United States of America*, *18*, 7432–7437.

Olson, E. N. (2014). MicroRNAs as therapeutic targets and biomarkers of cardiovascular disease. *Science Translational Medicine*, *239*, 239ps3.

Owen, B. M., Bookout, A. L., Ding, X., Lin, V. Y., Atkin, S. D., Gautron, L., et al. (2013). FGF21 contributes to neuroendocrine control of female reproduction. *Nature Medicine*, *9*, 1153–1156.

Owen, B. M., Mangelsdorf, D. J., & Kliewer, S. A. (2015). Tissue-specific actions of the metabolic hormones FGF15/19 and FGF21. *Trends in Endocrinology and Metabolism*, *1*, 22–29.

Perez, L. M., Bernal, A., San Martin, N., Lorenzo, M., Fernandez-Veledo, S., & Galvez, B. G. (2013). Metabolic rescue of obese adipose-derived stem cells by Lin28/Let7 pathway. *Diabetes*, *7*, 2368–2379.

Pineau, P., Volinia, S., McJunkin, K., Marchio, A., Battiston, C., et al. (2010). miR-221 overexpression contributes to liver tumorigenesis. *Proceedings of the National Academy of Sciences of the United States of America*, *107*(1), 264–269.

Pogribny, I. P., Starlard-Davenport, A., Tryndyak, V. P., Han, T., Ross, S. A., Rusyn, I., et al. (2010). Difference in expression of hepatic microRNAs miR-29c, miR-34a, miR-155, and miR-200b is associated with strain-specific susceptibility to dietary nonalcoholic steatohepatitis in mice. *Laboratory Investigation*, *10*, 1437–1446.

Potthoff, M. J., Kliewer, S. A., & Mangelsdorf, D. J. (2012). Endocrine fibroblast growth factors 15/19 and 21: From feast to famine. *Genes & Development, 4*, 312–324.

Poy, M. N., Eliasson, L., Krutzfeldt, J., Kuwajima, S., Ma, X., Macdonald, P. E., et al. (2004). A pancreatic islet-specific microRNA regulates insulin secretion. *Nature, 7014*, 226–230.

Rayner, K. J., Esau, C. C., Hussain, F. N., McDaniel, A. L., Marshall, S. M., van Gils, J. M., et al. (2011). Inhibition of miR-33a/b in non-human primates raises plasma HDL and lowers VLDL triglycerides. *Nature, 7369*, 404–407.

Rayner, K. J., Suarez, Y., Davalos, A., Parathath, S., Fitzgerald, M. L., Tamehiro, N., et al. (2010). MiR-33 contributes to the regulation of cholesterol homeostasis. *Science, 5985*, 1570–1573.

Rottiers, V., & Naar, A. M. (2012). MicroRNAs in metabolism and metabolic disorders. *Nature Reviews. Molecular Cell Biology, 4*, 239–250.

Rottiers, V., Najafi-Shoushtari, S. H., Kristo, F., Gurumurthy, S., Zhong, L., Li, Y., et al. (2011). MicroRNAs in metabolism and metabolic diseases. *Cold Spring Harbor Symposia on Quantitative Biology, 76*, 225–233.

Rottiers, V., Obad, S., Petri, A., McGarrah, R., Lindholm, M. W., et al. (2013). Pharmacological inhibition of a microRNA family in non-human primates by a seed-targeting 8-mer antimiR. *Science Translational Medicine, 5*(212), 212ra162.

Schreuder, T. C., Marsman, H. A., Lenicek, M., van Werven, J. R., Nederveen, A. J., Jansen, P. L., et al. (2010). The hepatic response to FGF19 is impaired in patients with nonalcoholic fatty liver disease and insulin resistance. *American Journal of Physiology. Gastrointestinal and Liver Physiology, 3*, G440–G445.

So, W. Y., Cheng, Q., Chen, L., Evans-Molina, C., Xu, A., Lam, K. S., et al. (2013). High glucose represses beta-klotho expression and impairs fibroblast growth factor 21 action in mouse pancreatic islets: Involvement of peroxisome proliferator-activated receptor gamma signaling. *Diabetes, 11*, 3751–3759.

Spiegelman, B. M., & Flier, J. S. (2001). Obesity and the regulation of energy balance. *Cell, 4*, 531–543.

Sun, L., & Trajkovski, M. (2014). MiR-27 orchestrates the transcriptional regulation of brown adipogenesis. *Metabolism, 2*, 272–282.

Trajkovski, M., Ahmed, K., Esau, C. C., & Stoffel, M. (2012). MyomiR-133 regulates brown fat differentiation through Prdm16. *Nature Cell Biology, 12*, 1330–1335.

Trajkovski, M., Hausser, J., Soutschek, J., Bhat, B., Akin, A., Zavolan, M., et al. (2011). MicroRNAs 103 and 107 regulate insulin sensitivity. *Nature, 7353*, 649–653.

Tsai, W.-C., Hsu, P. W., Lai, T.-C., Chau, G.-Y., Lin, C.-W., et al. (2009). MicroRNA-122, a tumor suppressor microRNA that regulates intrahepatic metastasis of hepatocellular carcinoma. Hepatology, 49(5), 1571–1582.

Wei, W., Dutchak, P. A., Wang, X., Ding, X., Wang, X., Bookout, A. L., et al. (2012). Fibroblast growth factor 21 promotes bone loss by potentiating the effects of peroxisome proliferator-activated receptor gamma. *Proceedings of the National Academy of Sciences of the United States of America, 8*, 3143–3148.

Wente, W., Efanov, A. M., Brenner, M., Kharitonenkov, A., Koster, A., Sandusky, G. E., et al. (2006). Fibroblast growth factor-21 improves pancreatic beta-cell function and survival by activation of extracellular signal-regulated kinase 1/2 and Akt signaling pathways. *Diabetes, 9*, 2470–2478.

Wu, X., Ge, H., Lemon, B., Vonderfecht, S., Baribault, H., Weiszmann, J., et al. (2010). Separating mitogenic and metabolic activities of fibroblast growth factor 19 (FGF19). *Proceedings of the National Academy of Sciences of the United States of America, 32*, 14158–14163.

Wu, A. L., Kolumam, G., Stawicki, S., Chen, Y., Li, J., Zavala-Solorio, J., et al. (2011). Amelioration of type 2 diabetes by antibody-mediated activation of fibroblast growth factor receptor 1. *Science Translational Medicine, 113*, 113–126.

Xu, Y., Zalzala, M., Xu, J., Li, Y., Yin, L., & Zhang, Y. (2015). A metabolic stress-inducible miR-34a-HNF4alpha pathway regulates lipid and lipoprotein metabolism. *Nature Communications*, *6*, 7466.

Yamakuchi, M., Ferlito, M., & Lowenstein, C. J. (2008). miR-34a repression of SIRT1 regulates apoptosis. *Proceedings of the National Academy of Sciences of the United States of America*, *36*, 13421–13426.

Yamauchi, T., & Kadowaki, T. (2008). Physiological and pathophysiological roles of adiponectin and adiponectin receptors in the integrated regulation of metabolic and cardiovascular diseases. *International Journal of Obesity (London)*, *32*, S13–S18.

Zhang, Y., Yang, P., Sun, T., Li, D., Xu, X., et al. (2013). miR-126 and miR-126* repress recruitment of mesenchymal stem cells and inflammatory monocytes to inhibit breast cancer metastasis. Nature Cell Biology, 15(3), 284–294.

CHAPTER EIGHT

The Role of Alpha-Klotho as a Universal Tumor Suppressor

T. Rubinek*,[1], I. Wolf*,†

*Institute of Oncology, Tel Aviv Sourasky Medical Center, Tel Aviv, Israel
†Sackler Faculty of Medicine, Tel Aviv University, Tel Aviv, Israel
[1]Corresponding author: e-mail address: rubinekt@gmail.com

Contents

Abstract

The klotho gene is implicated in many physiological activities, among them aging, glucose metabolism, and phosphate and calcium metabolism. Many cellular activities of klotho were implicated in promoting these activities. Two of them, inhibition of the insulin-like growth factor-1 pathway and of the Wnt signaling pathway, are also major pathways associated with cancer development and progression. These discoveries prompted a surge of research aiming to elucidate the role of klotho in cancer. Studies show that klotho is universally silenced in a wide array of malignancies, including breast, pancreatic, ovarian, lung, colorectal, and melanoma, and that klotho's expression can serve as an invaluable prognostic marker. Epigenetic mechanisms, ie, promoter hypermethylation and histone deacetylation, are mainly associated with klotho's

Vitamins and Hormones, Volume 101
ISSN 0083-6729
http://dx.doi.org/10.1016/bs.vh.2016.03.001

silencing; however, different micro-RNAs were also demonstrated to be involved in the process. The activity of klotho on cancer cells growth was also widely investigated, and accumulating data suggest that klotho forced expression or treatment with the soluble protein can inhibit cancer development and progression. Moreover, studies now aim to reveal the specific region in klotho protein that underlies this anticancer activity in order to develop efficient and safe klotho-based medications.

1. KLOTHO STRUCTURE, FUNCTIONS, AND ACTIVITIES

In Greek mythology Klotho is one of the three Fates that determine longevity. The α-klotho gene (*kl*), named after her, was discovered in 1997 as an aging suppressor gene (Kuro-o et al., 1997). Mice that do not express the α-klotho (which will be referred from here on as klotho) exhibit symptoms resembling human premature aging, including soft tissue calcification, arteriosclerosis, skin atrophy, gonadal dysplasia, infertility, hypoglycemia, severe hyperphosphatemia, osteoporosis, emphysema, and an overall shorter life span (Kuro-o et al., 1997). On the other hand, overexpression of the klotho gene in mice slowed the aging process and extended the life span by 20–30% (Kurosu et al., 2005).

Human klotho is a 1012-amino acid single-pass transmembrane protein. It is composed of a short intracellular domain of about 10 amino acids, a short transmembrane domain, and the major part of the protein, the extracellular domain. The extracellular domain of klotho is composed of two internal repeats, KL1 and KL2 that share 28% amino acid sequence homology to each other (Kuro-o et al., 1997; Kurosu et al., 2005; Matsumura et al., 1998). Each domain shares amino acid sequence homology with β-glucosidase (between 32% and 41%) but lack glucosidase activity. However, glucoronidase and sialidase activities of klotho were described (Cha et al., 2008; Chang et al., 2005; Tohyama et al., 2004). The klotho gene is highly conserved in humans, mice, and rats and is also found in Zebra fish and nematodes. The main product of the klotho gene (*KL*) is the full-length protein; however, a splice variant, a product of alternative splicing, that translates only the KL1 domain, as a soluble, secreted protein, was also described (Kato et al., 2000; Matsumura et al., 1998).

The klotho protein can be detected in several compartments—cellular and acellular. In cells it is found mainly in the cytoplasm and cell membrane (Kato et al., 2000). Klotho can also be cleaved and shed from plasma membrane by different sheddases and can be detected in blood, cerebrospinal fluid, and urine (Hu et al., 2011; Imura et al., 2004; Yamazaki et al., 2010). The main proteases that participate in klotho's shedding are ADAM10 and ADAM17 (Chen, Podvin, Gillespie, Leeman, & Abraham, 2007; Chen et al., 2014). The action of these sheddases yields secreted klotho that is composed of the entire extracellular domain, KL1 and KL2, termed soluble klotho (sKlotho) as well as the proteins KL1 and KL2, as separate polypeptides.

Klotho is widely expressed; however, the highest expression of klotho was observed in the distal convoluted renal tubules of the kidney and in the choroid plexus in the brain (Kuro-o et al., 1997). However, it is expressed, although at lower levels, in other tissues like various exocrine and endocrine tissues, including the breast, pancreas, and ovaries (Abramovitz et al., 2011; Lojkin et al., 2015; Wolf et al., 2008).

Several activities had been attributed to klotho. (1) Klotho is an important regulator of the fibroblast growth factor (FGF) pathway and is a mandatory cofactor for the interaction of FGF23 with FGFRs 1, 3, or 4 (Kurosu et al., 2006; Urakawa et al., 2006). (2) It is an inhibitor of the insulin and of insulin-like growth factor (IGF)-1 pathways (Kurosu et al., 2005; Wolf et al., 2008). (3) Klotho can inhibit the Wnt signaling pathway by binding the ligand Wnt (Liu et al., 2007). (4) Klotho enhances the activity of the calcium channels TRPV by recruiting these channels to the plasma membrane (Cha et al., 2008; Chang et al., 2005).

In humans the function of klotho is not clear. Klotho is associated with a number of pathological states like chronic renal failure (Koh et al., 2001), hypertension (Kamari et al., 2015; Kuwahara et al., 2008; Oguro et al., 2010; Saito et al., 2000; Wang & Sun, 2009), and aging (Arking et al., 2002; Semba et al., 2011; Yamazaki et al., 2010). One important aspect of klotho's activity is the role it plays in cancer.

2. KLOTHO AND CANCER

Many of klotho's activities and the pathways it regulates (as described earlier) play an important role in cancer development and progression (Hanahan & Weinberg, 2011). Indeed, many studies show that klotho has an important role in cancer development, progression, invasion, and prognosis. These studies focused on klotho expression in malignant tissue as compared to the normal tissue; the correlation between klotho expression in malignant cells and prognosis; the therapeutic effect of klotho on cancer cell proliferation, apoptosis, and survival; signaling pathways activated by klotho in cancer cells.

2.1 Expression of Klotho in Malignant Tissue Compared to Normal Tissue

One of the hallmarks of cancer is silencing of tumor suppressor genes in the tumor tissue compared to the normal tissue (Hanahan & Weinberg, 2011). The expression of klotho was studied in many tumor tissues, including, breast, cervix, ovary, prostate, melanoma, colon, gastric, pancreatic, esophageal, liver, and several types of lung cancer. In all tumor types klotho expression was reduced in the tumor tissue compared to the normal adjacent one (Abramovitz et al., 2011; Camilli et al., 2011; Jiang, Gu, & Chen, 2014; Lee et al., 2010; Lojkin et al., 2015; Pan et al., 2011; Rubinek et al., 2012; Tang et al., 2014; Wang et al., 2011; Wolf et al., 2008; Xie, Zhou, Yuan, et al., 2013). In certain tumor types, like breast, pancreatic, and cervical cancers, decreased klotho expression is observed in the majority of tumor samples (about 80% of samples; Abramovitz et al., 2011; Lee et al., 2010; Rubinek et al., 2012; Wolf et al., 2008) (Fig. 1). On the other hand, in

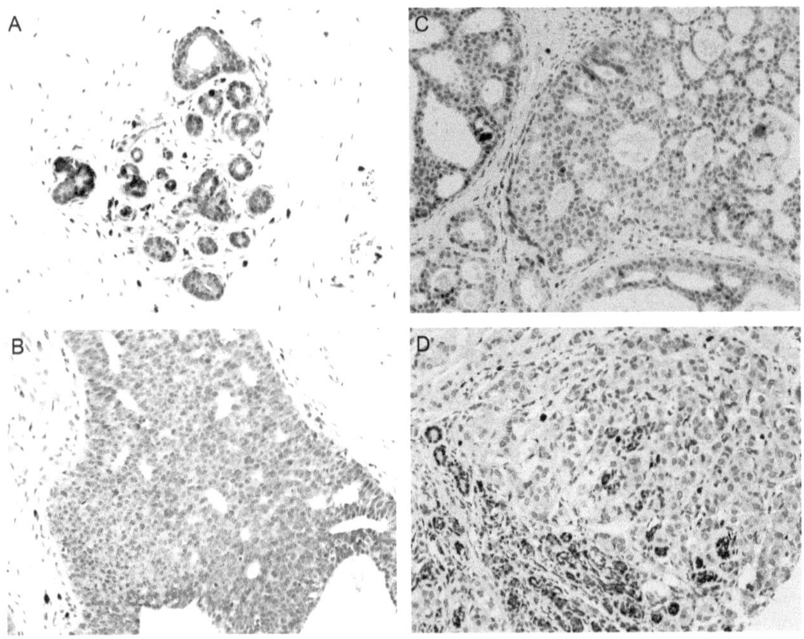

Fig. 1 Klotho expression is downregulated in early stages of the malignant process. Immunohistochemical staining of klotho in mammary tissues. Depicted high klotho expression in normal mammary tissue (A) and reduced expression is observed already in hyperplastic tissue (B), as well as in ductal carcinoma in situ (C) and invasive carcinoma (D). (See the color plate.)

certain tumors, like ovarian and certain lung cancers, about 50% reduction in klotho expression is observed (Lojkin et al., 2015; Usuda et al., 2011b). However, a minimal reduction in klotho expression was noted in cervical cancer (Aviel-Ronen et al., 2016).

2.2 Correlation Between Tumor Aggressiveness and Klotho Expression

In breast cancer, klotho expression was studied in over 100 breast cancer samples and in the normal adjacent tissue (Rubinek et al., 2012). Both klotho's mRNA and protein are highly expressed in normal ductal epithelial cells of the mammary gland, but its expression significantly reduced in the tumor cells. Examination of klotho expression in earlier, less malignant processes, revealed that in about third of moderate hyperplastic tissues klotho expression was silenced; however, in the more malignant cells of atypical hyperplasia, klotho was not detected in any of the samples (Fig. 1) (Rubinek et al., 2012). Expression of klotho is also reduced in esophageal carcinoma, and its expression negatively correlated with adverse clinical staging, histological grade, lymph node metastasis, and invasion depth (Tang et al., 2014). Patients with klotho-positive tumors had longer survival than those with klotho-negative tumors and klotho expression was an important factor indicating good prognosis. Similarly, in hepatocellular carcinoma (HCC), klotho staining was inversely correlated with clinical stage and histologic grade (Xie, Zhou, Yuan, et al., 2013). Patients with klotho-expressing tumors had longer survival periods than did those with klotho-negative tumors. In large-cell neuroendocrine carcinoma of the lung the expression of klotho was significantly associated with a good outcome of resected patients (Usuda et al., 2011a). Additionally, survival benefit of klotho expression in the primary tumor was significantly higher, compared with that of patients without klotho expression. Likewise, klotho expression in small cell lung cancer (SCLC) is reduced and its expression is significantly associated with a better overall survival (Usuda et al., 2011b).

Klotho expression was evaluated in tissue and serum of clear cells (cc) renal cell carcinoma (RCC) patients and this was correlated with disease progression (Gigante et al., 2015). Klotho protein levels were significantly decreased in RCC tissues compared with normal tissues and the more advanced the disease, the more evident the downregulation. This trend was also observed in serum samples, thus, statistically significant differences resulted between serum klotho levels and tumor size, Fuhrman grade, and clinical stage. Cancer-specific survival (CSS) and progression-free survival (PFS) were significantly shorter in patients with lower levels of klotho. Multivariate analysis showed that low serum levels of klotho were independent adverse prognostic factors for CSS and PFS. These results indicate that decreased klotho expression correlates with RCC progression and suggest a key role of declining klotho in the onset of cancer metastasis.

Klotho expression in tumor cells can serve as a biomarker. In cervical cancer klotho was silenced in a small fraction of cervical cancer. However, differential expression was observed between squamous cell carcinoma and adenocarcinoma, with reduced expression mainly in adenocarcinoma (Aviel-Ronen et al., 2016).

Association of klotho expression with cancer-related genes: Klotho expression was analyzed in over 150 samples of ovarian cancer and lower expression of klotho was observed in third of the samples (Lojkin et al., 2015). A correlation between reduced klotho expression and wild-type BRCA was observed. In pancreatic cancer a reduction of klotho expression was observed in about 75% of tumors. A negative correlation between expression of klotho and miR504 was observed, and reduced expression of klotho was associated with worse prognosis (Jiang et al., 2014). Klotho expression, as assessed by mRNA expression, was reduced in gastric cancer in the majority of samples (Wang et al., 2011). Additionally, its promoter was found to be hypermethylated in over 80% of these samples and methylation was found to be associated with poor prognosis. In melanoma, klotho expression was decreased in about 20% of metastasis, and this decrease negatively correlated with Wnt5a expression (Camilli et al., 2011). In about 60% of colon cancer samples, promoter hypermethylation was observed (Pan et al., 2011).

2.3 Means of Silencing

The silencing of klotho in cancer tissues is a nearly universal event. However, the precise mode of silencing has been only partially elucidated so far. Epigenetic alterations, namely, DNA hypermethylation at promoter sites and histone modifications, play an important part in silencing of tumor suppressor genes (Sharma, Kelly, & Jones, 2010). It was demonstrated that these mechanisms play an important role in klotho expression in tumor tissues. A 1439-bp long CpG island is located in the promoter region of klotho and extensive hypermethylation of this region was observed in cancer cell lines and in patient's samples (Rubinek et al., 2012). Indeed, demethylation of the DNA restored klotho's expression. DNA methylation is often associated with specific histone modifications that can cooperatively affect chromatin structure to silence gene expression. It has been shown that in addition to hypermethylation of klotho's promoter in cancer cells, the histones at these regions are deacetylated (Abramovitz et al., 2011; Lee et al., 2010; Pan et al., 2011; Rubinek et al., 2012; Wang et al., 2011; Xie, Zhou, Shu, et al., 2013; Xie, Zhou, Yuan, et al., 2013).

An important mechanism for tumor suppressor silencing is through micro-RNA (miR). Several miRs were shown to be involved in silencing of klotho expression in cancer. In PDAC miR504 was found to negatively correlate with klotho expression (Jiang et al., 2014). In gastric cancer miR-199a-5p seems to be involved in the silencing of klotho, as well as miR-10b in NSCLC, which its expression levels were higher in tumor cells as compared with an adjacent normal tissue (He et al., 2014; Huang, Sun, Wang, He, & Li, 2015). Furthermore, overexpression of miR-10b reduced klotho expression, whereas its silencing elevated klotho expression (Huang et al., 2015).

In colorectal cancer patients' high plasma miR-15b level and negative klotho expression in tumor tissues significantly correlated with poor survival (Li et al., 2016). Whereas the opposite was observed when miR-15b was silenced, ie, silencing led to increased klotho protein expression and subsequently decreased colony formation ability, invasion, and migration of HCT116 cells in vitro and liver metastasis of HCT116 tumors in vivo.

2.4 Klotho Single-Nucleotide Polymorphisms

Although no silencing or inactivating mutations of klotho were detected in tumor cells, klotho SNPs (single-nucleotide polymorphisms) were associated with cancer susceptibility. A functional variant of *klotho*, termed KL-VS, contains six sequence variants in complete linkage disequilibrium, two of which result in amino acid substitutions, F352V and C370S. It was found that homozygosity is associated with reduced longevity (Arking et al., 2002). In breast cancer patients, among BRCA1 carriers, heterozygosity for the KL-VS allele was associated with increased breast and ovarian cancer risk and younger age at breast cancer diagnosis (Wolf et al., 2010). Moreover, this SNP exhibits diminished antitumor effects, as assessed by reduced capacity to inhibit breast cancer cells tumor formation. Another polymorphism, G-395A of *klotho* gene, was shown to increase the risk of CRC (Liu et al., 2015). In prostate cancer it was shown that the C1548T polymorphism of *klotho* gene is associated with a risk factor for the development of prostate cancer (Kim et al., 2014).

3. THERAPEUTIC EFFECT OF KLOTHO ON CANCER DEVELOPMENT AND PROGRESSION

3.1 Effects of Klotho on Tumor Cells In Vitro

As klotho is silenced in a wide array of malignant tissues, a question was raised as to its (potential) therapeutic activity in cancer. Klotho is expressed in different tissues; however, klotho is also a circulating hormone with systemic effects. Thus the effect of klotho on cancer cell proliferation, survival, and invasiveness was evaluated by its reexpression in tumor cells and by treatment of the cells with sKlotho. The first tumor where klotho effects on the tumorigenicity were studied was breast cancer (Wolf et al., 2008). Overexpression of klotho in different types of breast cancer cell lines reduced tumorigenicity of cells as assessed by colony formation assays and viability assays. Similarly, klotho is expressed in breast cancer cell lines, though to a lesser extent than in normal mammary gland. Therefore, further silencing of klotho was conducted, and this led to an increase in breast cancer cell proliferation. Inhibition of proliferation was evident also by elevation in the tumor suppressor p53. Apoptosis by klotho in breast cancer cells was evidenced by cleavage of PARP, a known indicator for apoptosis induction. As mentioned, klotho is a hormone present in the circulation at nanomolar concentrations. Indeed, there are no data regarding klotho levels in cancer patients; however, the effect of the soluble protein in cancer cells in vitro and in vivo (see later) was studied. In pancreatic cancer cells it was shown that sKlotho inhibits cell proliferation and invasiveness (Abramovitz et al., 2011). Furthermore, klotho also enhanced the effect of gemcitabine and 5-fluorouracil (5FU) on the survival of these cells.

A similar effect of klotho was observed in other tumor types. Expression of klotho in several types of colon cancer cells inhibited their proliferation, induced cell cycle arrest and apoptosis, as well as inhibited their invasive capacity (Li et al., 2014; Pan et al., 2011). In gastric cancer overexpression of klotho inhibited cell proliferation, colony formation, while it induced apoptosis and autophagy (Wang et al., 2011; Xie, Zhou, Shu, et al., 2013). In hepatocellular cells klotho overexpression inhibited cell proliferation and induced apoptosis, whereas klotho silencing had the opposite effect (Shu et al., 2013; Sun et al., 2015; Tang et al., 2016).

Overexpression of klotho reduced the proliferation of lung cancer A549 cells, whereas klotho silencing in these cells enhanced proliferation. Overexpression of klotho promoted also apoptosis of these cells. Thus, overexpression of klotho led to upregulation of the proapoptotic gene bax, whereas the antiapoptotic gene bcl-2 was found to be downregulated. In contrast, bax and bcl-2 were found to be downregulated and upregulated, respectively, upon klotho silencing (Chen, Ma, Liu, Zhao, & Wu, 2012; Chen, Wang, Zhao, & Wu, 2010). In melanoma, klotho inhibits invasiveness (Camilli et al., 2011). Klotho did not show any effects on noncancerous HEK-293 cells (Wolf et al., 2008).

3.2 Klotho Anticancer Activity In Vivo

The antitumor effect of klotho was demonstrated using several in vivo models. The most thoroughly studied model is the xenograft model, where cancer cells are transplanted into the flanks of athymic nude mice (Tomayko & Reynolds, 1989). In this model, following cell transplantation, the mice are treated with klotho. Klotho treatment can commence when a palpable tumor is established, and in this case klotho's curative effect is tested. Alternatively, the preventive effect of klotho can be tested when it is administered at the same time of cells' transplantation. Klotho was shown to possess curative effects on colorectal cancer tumors, using intratumoral injections of klotho-expressing lentivirus, as assessed by reduced tumor weight and size (Li et al., 2014). A similar effect of klotho was demonstrated in HCC cells (Tang et al., 2016). Klotho also exhibits a preventive effect on cancer development. Treatment of mice-bearing pancreatic (Abramovitz et al., 2011), breast (Ligumsky et al., 2015), and lung (Doi et al., 2011) xenograft tumors with sKlotho reduced tumor size and weight. Breast tumors from the klotho-treated group express less vimentin, Ki67, and cytokeratin AE1/AE3, all indicative of reduced proliferative properties of klotho and elevated differentiation (Ligumsky et al., 2015).

The second in vivo model used is a transgenic mouse model with klotho overexpression. These mice are injected with cancer cells and metastasis formation and mice survival rate is compared with wild-type mice. It was found that the human cancer cell line 3LL (Lewis Lung cancer) metastasized less in klotho-overexpressing mice compared to control mice and mice survival was improved (Doi et al., 2011). Importantly, klotho serum levels are higher in these mice; however, these mice differ in other aspects, like glucose levels, a factor that might affect tumor growth as well.

4. KLOTHO AND CHEMOTHERAPIES

4.1 Cisplatin

Cisplatin is an intercalating agent that induces apoptosis of cancer cells, and it has been the most widely used drug in the first-line chemotherapy.

However, the increase in cisplatin-resistant cancer cells has become a major obstacle in clinical management of cancers (Amable, 2016). Klotho was shown to attenuate the resistance of lung cancer to cisplatin-based chemotherapy and klotho overexpression markedly restored apoptosis of the resistant cells (Wang et al., 2013). In accordance, klotho knockdown cells showed enhanced resistance to chemotherapy. Further analysis showed that the cisplatin resistance obtained following klotho silencing was abrogated by inhibition of PI3K/Akt pathway with a specific inhibitor (LY294002). Moreover, klotho modulated the resistance to cisplatin in a xenograft nude mice model. These observations suggested that klotho could improve the resistance of lung cancer cells to chemotherapy and may serve as a potential target for the gene therapy of lung cancers resistant to cisplatin-based chemotherapy.

Cisplatin is a mainstay of ovarian cancer therapy. We found that klotho treatment of ovarian cancer cell lines sensitizes some of them to cisplatin (Lojkin et al., 2015). It was shown that inhibition of the PI3K pathway sensitizes cells to cisplatin, and it is possible that the ability of klotho to inhibit this pathway confers increased sensitivity to this drug.

4.2 5FU and Gemcitabine

5FU and gemcitabine are commonly used for the treatment of pancreatic adenocarcinoma. 5FU is an irreversible inhibitor of thymidylate synthase and gemcitabine is a nucleoside analog. We showed that treatment with klotho of several pancreatic cancer cell lines leads to increased sensitivity to each of these chemotherapies (Abramovitz et al., 2011). This may stem from the different pathways utilized by klotho and the chemotherapies to inhibit cell growth and induce cell death. In line with these discoveries, in SCLC patients, administration of perioperative chemotherapy had no significant effect in improving the survival. However, patients showing klotho expression at stage I and II benefited from perioperative chemotherapy (Usuda et al., 2011b).

5. CIRCULATING KLOTHO IN CANCER

Klotho is present in the circulation, and its levels in different pathologies were evaluated. Only three studies so far aimed to determine klotho serum levels in cancer states. In HCC patients serum levels of sKlotho were significantly increased compared to healthy controls (Tang et al., 2016). Similarly, in esophageal cancer, the level of serum klotho was markedly higher than control group (Tang et al., 2014). This result is counterintuitive as klotho inhibits growth of cancer cells in vitro and in vivo. The factors that lead to elevated levels of circulating klotho in these cancer patients, and the reason for the lack of effect on cancer cell proliferation, are still unknown and more studies are required in order to corroborate these results.

On the other hand, in RCC patients it was shown that klotho serum levels decreased in correlation with elevated tumor size, Fuhrman grade, and clinical stage (Gigante et al., 2015).

6. CELLULAR TARGETS OF KLOTHO IN CANCER CELLS

6.1 IGF-1/PI3K/AKT

Klotho can inhibit the IGF-1 pathway in hepatocytes and in muscle cells (Kurosu et al., 2005) (Fig. 2). The IGF-1 pathway plays an important role in the tumorigenic process in many cancer types (Werner, 2012). The downstream effectors of IGF-1 are PI3K and AKT, both play a crucial role in many cancer types and are considered as major targets for therapy development (Hanahan & Weinberg, 2011). Evidently, klotho's effect on this pathway in cancer cells was evaluated. Indeed, a tight correlation between klotho antitumor effects and inhibition of the IGF-1 pathway was observed in several cancers. In breast, ovarian, colon, hepatocellular, and pancreatic cancer cell lines it was shown that klotho inhibits IGF-1 signaling as evidenced by inhibition of IGF-1R phosphorylation and the downstream signaling (Abramovitz et al., 2011; Chen et al., 2010; Li et al., 2014; Ligumsky et al., 2015; Lojkin et al., 2015; Shu et al., 2013; Wolf et al., 2008; Xie, Zhou, Shu, et al., 2013). Klotho also abrogated IGF-1-induced proliferation in breast cancer cells (Chen et al., 2010; Wolf et al., 2008). Another association between klotho and the IGF-1 pathway in cancer is the finding that in pancreatic cancer klotho expression is negatively correlated with phosphorylated IGF-1R (Jiang et al., 2014).

6.2 Wnt/β-Catenin

The canonical Wnt pathway is implicated with the development of several malignant tumors like breast cancer, colorectal cancer, melanoma, prostate cancer, lung cancer, and several other cancer types (Polakis, 2000). Wnt activation leads to elevated levels of β-catenin in the nucleus and/or cytoplasm. Nuclear β-catenin induces the transcriptional activity of TCF/LEF, which subsequently promotes the synthesis of genes which trigger cancer cell

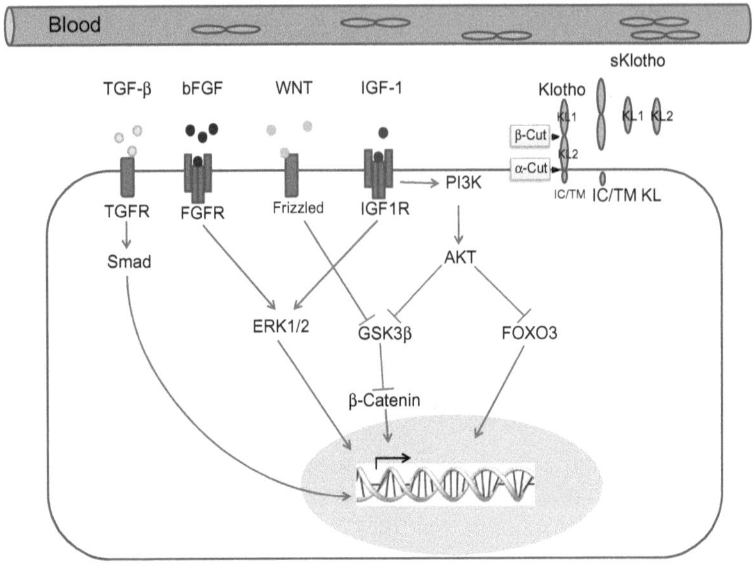

Fig. 2 Signaling pathways affected by klotho in cancer cells. Cancer cells are regulated by multiple pathways, and klotho may modulate several of these pathways. Klotho was shown to inhibit the IGF-1 pathway and the Wnt pathway. These two pathways may converge at the GSK3β signaling molecule. Thus, the IGF-1 pathway inhibits GSK3β. Similarly, activated Wnt pathway also inhibits GSK3β. Therefore, inhibition of both pathways by klotho promotes GSK3β activity and reduces cell viability. Furthermore, klotho modulates the activity of bFGF, which activates ERK1/2. Similarly, both IGF-1 and bFGF may elicit ERK1/2 phosphorylation. Thus, although klotho seems to operate through many, unrelated pathways, perhaps some of these activities are connected. TGF-β was also shown to activate the ERK1/2 pathway. Klotho may access cells through the circulation, as it is a hormone. However, although it is silenced in many cancer types, it is expressed, at least to some extent, in some cancer cells. It may operate in the cytoplasm, or as a membrane-bound protein. However, it may be cleaved through ADAM10 and ADAM17 sheddasses and act from outside the cell on the different receptors. Klotho is cleaved to the entire extracellular domain, and also into the subdomains KL1 and KL2. KL1 can effectively inhibit cancer cell growth, and inhibit the IGF-1, bFGF, and Wnt pathways. (See the color plate.)

growth, motility, and invasiveness (Klaus & Birchmeier, 2008). Klotho was shown to inhibit the Wnt pathway by binding the Wnt ligands (mostly Wnt3A and Wnt5A), thus leading to reduced binding of Wnt to their membrane receptors (Liu et al., 2007). In cervical, esophageal, lung, and HCC cancers and in melanoma it was shown that klotho inhibits activation of Wnt-TCF/β-catenin signaling pathway and inhibited expression of target genes of the pathway like *c-MYC* and *CCND1* (Camilli et al., 2011; Chen et al., 2012; Lee et al., 2010; Sun et al., 2015; Tang et al., 2014, 2016).

6.3 TGF-β1

TGF-β1 has been identified as a potent and universal growth factor that can independently induce epithelial-to-mesenchymal transition (EMT) in various types of cells (Joshi & Cao, 2010). TGF-β1 inhibits expression of the genes essential for an epithelial phenotype (eg, E-cadherin) and induces expression of the genes that confer a mesenchymal phenotype (eg vimentin, collagen-1, N-cadherin) and invasiveness (eg metalloproteases). Thus, it confers cancer cells the ability to migrate and metastasize. sKlotho treatment inhibited TGF-β1 signaling and suppressed EMT in A549 lung cancer cells (Doi et al., 2011). This was accompanied by inhibition of TGF-β1 signaling pathway as assessed by reduced phosphorylation of Smad3, reduced Smad transcriptional activity, and reduction in TGF-β1 binding to cell surface. While TGF-β1 treatment decreased epithelial marker expression, increased mesenchymal marker expression, and increased migration of A549 cells, sKlotho treatment attenuated these responses.

7. KLOTHO ACTIVE FRAGMENT

Klotho activity can be attributed, grossly, to the membrane-bound klotho, the entire membrane-free sKlotho (KL1 + KL2) or to the subdomains KL1 or KL2 (Fig. 2). While only the membrane-bound klotho functions as a cofactor for FGF23 activity, the IGF-1 and Wnt inhibitory functions can be carried out by the membrane-bound, sKlotho or KL1 (Abramovitz et al., 2011; Kurosu et al., 2006; Liu et al., 2007). Many efforts to identify the antitumor active domain of klotho were invested. It was found that the membrane-bound klotho as well as sKlotho possess antitumor activities. Thus, overexpression of klotho or treatment with the sKlotho inhibited growth, motility, and invasiveness in most of the cancers studied (see above). The efficacy of sKlotho was demonstrated in vivo as treatment of mice bearing various tumors (pancreas, lung, and colon) with soluble protein inhibited tumor growth. Following these findings, the activity of KL1 and KL2 domain was studied separately. It was found that while KL1 inhibits cancer cell proliferation, to a similar extent as full-length klotho, KL2 is devoid of this activity (Ligumsky et al., 2015). Moreover, administration of KL1 to mice with xenograft tumors of breast and pancreatic cancer inhibited tumor growth in vivo as well (Abramovitz et al., 2011; Ligumsky et al., 2015). This differential activity of klotho has mechanistic as well as therapeutic implications. The klotho proteins that possess anticancer activity (membrane-bound klotho, sKlotho, and KL1) share mechanistic similarities as they can inhibit the IGF-1 and Wnt pathways. In contrast, only the membrane-bound klotho can function as a cofactor for FGF23. On the other hand, KL2 does not inhibit cancer cells, as well as any of the above functions. These mechanistic findings reflect on the therapeutic potential of klotho. More studies, aiming to elucidate the mechanism of action of klotho in cancer, as well as the adverse effects of klotho treatment, may facilitate its usage as a medication for cancer.

REFERENCES

Abramovitz, L., Rubinek, T., Ligumsky, H., Bose, S., Barshack, I., Avivi, C., … Wolf, I. (2011). KL1 internal repeat mediates klotho tumor suppressor activities and inhibits bFGF and IGF-I signaling in pancreatic cancer. *Clinical Cancer Research*, *17*(13), 4254–4266. http://dx.doi.org/10.1158/1078-0432.ccr-10-2749.

Amable, L. (2016). Cisplatin resistance and opportunities for precision medicine. *Pharmacological Research*, *106*, 27–36. http://dx.doi.org/10.1016/j.phrs.2016.01.001.

Arking, D. E., Krebsova, A., Macek, M., Macek, S., Milan, J., Arking, A., Mian, I. S., … Dietz, H. C. (2002). Association of human aging with a functional variant of klotho. *Proceedings of the National Academy of Sciences of the United States of America*, *99*(2), 856–861.

Aviel-Ronen, S., Rubinek, T., Zadok, O., Vituri, A., Avivi, C., Wolf, I., & Barshack, I. (2016). Klotho expression in cervical cancer: Differential expression in adenocarcinoma and squamous cell carcinoma. *Journal of Clinical Pathology*, *69*(1), 53–57. http://dx.doi.org/10.1136/jclinpath-2015-202929.

Camilli, T. C., Xu, M., O'Connell, M. P., Chien, B., Frank, B. P., Subaran, S., … Weeraratna, A. T. (2011). Loss of Klotho during melanoma progression leads to increased filamin cleavage, increased Wnt5A expression, and enhanced melanoma cell motility. *Pigment Cell & Melanoma Research*, *24*(1), 175–186. http://dx.doi.org/10.1111/j.1755-148X.2010.00792.x.

Cha, S.-K., Ortega, B., Kurosu, H., Rosenblatt, K. P., Kuro-o, M., & Huang, C.-L. (2008). Removal of sialic acid involving Klotho causes cell-surface retention of TRPV5 channel via binding to galectin-1. *Proceedings of the National Academy of Sciences of the United States of America*, *105*, 9805–9810.

Chang, Q., Hoefs, S., van der Kemp, A. W., Topala, C. N., Bindels, R. J., & Hoenderop, J. G. (2005). The b-glucuronidase klotho hydrolyzes and activates the TRPV5 channel. *Science*, *310*(5747), 490–493.

Chen, B., Ma, X., Liu, S., Zhao, W., & Wu, J. (2012). Inhibition of lung cancer cells growth, motility and induction of apoptosis by Klotho, a novel secreted Wnt antagonist, in a dose-dependent manner. *Cancer Biology & Therapy*, *13*(12), 1221–1228. http://dx.doi.org/10.4161/cbt.21420.

Chen, C.-D., Podvin, S., Gillespie, E., Leeman, S. E., & Abraham, C. R. (2007). Insulin stimulates the cleavage and release of the extracellular domain of Klotho by ADAM10 and ADAM17. *Proceedings of the National Academy of Sciences of the United States of America*, *104*(50), 19796–19801.

Chen, C. D., Tung, T. Y., Liang, J., Zeldich, E., Tucker Zhou, T. B., Turk, B. E., & Abraham, C. R. (2014). Identification of cleavage sites leading to the shed form of the anti-aging protein klotho. *Biochemistry*, *53*(34), 5579–5587. http://dx.doi.org/10.1021/bi500409n.

Chen, B., Wang, X., Zhao, W., & Wu, J. (2010). Klotho inhibits growth and promotes apoptosis in human lung cancer cell line A549. *Journal of Experimental & Clinical Cancer Research*, *29*, 99. http://dx.doi.org/10.1186/1756-9966-29-99. 1756-9966-29-99 [pii].

Doi, S., Zou, Y., Togao, O., Pastor, J. V., John, G. B., Wang, L., ... Kuro, O. M. (2011). Klotho inhibits transforming growth factor-{beta}1 (TGF-{beta}1) signaling and suppresses renal fibrosis and cancer metastasis in mice. *The Journal of Biological Chemistry*, *286*(10), 8655–8665. http://dx.doi.org/10.1074/jbc.M110.174037. M110.174037 [pii].

Gigante, M., Lucarelli, G., Divella, C., Netti, G. S., Pontrelli, P., Cafiero, C., ... Ranieri, E. (2015). Soluble serum alphaKlotho is a potential predictive marker of disease progression in clear cell renal cell carcinoma. *Medicine (Baltimore)*, *94*(45), e1917. http://dx.doi.org/10.1097/md.0000000000001917.

Hanahan, D., & Weinberg, R. A. (2011). Hallmarks of cancer: The next generation. *Cell*, *144*(5), 646–674.

He, X. J., Ma, Y. Y., Yu, S., Jiang, X. T., Lu, Y. D., Tao, L., ... Tao, H. Q. (2014). Upregulated miR-199a-5p in gastric cancer functions as an oncogene and targets klotho. *BMC Cancer*, *14*, 218. http://dx.doi.org/10.1186/1471-2407-14-218.

Hu, M. C., Shi, M., Zhang, J., Quinones, H., Griffith, C., Kuro-o, M., & Moe, O. W. (2011). Klotho deficiency causes vascular calcification in chronic kidney disease. *Journal of the American Society of Nephrology*, *22*(1), 124–136. http://dx.doi.org/10.1681/asn.2009121311.

Huang, J., Sun, C., Wang, S., He, Q., & Li, D. (2015). microRNA miR-10b inhibition reduces cell proliferation and promotes apoptosis in non-small cell lung cancer (NSCLC) cells. *Molecular BioSystems*, *11*(7), 2051–2059. http://dx.doi.org/10.1039/c4mb00752b.

Imura, A., Iwano, A., Tohyama, O., Tsuji, Y., Nozaki, K., Hashimoto, N., ... Nabeshima, Y.-I. (2004). Secreted Klotho protein in sera and CSF: Implication for post-translational cleavage in release of Klotho protein from cell membrane. *FEBS Letters*, *565*(1–3), 143–147.

Jiang, B., Gu, Y., & Chen, Y. (2014). Identification of novel predictive markers for the prognosis of pancreatic ductal adenocarcinoma. *Cancer*, *32*(6), 218–225. http://dx.doi.org/10.3109/07357907.07352014.07905586. Epub 07352014 Apr 07357918.

Joshi, A., & Cao, D. (2010). TGF-beta signaling, tumor microenvironment and tumor progression: The butterfly effect. *Frontiers in Bioscience (Landmark Edition)*, *15*, 180–194.

Kamari, Y., Fingrut, O., Shaish, A., Almog, T., Kandel-Kfir, M., Harats, D., ... Wolf, I. (2015). The effect of klotho treatment on atherogenesis, blood pressure, and metabolic parameters in experimental rodent models. *Hormone and Metabolic Research*, *48*(3), 196–200. http://dx.doi.org/10.1055/s-0035-1549879.

Kato, Y., Arakawa, E., Kinoshita, S., Shirai, A., Furuya, A., Yamano, K., ... Nabeshima, Y.-I. (2000). Establishment of the anti-klotho monoclonal antibodies and detection of klotho protein in kidneys. *Biochemical and Biophysical Research Communications*, *267*(2), 597–602.

Kim, H. J., Lee, J., Lee, S. Y., Cheong, H. S., Kye, Y. S., Kim, W., ... Myung, S. C. (2014). The association between KL polymorphism and prostate cancer risk in Korean patients. *Molecular Biology Reports*, *41*(11), 7595–7606. http://dx.doi.org/10.1007/s11033-014-3647-y.

Klaus, A., & Birchmeier, W. (2008). Wnt signalling and its impact on development and cancer. *Nature Reviews Cancer*, *8*(5), 387–398. http://www.nature.com/nrc/journal/v8/n5/suppinfo/nrc2389_S1.html.

Koh, N., Fujimori, T., Nishiguchi, S., Tamori, A., Shiomi, S., Nakatani, T., ... Nabeshima, Y. (2001). Severely reduced production of klotho in human chronic renal failure kidney. *Biochemical and Biophysical Research Communications*, *280*(4), 1015–1020. http://dx.doi.org/10.1006/bbrc.2000.4226.

Kuro-o, M., Matsumura, Y., Aizawa, H., Kawaguchi, H., Suga, T., Utsugi, T., … Nabeshima, Y.-I. (1997). Mutation of the mouse klotho gene leads to a syndrome resembling ageing. *Nature*, *390*(6655), 45–51.

Kurosu, H., Ogawa, Y., Miyoshi, M., Yamamoto, M., Nandi, A., Rosenblatt, K. P., … Kuro-o, M. (2006). Regulation of fibroblast growth factor-23 signaling by klotho. *The Journal of Biological Chemistry*, *281*(10), 6120–6123.

Kurosu, H., Yamamoto, M., Clark, J. D., Pastor, J. V., Nandi, A., Gurnani, P., … Kuro-o, M. (2005). Suppression of aging in mice by the hormone klotho. *Science*, *309*(5742), 1829–1833.

Kuwahara, N., Sasaki, S., Kobara, M., Nakata, T., Tatsumi, T., Irie, H., … Hushiki, S. (2008). HMG-CoA reductase inhibition improves anti-aging klotho protein expression and arteriosclerosis in rats with chronic inhibition of nitric oxide synthesis. *International Journal of Cardiology*, *123*(2), 84–90. Epub 2007 Apr 2016.

Lee, J., Jeong, D.-J., Kim, J., Lee, S., Park, J.-H., Chang, B., … Lee, M.-S. (2010). The anti-aging gene KLOTHO is a novel target for epigenetic silencing in human cervical carcinoma. *Molecular Cancer*, *9*(1), 109–119.

Li, J., Chen, Y., Guo, X., Zhou, L., Jia, Z., Tang, Y., … Ren, C. (2016). Inhibition of miR-15b decreases cell migration and metastasis in colorectal cancer. *Tumour Biology*. http://dx.doi.org/10.1007/s13277-015-4396-9.

Li, X. X., Huang, L. Y., Peng, J. J., Liang, L., Shi, D. B., Zheng, H. T., & Cai, S. J. (2014). Klotho suppresses growth and invasion of colon cancer cells through inhibition of IGF1R-mediated PI3K/AKT pathway. *International Journal of Oncology*, *45*(2), 611–618. http://dx.doi.org/10.3892/ijo.2014.2430. Epub 2014 May 3899.

Ligumsky, H., Rubinek, T., Merenbakh-Lamin, K., Yeheskel, A., Sertchook, R., Shahmoon, S., … Wolf, I. (2015). Tumor suppressor activity of klotho in breast cancer is revealed by structure-function analysis. *Molecular Cancer Research*, *13*(10), 1398–1407. http://dx.doi.org/10.1158/1541-7786.mcr-15-0141.

Liu, C., Cui, W., Wang, L., Yan, L., Ruan, X., Liu, Y., … Zhang, X. (2015). Klotho gene polymorphisms are related to colorectal cancer susceptibility. *International Journal of Clinical and Experimental Pathology*, *8*(6), 7446–7449.

Liu, H., Fergusson, M. M., Castilho, R. M., Liu, J., Cao, L., Chen, J., … Finkel, T. (2007). Augmented Wnt signaling in a mammalian model of accelerated aging. *Science*, *317*(5839), 803–806.

Lojkin, I., Rubinek, T., Orsulic, S., Schwarzmann, O., Karlan, B. Y., Bose, S., & Wolf, I. (2015). Reduced expression and growth inhibitory activity of the aging suppressor klotho in epithelial ovarian cancer. *Cancer Letters*, *362*(2), 149–157. http://dx.doi.org/10.1016/j.canlet.2015.03.035.

Matsumura, Y., Aizawa, H., Shiraki-Iida, T., Nagai, R., Kuro-o, M., & Nabeshima, Y.-I. (1998). Identification of the human klotho gene and its two transcripts encoding membrane and secreted klotho protein. *Biochemical and Biophysical Research Communications*, *242*(3), 626–630.

Oguro, R., Kamide, K., Kokubo, Y., Shimaoka, I., Congrains, A., Horio, T., … Rakugi, H. (2010). Association of carotid atherosclerosis with genetic polymorphisms of the klotho gene in patients with hypertension. *Geriatrics & Gerontology International*, *10*(4), 311–318.

Pan, J., Zhong, J., Gan, L. H., Chen, S. J., Jin, H. C., Wang, X., & Wang, L. J. (2011). Klotho, an anti-senescence related gene, is frequently inactivated through promoter hypermethylation in colorectal cancer. *Tumour Biology*, *32*(4), 729–735. http://dx.doi.org/10.1007/s13277-011-0174-5.

Polakis, P. (2000). Wnt signaling and cancer. *Genes & Development*, *14*(15), 1837–1851.

Rubinek, T., Shulman, M., Israeli, S., Bose, S., Avraham, A., Zundelevich, A., … Wolf, I. (2012). Epigenetic silencing of the tumor suppressor klotho in human breast cancer.

Breast Cancer Research and Treatment, 133, 649–657. http://dx.doi.org/10.1007/s10549-011-1824-4.

Saito, Y., Nakamura, T., Ohyama, Y., Suzuki, T., Iida, A., Shiraki-Iida, T., … Nagai, R. (2000). In vivo klotho gene delivery protects against endothelial dysfunction in multiple risk factor syndrome. *Biochemical and Biophysical Research Communications, 276*(2), 767–772.

Semba, R. D., Cappola, A. R., Sun, K., Bandinelli, S., Dalal, M., Crasto, C., … Ferrucci, L. (2011). Plasma klotho and mortality risk in older community-dwelling adults. *The Journals of Gerontology Series A, Biological Sciences and Medical Sciences, 66*(7), 794–800.

Sharma, S., Kelly, T. K., & Jones, P. A. (2010). Epigenetics in cancer. *Carcinogenesis, 31*(1), 27–36. http://dx.doi.org/10.1093/carcin/bgp220.

Shu, G., Xie, B., Ren, F., Liu, D. C., Zhou, J., Li, Q., … Zhou, J. (2013). Restoration of klotho expression induces apoptosis and autophagy in hepatocellular carcinoma cells. *Cell Oncology (Dordrecht), 36*(2), 121–129. http://dx.doi.org/10.1007/s13402-012-0118-0.

Sun, H., Gao, Y., Lu, K., Zhao, G., Li, X., Li, Z., & Chang, H. (2015). Overexpression of Klotho suppresses liver cancer progression and induces cell apoptosis by negatively regulating wnt/beta-catenin signaling pathway. *World Journal of Surgical Oncology, 13*, 307. http://dx.doi.org/10.1186/s12957-015-0717-0.

Tang, X., Fan, Z., Wang, Y., Ji, G., Wang, M., Lin, J., & Huang, S. (2014). Expression of klotho and beta-catenin in esophageal squamous cell carcinoma, and their clinicopathological and prognostic significance. *Diseases of the Esophagus*. http://dx.doi.org/10.1111/dote.12289.

Tang, X., Wang, Y., Fan, Z., Ji, G., Wang, M., Lin, J., … Meltzer, S. J. (2016). Klotho: A tumor suppressor and modulator of the Wnt/beta-catenin pathway in human hepatocellular carcinoma. *Laboratory Investigation, 96*(2), 197–205. http://dx.doi.org/10.1038/labinvest.2015.86.

Tohyama, O., Imura, A., Iwano, A., Freund, J. N., Henrissat, B., Fujimori, T., & Nabeshima, Y. (2004). Klotho is a novel beta-glucuronidase capable of hydrolyzing steroid beta-glucuronides. *Journal of Biological Chemistry, 279*(11), 9777–9784. http://dx.doi.org/10.1074/jbc.M312392200.

Tomayko, M. M., & Reynolds, C. P. (1989). Determination of subcutaneous tumor size in athymic (nude) mice. *Cancer Chemotherapy and Pharmacology, 24*(3), 148–154.

Urakawa, I., Yamazaki, Y., Shimada, T., Iijima, K., Hasegawa, H., Okawa, K., … Yamashita, T. (2006). Klotho converts canonical FGF receptor into a specific receptor for FGF23. *Nature, 444*(7120), 770–774.

Usuda, J., Ichinose, S., Ishizumi, T., Ohtani, K., Inoue, T., Saji, H., … Ikeda, N. (2011a). Klotho is a novel biomarker for good survival in resected large cell neuroendocrine carcinoma of the lung. *Lung Cancer, 72*(3), 355–359. http://dx.doi.org/10.1016/j.lungcan.2010.10.008.

Usuda, J., Ichinose, S., Ishizumi, T., Ohtani, K., Inoue, T., Saji, H., … Ikeda, N. (2011b). Klotho predicts good clinical outcome in patients with limited-disease small cell lung cancer who received surgery. *Lung Cancer (Amsterdam, Netherlands), 74*(2), 332–337.

Wang, Y., Chen, L., Huang, G., He, D., He, J., Xu, W., … Wu, J. (2013). Klotho sensitizes human lung cancer cell line to cisplatin via PI3k/Akt pathway. *PloS One, 8*(2), 21.

Wang, Y., & Sun, Z. (2009). Klotho gene delivery prevents the progression of spontaneous hypertension and renal damage. *Hypertension, 54*(4), 810–817. http://dx.doi.org/10.1161/HYPERTENSIONAHA.109.134320. HYPERTENSIONAHA.109.134320 [pii].

Wang, L., Wang, X., Jie, P., Lu, H., Zhang, S., Lin, X., … Jin, H. (2011). Klotho is silenced through promoter hypermethylation in gastric cancer. *American Journal of Cancer Research, 1*(1), 111–119.

Werner, H. (2012). Tumor suppressors govern insulin-like growth factor signaling pathways: Implications in metabolism and cancer. *Oncogene*, *31*(22), 2703–2714. http://dx.doi.org/10.1038/onc.2011.447.

Wolf, I., Laitman, Y., Rubinek, T., Abramovitz, L., Novikov, I., Beeri, R., ... Kaufman, B. (2010). Functional variant of KLOTHO: A breast cancer risk modifier among BRCA1 mutation carriers of Ashkenazi origin. *Oncogene*, *29*, 26–33. http://dx.doi.org/10.1038/onc.2009.301. onc2009301 [pii].

Wolf, I., Levanon-Cohen, S., Bose, S., Ligumsky, H., Sredni, B., Kanety, H., ... Rubinek, T. (2008). Klotho: A tumor suppressor and a modulator of the IGF-1 and FGF pathways in human breast cancer. *Oncogene*, *27*(56), 7094–7105.

Xie, B., Zhou, J., Shu, G., Liu, D. C., Chen, J., & Yuan, L. (2013a). Restoration of klotho gene expression induces apoptosis and autophagy in gastric cancer cells: Tumor suppressive role of klotho in gastric cancer. *Cancer Cell International*, *13*(1), 1475–2867.

Xie, B., Zhou, J., Yuan, L., Ren, F., Liu, D. C., Li, Q., & Shu, G. (2013b). Epigenetic silencing of Klotho expression correlates with poor prognosis of human hepatocellular carcinoma. *Human Pathology*, *44*(5), 795–801. http://dx.doi.org/10.1016/j.humpath.2012.07.023.

Yamazaki, Y., Imura, A., Urakawa, I., Shimada, T., Murakami, J., Aono, Y., ... Nabeshima, Y.-I. (2010). Establishment of sandwich ELISA for soluble alpha-Klotho measurement: Age-dependent change of soluble alpha-Klotho levels in healthy subjects. *Biochemical and Biophysical Research Communications*, *398*(3), 513–518.

CHAPTER NINE

Klotho Is a Neuroprotective and Cognition-Enhancing Protein

C.R. Abraham[1], P.C. Mullen, T. Tucker-Zhou, C.D. Chen, E. Zeldich
Boston University School of Medicine, Boston, MA, United States
[1]Corresponding author: e-mail address: cabraham@bu.edu

Contents

Abstract

In this chapter, we will describe what has been learned about Klotho and its potential functions in the brain. Klotho is localized in the choroid plexus and, to a lesser extent, in hippocampal neurons. Cognitive decline is a common issue in human aging affecting over 50% of the population. This cognitive decline can also be seen in animal models such as the Rhesus monkey. A long-term study undertaken by our lab demonstrated that normal brain aging in rhesus monkeys and other animal models is associated with a significant downregulation of Klotho expression. This observation substantiates data from other laboratories that have reported that loss of Klotho accelerates the development of aging-like phenotypes, including cognitive deficits, whereas Klotho overexpression extends life span and enhances cognition in mice and humans. Klotho is a type 1 transmembrane pleiotropic protein predominantly expressed in kidney and brain and shed by ADAM 10 and 17 into the blood and cerebral spinal fluid, respectively. While the renal functions of Klotho are well known, its roles in the brain remain to be fully elucidated. We recently demonstrated that Klotho protects hippocampal neurons from amyloid and glutamate toxicity via the activation of an antioxidant enzymatic system suggesting Klotho is a neuroprotective protein. Furthermore, Klotho is necessary

Vitamins and Hormones, Volume 101
ISSN 0083-6729
http://dx.doi.org/10.1016/bs.vh.2016.02.004

for oligodendrocyte maturation and myelin integrity. Through its diverse roles in the brain, Klotho has become a new therapeutic target for neurodegenerative diseases such as Alzheimer's disease and demyelinating diseases like multiple sclerosis. Discovery of small molecule Klotho enhancers may lead to novel treatments for these incurable disorders.

1. INTRODUCTION

α-Klotho (Klotho) is an antiaging protein named after the Greek goddess Clotho, who spins the thread of life (Kuro-o et al., 1997). The Klotho gene was discovered by Kuro-o and colleagues while attempting to develop mice overexpressing the rabbit type I sodium-proton exchanger (NHE-1). Insertion of the NHE-1 transgene into the Klotho promoter resulted in a striking phenotype reminiscent of premature aging in humans. Subsequent analysis of Klotho-deficient (Klotho −/−) mice revealed systemic age-related abnormalities including gait disturbance, emphysema, osteoporosis, arteriosclerosis, hypomyelination, hippocampal neurodegeneration, and cognitive deficits (Chen et al., 2013; Kuro-o et al., 1997; Li et al., 2004; Nagai et al., 2003; Shiozaki et al., 2008). Klotho −/− mice begin to develop symptoms at 3–4 weeks and die prematurely at 2–3 months. Subsequent experiments demonstrated that Klotho overexpression confers resistance to oxidative stress, enhances cognitive performance, and extends lifespan by 20% and 30% in female and male mice, respectively (Brobey et al., 2015; Dubal et al., 2014; Kurosu et al., 2005; Zeldich et al., 2014).

A number of observations from the first report on Klotho −/− mice suggest that Klotho acts as a circulating hormone. For example, although Klotho −/− mice develop normally, several organs that do not express Klotho, including the thymus, skin, and lungs, are severely affected by Klotho depletion. Furthermore, exogenous expression of Klotho cDNA in the brain and testes of knockout mice improves aging phenotypes across all organ systems, indicating that Klotho acts in a noncell-autonomous manner. Findings from the initial study on Klotho were groundbreaking not only due to the discovery of the Klotho gene but also because the authors identified the possibility that Klotho acts as a hormone with pleiotropic and beneficial functions throughout the body (Kuro-o et al., 1997).

In humans, the Klotho gene is located on chromosome 13 and encodes a single-pass type I transmembrane protein that shares sequence homology with β-glycosidases (Tohyama et al., 2004). Klotho contains a 130-kDa extracellular domain, a transmembrane domain, and a very short 8 or 9 amino acid intracellular domain. The large extracellular domain is shed from the cell surface by ADAM10, ADAM17, and BACE1 (Bloch et al., 2009; Chen, Podvin, Gillespie, Leeman, & Abraham, 2007). The shed ectodomain of Klotho (sKlotho) is detectable in the blood and cerebrospinal fluid (CSF), where it likely functions as a humoral factor (Imura et al., 2004; Kurosu et al., 2005). While the membrane bound form of Klotho is a coreceptor for FGF23 (Urakawa et al., 2006), sKlotho modulates ion channels (Cha et al., 2009, 2008; Chang et al., 2005), acts as an antioxidant, inhibits insulin/IGF1 signaling, and enhances oligodendrocyte maturation (Chen et al., 2013; Kurosu et al., 2005; Zeldich et al., 2014).

Klotho is predominantly found in the kidney, but is also expressed in the brain, lung, skeletal muscle, urinary bladder, testes, and ovaries (Kuro-o et al., 1997). Several groups have identified a relationship between Klotho and cognitive performance (Degaspari et al., 2015; Dubal et al., 2014; Nagai et al., 2003; Park et al., 2013), but very little is known regarding the molecular mechanisms of Klotho in the brain and a putative receptor has not been identified. Much of our mechanistic understanding comes from its function in the kidney, where Klotho regulates vitamin D, calcium, and phosphate levels by acting as a coreceptor for FGF23 (Kuro-o, 2012; Kurosu et al., 2006; Urakawa et al., 2006). Although Klotho undoubtedly plays a vital role in the kidney, investigating its neuronal function is especially important given that Klotho expression decreases with age in white matter, and sKlotho is reduced in the CSF of patients with multiple sclerosis (MS) and Alzheimer's disease (Duce et al., 2008; Emami Aleagha et al., 2015; Semba et al., 2014).

Recently, a shorter, 70 kDa secreted Klotho protein resulting from differential splicing has been identified as the major isoform in the brain with the use of specific antibodies. As we have shown previously with the long Klotho isoform (Duce et al., 2008), the shorter form also diminishes with aging and in an AD mouse model (Massó et al., 2015). Another interesting recent finding is that application of secreted recombinant Klotho together with FGF23 to hippocampal cells impacts neuronal morphology and synaptic density via the activation of the Akt pathway (Hensel et al., 2016).

2. THE EXPRESSION AND LOCALIZATION OF KLOTHO IN THE BRAIN

Klotho expression varies with age, species, and cell type. Expression in the brain is widespread throughout postnatal development with Klotho mRNA detectable as early as postnatal day 1 (P1). The most thorough analysis of changes in Klotho expression during development was undertaken using the rat brain as a model. In most areas of the developing rat brain, Klotho mRNA levels are high early in development, decline sharply in week 2 of postnatal development then slowly climb back to adult levels (P75) that are similar to levels seen in early development. These changes were more pronounced in areas of the amygdala, while mRNA levels in the hippocampus remained relatively constant (Clinton et al., 2013). Klotho levels also fluctuate with advancing age. In aged rhesus monkeys, Klotho is reduced at both the mRNA and protein level in white matter from the corpus callosum. Analysis of white matter from mice and rats confirmed these findings indicating that reduced Klotho in the brain is common among aged mammals (Duce et al., 2008). This decline with age is thought to be due to hypermethylation of the promoter region (Duce et al., 2008; King, Rosene, & Abraham, 2012). Hypermethylation of the Klotho promoter not only increases with age, but in vitro methyl modification of the identified methylation sites was capable of reducing gene transcription. These findings are particularly significant given the recent evidence showing that age-related decreases in Klotho expression may be linked to neurodegeneration (Semba et al., 2014).

Measurements of Klotho mRNA expression by RT-PCR in mice show that Klotho is expressed in the kidney, choroid plexus, pituitary gland, skeletal muscles, urinary bladder, and reproductive organs (Kuro-o et al., 1997). Although Klotho clearly has an important function in the brain, its exact distribution in the central nervous system (CNS) remains unclear. Studies using in situ hybridization and immunohistochemistry (IHC) in mice and rats demonstrate that Klotho expression in the CNS is highest in the choroid plexus, where it is localized to the apical membrane of ependymal cells (Li et al., 2004). Soluble Klotho is found in the CSF of humans and mice, indicating that the ectodomain is released into the CSF following proteolytic cleavage from ependymal cells of the choroid plexus (Imura et al., 2004). It is thought that cleavage into the CSF and blood by ADAM 10 and 17 mediates the hormonal activity of sKlotho throughout the brain and body (Bloch et al., 2009; Chen et al., 2007).

Although expression in the CNS is highest in the choroid plexus, Klotho is also found at lower levels in other brain regions. Klotho has been identified in the cerebellar Purkinje cells of mice, where it is localized to the plasma membrane and nuclear envelope (German, Khobahy, Pastor, Kuro, & Liu, 2012). IHC in rats found that Klotho is present at the membrane of cortical and hippocampal neurons, which may explain Klotho's influence on cognition. In this study, researchers confirmed IHC results and extended findings using in situ hybridization. Klotho mRNA was detected throughout the cortex and hippocampus, as well as various other limbic areas including amygdalar, thalamic, and hypothalamic nuclei. Outside of the choroid plexus, Klotho mRNA levels were highest in the medial preoptic area, median preoptic nucleus, ventromedial hypothalamus, and basolateral amygdala (Clinton et al., 2013). Compared to these regions, expression in cortical and hippocampal areas was slightly lower. Characterization of Klotho expression in vivo in glial cells showed that Klotho is not present in microglia or astrocytes, but is expressed in oligodendrocytes, which may be important for Klotho's influence on myelination (Clinton et al., 2013). However, primary cultured astrocytes do express Klotho (Abraham, Chen, Cuny, Glicksman, & Zeldich, 2012). In sum, results from localization studies suggest that a key function of Klotho in the brain is to act as a secreted humoral factor in the CSF, but more detailed functional analysis indicates that Klotho has varied region-specific effects throughout the CNS. For example, expression in hippocampal neurons likely influences spatial memory, while amygdalar expression may relate to fear learning. At this time a potential role of transmembrane Klotho in neurons in cell-to-cell communication with oligodendrocytes or other neurons via FGF signaling cannot be excluded and requires further investigation. With the identification of the exact cleavage site that releases sKlotho, researchers can now produce knockin mice that express only the transmembrane or the shed forms of Klotho in selected brain cells to determine the functions of each isoform (Chen et al., 2014).

3. KLOTHO AND COGNITION

A growing body of evidence points to a connection between Klotho and cognition. Mouse models demonstrate that Klotho deficiency impairs performance on tests of learning and memory, while overexpression improves cognitive function (Dubal et al., 2014; Nagai et al., 2003). Interestingly, cognition is enhanced in humans carrying a single copy of the Klotho-VS polymorphism, while homozygous Klotho-VS individuals exhibit a lower IQ from a young age (Deary et al., 2005). Biochemical analysis of Klotho mutant mice and characterization of the Klotho-VS polymorphism have led to the identification of candidate cellular and molecular mechanisms underlying Klotho's influence on cognition.

3.1 Antioxidant and Neuroprotective Role of Klotho in the Hippocampus

Klotho's impact on cognition likely involves its activity in the hippocampus and/or cerebral cortex, as these regions are crucial for the consolidation and storage of long-term memories (Huang, Nguyen, Abel, & Kandel, 1996; Scoville & Milner, 2000). Klotho −/− mice exhibit decreased synaptophysin immunoreactivity in the hippocampus, indicating a reduction in synaptic vesicles. This effect is most profound at the stratum lucidum, the tract of mossy fiber projections connecting granule cells of the dentate gyrus with pyramidal cells in the CA3 field. This pathway facilitates normal hippocampal function, which is vital for acquiring and consolidating new memories (Scoville & Milner, 2000). When researchers assessed the morphology of single neurons from the stratum lucidum of Klotho −/− mice, they found that individual axons were normal in size and number of vesicles, but the total number of synapses at the stratum lucidum was markedly decreased (Li et al., 2004). Subsequent reports have confirmed the presence of structural abnormalities in the hippocampus of Klotho −/− mice (Shiozaki et al., 2008).

Although the exact mechanism is unknown, it appears that Klotho facilitates normal hippocampal structure and function. One possible mechanism is that Klotho acts as part of the endogenous antioxidant defense system, which has an important neuroprotective role in the hippocampus. In support of this hypothesis, markers of DNA and lipid oxidation are elevated in the hippocampus of Klotho-deficient mice, and expression of the proapoptotic protein bax is increased, while expression of antiapoptotic blc-2 and bcl-Xl is decreased (Nagai et al., 2003; Shiozaki et al., 2008). Consistent with findings in mice lacking Klotho, mice that overexpress Klotho show a 50% reduction in DNA oxidation, as measured by urinary 8-OHdG levels. In CHO and HeLa cells, treatment with Klotho suppresses paraquat-induced lipid oxidation and apoptosis, via inhibition of insulin/IGF-1 signaling, activation of FOXO1, and subsequent transcription of SOD2. Treatment of mice overexpressing Klotho with paraquat revealed that Klotho is also protective against oxidative stress in vivo (Yamamoto et al., 2005). Recently, the importance of Klotho's antioxidant function for its neuroprotective role was confirmed using a hippocampus-derived cell line (HT22) and primary hippocampal neurons (Zeldich et al., 2014). The addition of sKlotho to these cells significantly enhanced the expression of the thioredoxin/peroxiredoxin (Trx/Prx) redox system. The greatest effect was seen on peroxiredoxin-2 (Prx-2), an antioxidant enzyme, that showed an increase in expression at both mRNA and protein levels.

Because the hippocampus is integrally involved in the acquisition and consolidation of episodic memories (Scoville & Milner, 2000), and Klotho −/− mice have altered hippocampal morphology and impaired hippocampal long-term potentiation (LTP) (Park et al., 2013), it is unsurprising that Klotho −/− mice score lower than wild-type (WT) littermates on tests of long-term memory (LTM). Klotho −/− mice perform more poorly on novel object recognition, as well as context-cued and tone-cued fear conditioning tasks—implicating both hippocampal and amygdalar dysfunction. Notably, researchers found that memory impairments observed in Klotho −/− mice could be ameliorated by treatment with the potent antioxidant α-tocopherol, which further implicates the importance of Klotho's antioxidant function in its neuroprotective role (Nagai et al., 2003).

It is important to point out that memory impairments observed in Klotho −/− mice are age dependent. WT mice perform better than Klotho −/− mice when assessed at 7 weeks of age, but not at 6 weeks. This effect may relate to the accumulation of oxidative cellular damage over time in the Klotho-deficient mice (Nagai et al., 2003). In addition, deficits were seen in LTM but not short-term memory (STM); mice lacking Klotho performed worse when tested 24 hours, but not 1-h posttraining, implying that Klotho depletion causes dysfunction of the hippocampal circuitry involved in LTM consolidation, but not prefrontal circuitry responsible for STM. Conversely, experiments with Klotho-overexpressing mice (Klotho-OE) show that Klotho overexpression enhances both STM and LTM in an age-independent manner. Young- and middle-aged Klotho-OE mice perform better on the Morris water maze task, a hippocampal dependent test of spatial memory, as well as in context-cued fear conditioning. Klotho-OE mice also score higher on the Y maze task, a paradigm that tests working memory, which is a type of STM (Dubal et al., 2014).

3.2 Klotho and LTP

To better understand the effect of Klotho on cognition at the molecular level, researchers examined the relative amounts of GluN2A, GluN2B, and GluN2C, *N*-methyl-D-aspartate receptor (NMDAR) subunits in the hippocampus and prefrontal cortex (PFC) of Klotho-OE mice. Hippocampal NMDARs are involved in the induction of LTP, which is thought to be the cellular correlate of LTM (Bliss & Collingridge, 1993; Malenka & Nicoll, 1993). LTP is triggered when glutamate binds NMDARs and calcium enters the cell—initiating a complex cascade of intracellular signaling pathways. Ca^{2+} activates calcium-sensitive proteins such as protein kinase C and calcium/calmodulin-dependent protein kinases (CaMKII and CaMKIV). These kinases and many other molecules stimulate diverse biochemical pathways that ultimately alter the strength of synapses. Modulation of glutamate receptors, de novo protein synthesis, and changes in neuronal morphology are three well-characterized mechanisms that result in enhanced synaptic connectivity. For example, calcium influx stimulates CaMKII, which then modulates NMDARs by binding to GluN2B subunits (Barria & Malinow, 2005).

The NMDAR is required for many forms of LTP, and NMDAR-dependent LTP in the hippocampus is necessary for spatial memory, temporal memory, and multimodal associative learning, among other forms of memory (Huerta, Sun, Wilson, & Tonegawa, 2000; Morris, Anderson, Lynch, & Baudry, 1986; Rondi-Reig, Libbey, Eichenbaum, & Tonegawa, 2001). In Klotho-OE mice, the GluN2B subunit is enriched at the postsynaptic density in the hippocampus and PFC at the protein level, but not at the mRNA level, implicating the involvement of posttranslational modifications (Dubal et al., 2014). GluN2B currents are involved in LTP (Smith, Smith, Bredemann, & McMahon, 2015) and this subunit has an important impact on learning and memory. Depletion with age is linked to cognitive decline, and overexpression rescues these deficits (Brim et al., 2013). Because Klotho overexpression upregulates the GluN2B subunit in the hippocampus and PFC, researchers believe modulation of this subunit is involved in spatial and working memory improvements observed in Klotho-OE mice. Consistent with this hypothesis, Klotho overexpression results in enhanced hippocampal LTP, which likely mediates Klotho's impact on cognition. To highlight the involvement of the GluN2B subunit, Klotho-OE and WT mice were treated with a low dose of ifenprodil, a selective inhibitor of NMDARs containing the N2B subunit. Treatment with ifenprodil did not affect WT mice, but abolished memory improvements seen in Klotho-OE mice, revealing a vital role for the GluN2B subunit in Klotho induced cognitive enhancement (Dubal et al., 2014).

Although glutamate is the primary excitatory neurotransmitter involved in hippocampal LTP, this process is also subject to modulation by a number of other neurotransmitters and hormones including serotonin, acetylcholine (ACh), dopamine, norepinephrine, cortisol, and estrogens (Rodrigues, LeDoux, & Sapolsky, 2009; Sweatt, 2004). Modulatory inputs at the hippocampus influence learning and memory by impacting LTP, and it is possible that Klotho impacts cognition by interacting with these modulators. For example, impaired cognition and LTP in Klotho −/− mice may be related to decreased ACh-mediated JAK/STAT signaling (Park et al., 2013).

3.3 Klotho and the Neuroendocrine System

Recent evidence has raised the possibility that Klotho's impact on cognition is modulated by estradiol (E2). Like Klotho, E2 influences cognitive function, and postmenopausal decreases in E2 levels are associated with memory decrements in some postmenopausal women. E2 enhances hippocampal LTP (Foy et al., 1999; Sweatt, 2004; Woolley, Weiland, McEwen, & Schwartzkroin, 1997; Yun et al., 2007) and rescues memory deficits in ovariectomized (OVX) animals (Daniel, Hulst, & Berbling, 2006; Talboom, Williams, Baxley, West, & Bimonte-Nelson, 2008). Intriguingly, a recent study showed that E2-treated OVX rats exhibited ten times higher hippocampal Klotho expression, compared to vehicle-treated animals (Sarvari et al., 2015). This report is especially exciting because Klotho and E2 seem to act in similar ways by improving cognition and enhancing LTP by acting through GluN2B subunits (Smith & McMahon, 2006). These findings raise the possibility that Klotho and E2 may interact to enhance cognition—however, a definitive model requires additional analysis.

Since Klotho is a potent inhibitor of the IGF-I pathway, which negatively regulates the secretion of growth hormone (GH) from the pituitary gland, it was hypothesized that Klotho may enhance GH secretion. Recent studies indeed implicated Klotho, as a positive regulator of the secretion of the GH from normal somatotropes and pituitary adenomas through the possible involvement of IGF-I and bFGF pathways (Shahmoon et al., 2014). Measuring levels of soluble Klotho levels was also proposed as a marker in the follow-up of patients with acromegaly (Kohler et al., 2013).

3.4 The Klotho-VS Polymorphism and Human Cognition

In line with experiments linking Klotho and cognition in rodents, studies of human populations indicate that Klotho enhances learning and memory. Analysis of three independent cohorts—the Hillblom Aging Study, the Memory and Aging Project, and the Normal Aging Cohort—revealed that heterozygous carriers of the Klotho-VS polymorphism score significantly higher on neuropsychological tests of cognitive function (Dubal et al., 2014). These results were confirmed by meta-analysis of the three populations. The Klotho-VS polymorphism is a variant composed of six single-nucleotide polymorphisms (SNPs), with three located in exons and three within introns. Of the SNPs occurring in exons, two out of three lead to amino acid substitutions—F352V and C370S. Expression of this polymorphism in human embryonic kidney cells leads to increased formation of Klotho–FGF receptor complexes and higher levels of FGF23 signaling (Tucker Zhou, King, Chen, & Abraham, 2013). Heterozygous carriers of this variant also have higher sKlotho levels in the serum than noncarriers, which is thought to be a result of enhanced shedding that has been reported in HeLa cells transiently transfected with a truncated Klotho-VS construct (Arking et al., 2002). Increased Klotho levels in the serum of carriers and noncarriers tend to correlate with higher scores on tests of semantic fluency, category fluency, and modified trials (Dubal et al., 2014). On the other hand, women homozygous for the V/V genotype had lower nonverbal reasoning scores at age 79, after adjustment for cognitive ability at age 11 (Deary et al., 2005). Taken together, these results strongly suggest that increasing Klotho levels enhances cognition in both humans and rodents. Although the precise mechanism is not fully understood, two leading theories posit:

1. Klotho promotes memory formation and hippocampal function by acting as an antioxidant and neuroprotective agent.
2. Klotho influences memory formation by modulating LTP via upregulation of GluN2B NMDAR subunits, resulting in improved hippocampal and prefrontal dependent memory.

Given Klotho's pleiotropic function, it is likely that both of these processes contribute to Klotho's effect on learning and memory. In addition, modulators such as ACh and E2 may be related to Klotho's cognition-enhancing properties. Although significant progress has been made regarding our understanding of Klotho's impact on cognition, further work is required for a more precise and detailed characterization of the mechanisms and interactions involved.

3.5 The White Matter, Aging, and Cognition

In addition to modulating cognition through its antioxidant and LTP enhancing activity, it is possible that Klotho influences learning and memory by counteracting cognitive impairments associated with "normal" aging. Although no studies to date have specifically addressed Klotho's impact on normal cognitive aging, Klotho expression and levels decrease with age (Duce et al., 2008; Semba et al., 2014), as does cognitive function. These correlations may represent a relationship between aging, decreased Klotho, and cognitive decline, but causal evidence is needed to confirm and better understand this relationship.

Age-related cognitive decline (ARCD) is a well-documented phenomenon characterized by impairments in neural processing, executive function, and memory acquisition (Makris et al., 2007). Initially, neuronal death was thought to underlie ARCD, but as imaging methodologies improved, a number of reports were published describing no significant loss of neurons with age (Giannaris & Rosene, 2012). Studies have demonstrated that aged monkeys do not display a neurodegenerative phenotype in cognitively relevant areas such as the PFC or the hippocampal CA1 field (Peters, Leahu, Moss, & McNally, 1994; Peters, Nigro, & McNally, 1997; West, 1993). While significant loss of gray matter is not apparent in the aged brain of healthy animals, age-related white matter loss and myelin abnormalities have been observed in disease-free monkeys and humans. Ultrastructural analysis of white matter in the aged brain shows abnormal thickness, splitting, and redundancy of myelin sheaths, as well as reduced white matter volume and thinning in layer I of the PFC (Hinman & Abraham, 2007). These age-related white mater changes may yield severe functional consequences, as myelination facilitates normal neuronal communication, and permits high speed information processing required for normal brain function. Although age-related white matter changes are heterogeneous and complex, myelin breakdown and reduced white matter volume correlate with both age and cognitive decline (Peters, 1996).

White matter refers to the bundles of myelinated projections that connect various brain regions and was named because of its characteristic white appearance—the result of an abundance of myelinated axons. Myelin is a lipid- and protein-containing membrane that spirals around axons to form the myelin sheath, which provides electrical insulation for axons. In the CNS, oligodendrocytes are responsible for myelination, and this process results in markedly faster speeds of action potential (AP) propagation. By increasing the velocity of AP propagation, myelination facilitates the remarkable information processing capacity of the brain. When myelin is disrupted, normal AP conduction cannot occur, and neuronal communication is impaired. As a result, specific brain functions are impaired, depending on the regions and pathways affected. This process can be seen on a pathological level in MS, as demyelination and white matter lesions lead to neurological symptoms that depend on lesion location. For example, demyelination and cell death in pathways important for learning and memory results in symptoms related to cognition, while lesions in other brain areas may cause sensory, motor, and/or sexual dysfunction. Because demyelination occurs in both MS and normal brain aging, albeit to vastly different extents, it is possible that cognitive dysfunction in MS and ARCD both involve white matter changes and circuit level brain dysfunction despite differing underlying molecular mechanisms. White matter loss is common in neurodegenerative disorders such as MS, AD, and Parkinson's disease (PD), and may relate to cognitive dysfunction in these conditions. Although ubiquitous gray matter loss makes it difficult to study the cognitive consequence of white matter loss in AD, a recent study in humans diagnosed with PD shows that white matter loss correlates with cognitive deficits (Galantucci et al., 2015).

3.6 Klotho, Aging, and Neuroinflammation

It appears that during normal brain aging, the number of neurons does not decrease, but cellular structure and function are altered by aging processes. For example, increases in neuroinflammation with age are associated with the activation of microglia and astrocytes in white matter (Sloane, Hollander, Moss, Rosene, & Abraham, 1999; Sloane et al., 2000). When activated, these cells phagocytose myelin (Duce, Hollander, Jaffe, & Abraham, 2006) and release inflammatory and oxidant molecules, both of which are processes previously associated with aging. Microglial activation in aged monkeys correlates with cognitive deficits, and inflammation is associated with impaired cognition in chronic kidney disease (Madero, Gul, & Sarnak, 2008), suggesting a key role for inflammation in ARCD. Notably, Klotho inhibits the proinflammatory cytokine TNFα, and cognitively impaired nephrectomized rats show decreased levels of Klotho and increased levels of TNFα in the PFC (Degaspari et al., 2015). In parallel, Klotho was shown to be downregulated by TNFα via the transcription of nuclear factor kappa B (NFkB) in animal models of chronic kidney disease and colitis (Moreno et al., 2011; Thurston et al., 2010). These findings were confirmed in a study that demonstrated that Klotho was downregulated in the temporal lobe of epilepsy patients, TNFα upregulation in epilepsy patients is associated with chronic hippocampal inflammation and the expression of Klotho in the area was inversely correlated with the expression of NFkB and TNFα (Teocchi, Ferreira, da Luz de Oliveira, Tedeschi, & D'Souza-Li, 2013). We have demonstrated in vitro that endogenously secreted Klotho rescues cocultures of oligodendrocyte precursor cells and astrocytes from TNFα-induced cytotoxicity (Abraham et al., 2012).

4. KLOTHO, OLIGODENDROCYTE BIOLOGY, AND MYELINATION

A number of processes other than inflammation, such as oxidative stress and altered gene expression, also influence white matter degeneration and ARCD. Findings of decreased Klotho expression, white matter abnormalities, and cognitive deficits in both aged monkeys and Klotho-deficient mice support the hypothesis that Klotho is involved in ARCD and white matter degeneration in humans. Following the discovery of reduced Klotho expression in the white matter of aged mammals (Duce et al., 2008), recent efforts in our lab have been aimed at elucidating the role of Klotho in myelination, and the effect of Klotho on oligodendrocyte biology. We have found that Klotho impacts oligodendrocytes in vitro and myelination in vivo. Klotho activates Akt and Erk1/2 signaling, and enhances maturation of primary oligodendrocyte progenitor cells (OPCs) and the oligodendrocytic cell line, MO3.13 (Chen et al., 2015, 2013). Analysis of protein and mRNA levels shows that Klotho also enhances expression of myelin-associated proteins in OPCs and MO3.13 cells. In OPCs, treatment with Klotho upregulated the two major protein constituents of myelin, proteolipid protein (PLP), and myelin basic protein, as well as myelin-associated glycoprotein (MAG) and oligodendrocyte-specific protein. Analysis of Klotho-induced OPC maturation following treatment with the Akt inhibitor, LY294002, and the Erk-1 inhibitor, UO126, revealed a vital role for Akt signaling and a less important role for Erk-1 (Chen et al., 2013). Consistent with the involvement of Akt signaling, Klotho induced Akt phosphorylation at S473 in OPCs and MO3.13 cells (Chen et al., 2015, 2013).

Recent evidence from mouse models has provided further insights into the relationship between Klotho, oligodendrocytes, and myelination in vivo. Klotho-deficient mice have a markedly lower percentage of myelinated fibers in the corpus callosum and optic nerve, as well as fewer total and mature oligodendrocytes in the fimbria (Chen et al., 2013). As expected, analysis of Klotho −/− mice revealed a significant reduction in both protein and mRNA levels of myelin-related proteins MAG, PLP, and 2′,3′-cyclic-nucleotide 3′-phosphodiesterase. Klotho −/− mice also show impaired myelin at the node of Ranvier, as the paranodal region is diminished and the nodal region is enlarged. In sum, these findings strongly suggest that Klotho enhances myelination both in vitro and in vivo (Chen et al., 2013). In line with this hypothesis, a recent study using the cuprizone model of MS demonstrates that spontaneous remyelination is significantly enhanced in Klotho-OE mice (Zeldich, Chen, Avila, Medicetty, & Abraham, 2015). Further evidence for Klotho's therapeutic potential in MS comes from the finding that CSF levels are significantly lower in MS patients compared to healthy age-matched controls (Emami Aleagha et al., 2015), which may relate to the failure of remyelination mechanisms seen in MS patients. Given that Klotho enhances oligodendrocyte maturation and myelination and has antioxidant and neuroprotective properties, Klotho modulators represent powerful and promising therapeutics for demyelinating disorders such as MS. The ability of Klotho to stimulate remyelination may be especially important, because normal repair mechanisms are disrupted in MS patients (Franklin, 2002). Due to Klotho's therapeutic potential, development and optimization of small molecule Klotho modulators is a current focus in our laboratory (Abraham et al., 2012; King, Chen, et al., 2012).

5. KLOTHO, OXIDATIVE STRESS, AND NEURODEGENERATION

While upregulating Klotho to induce remyelination represents a promising treatment for MS (Zeldich et al., 2015), Klotho modulators may also be useful for treating other neurodegenerative diseases given Klotho's antioxidant and neuroprotective properties (Yamamoto et al., 2005; Zeldich et al., 2014). Cellular oxidative damage is thought to play a role in aging and has been implicated in numerous age-related brain diseases (Popa-Wagner, Mitran, Sivanesan, Chang, & Buga, 2013). Oxidative stress is mediated by reactive oxygen species (ROS) and reactive nitrogen species (RNS) that are generated as a result of normal cellular metabolism. These species include molecules such as superoxide, hydroxy radicals, hydrogen peroxide, nitric oxide, and peroxynitrite. ROS and RNS are not harmful at basal levels and some have functional roles as signaling molecules (Arancio et al., 1996). A suite of endogenous antioxidant enzymes including superoxide dismutase, glutathione, and catalase are responsible for neutralizing ROS and RNS as they are produced in order to prevent toxic accumulation of these species. In pathological conditions, ROS and RNS oxidize nucleic acids, proteins, and lipids causing cellular damage and eventually death. Because of the high-energy demands of neurons, the brain may be particularly susceptible to oxidative damage, and reactive oxidants are elevated in PD, amyotrophic lateral sclerosis (ALS), MS, and AD.

ROS levels are elevated during glutamate excitotoxicity, and these processes are thought to mediate neurodegeneration in disorders such as ALS, MS, and PD (Lau & Tymianski, 2010). In Klotho-treated HT22 cells and primary hippocampal neurons obtained from Klotho-OE mice, Klotho confers resistance to glutamate excitotoxicity and oxidative stress (Zeldich et al., 2014). Downregulation of the antioxidant enzyme Prx-2 via shRNA abolished the ability of Klotho to protect against glutamate toxicity, indicating a key role for Prx-2 in Klotho-mediated neuroprotection (Zeldich et al., 2014). Experiments in Klotho −/− mice have shown that lack of Klotho leads to decreased levels of Trx, which is correlated with higher p38 activation. In contrast, challenging the brains of Klotho transgenic mice with a neurotoxin, MPTP and analysis for residual neuron numbers and integrity in the substantia nigra pars compacta showed that Klotho overexpression significantly protects dopaminergic neurons against oxidative damage, partly by modulating p38 MAPK activation level. The data highlight the importance of ASK1/p38 MAPK pathway in the brain and identify Klotho as a stabilizer of the interaction of Trx with ASK1, leading to increased antioxidative resistance (Brobey et al., 2015). This recent study further confirms that Klotho regulates the Trx–Prx system to control oxidative stress. Remarkably, in cultures of human aortic vascular smooth muscle cells, Klotho-induced antioxidant defense is also associated with the increased expression of Prx (Prx-1), suggesting that the antioxidative mode of action of Klotho universally involves the regulation of this redox system (Rizzo et al., 2014).

If the brain is particularly vulnerable to oxidative stress, developing blood–brain barrier penetrant antioxidants may be especially valuable. A recent study in mice demonstrates the therapeutic benefit of reducing cellular oxidative damage by showing that antioxidant cerium oxide nanoparticles decrease ROS in the cerebellum and hippocampus and alleviate disease severity in the EAE model of MS (Heckman et al., 2013). Klotho may have a similar therapeutic mechanism, as it protects neurons from oxidative stress both in vitro and in vivo (Yamamoto et al., 2005; Zeldich et al., 2014), although there is some discrepancy with respect to the mechanism involved. In nonneuronal cell lines including COS, HeLa, and CHO cells, Klotho acts as an antioxidant by inhibiting Akt and FOXO1 resulting in increased SOD2 expression (Yamamoto et al., 2005). Conversely, in brain-derived cells such as the HT22 cell line, primary hippocampal neurons, and oligodendrocytes, Klotho induces phosphorylation of Akt (Chen et al., 2013; Zeldich et al., 2014), while in primary hippocampal neurons the Akt activation resulted in inhibitory phosphorylation of FOXO3a (Zeldich et al., 2014).

Because Klotho can reduce oxidative stress and protect against glutamate toxicity, Klotho boosters may be effective treatments for a number of neurodegenerative diseases. For example, substantial evidence suggests that excitotoxicity and oxidative damage play a role in ALS pathogenesis and motor neurons affected in ALS are highly susceptible to excitotoxicity (Barber, Mead, & Shaw, 2006; Foran & Trotti, 2009). Klotho −/− mice have fewer large anterior horn cells (AHCs) in the spinal cord, with concomitant increases in GFAP and neurofilament phosphorylation (Anamizu et al., 2005). Changes in AHCs of Klotho −/− mice resemble the loss of motor neurons seen in ALS, and degeneration in this region may account for altered gait in mice-lacking Klotho. However, motor abnormalities seen in Klotho −/− have been described as Parkinsonian, and may be the result of Purkinje cell degeneration in the cerebellum.

Purkinje cells are also sensitive to excitotoxic insult (Slemmer, De Zeeuw, & Weber, 2005), and neurodegeneration caused by the accumulation of neurofilaments (Tu et al., 1997). Klotho-deficient mice show increased phosphorylation and expression of neurofilament subunits in the cerebellum, and denser dendritic arbors in Purkinje cells, which may be due to decreased expression of MAP2, a microtubule-associated protein necessary for proper dendritic spacing. Purkinje cell bodies were reported to have abnormal inclusions similar to lipofuscin type granules and axons also showed signs of degeneration (Shiozaki et al., 2008). These findings suggest that Klotho protects Purkinje cells from excitotoxicity in wild-type mice, while Klotho-deficient mice lack this protective mechanism leading to Purkinje cell damage and death.

Klotho also appears to play a protective role in the retina by acting as an antioxidant. Cultured retinal pigment epithelial cells treated with Klotho show decreased markers of oxidative stress, and these cells play a critical role in retinal function (Kokkinaki et al., 2013). Although the retinas of Klotho −/− mice are normal throughout development, functional deficits arise around 7 weeks of age (Reish et al., 2013). In line with these findings, Klotho levels are elevated in a model of retinal degeneration, suggesting that Klotho is involved in retinal disease states, although it is unknown whether a causal relationship exists (Farinelli et al., 2013; Kokkinaki et al., 2013).

Klotho levels in the CSF decline with age and are further decreased in age-related neurodegenerative disorders such as MS and AD (Emami Aleagha et al., 2015; Semba et al., 2014). It can be difficult to interpret this type of correlative data because it is impossible to know if reduced Klotho is a cause or consequence of pathology. Regardless, a number of studies support the hypothesis that upregulating Klotho is useful for treating AD and MS. In line with this idea, Klotho overexpression provides therapeutic benefit in a transgenic mouse model of AD. Mice expressing human APP (hAPP) develop behavioral and neuropathological phenotypes that recapitulate features of AD such as neuritic plaques, impaired LTP, altered hippocampal spine morphology, and cognitive deficits. These mice also die prematurely, which may be a result of increased epileptiform activity in the cortex (Dubal et al., 2015). In transgenic mice overexpressing both hAPP and Klotho (hAPP × Klotho), the deleterious effects of hAPP expression on cognition and lifespan were ameliorated. Furthermore, hAPP mice had reduced levels of the GluN2A and GluN2B subunits in the hippocampus, while Klotho-OE and hAPP × Klotho mice had significantly higher levels of GluN2B when compared with either WT or hAPP mice. Consistent with results from Klotho-OE mice, LTP was enhanced in hAPP × Klotho mice. Although Klotho overexpression in hAPP mice does not alter the number or neuritic plaques, or the levels of phosphorylated tau, it does prevent loss of spinophillin staining in the hippocampus, demonstrating Klotho's neuroprotective role. These results indicate that although Klotho does not directly alter AD pathogenesis, enhancing Klotho levels in the brain may effectively counteract cognitive decline associated with dementia, either by enhancing LTP via upregulation of GluN2B or by providing neuroprotection through an antioxidant mechanism.

6. KLOTHO ENHANCERS AS A POTENTIAL TREATMENT FOR NEURODEGENERATION

Increasing Klotho appears to be a promising strategy for treating AD and MS, but due to Klotho's relatively large size, it cannot be directly used as a therapeutic. Instead, small molecule Klotho enhancers that can cross the BBB are being screened in vitro to identify lead compounds for testing in animal models of neurodegeneration (Abraham et al., 2012; King, Chen, et al., 2012).

Klotho is a large protein with antiaging and neuroprotective properties, which is downregulated in aging and Alzheimer's diseased brains. Klotho has been implicated in myelination processes, and a decline of Klotho is associated with cognitive and neuronal decline. Since Klotho is a large molecule and unlikely to cross the blood–brain barrier, it will be difficult to clinically administer Klotho to restore Klotho levels in the brain. Instead, a primary goal is to develop novel treatments based on Klotho-enhancing small molecule compounds, focusing on neurological improvement or protection. This goal aligns well in potential treatments for AD and MS, neither of which has a cure or long-term effective treatments.

Using a high throughput screen of 150,000 compounds, a number of Klotho-enhancing small molecule hits were identified (Abraham et al., 2012; King, Chen, et al., 2012). Medicinal chemistry optimization of the drug-like properties of these compounds has led to the recent identification of a lead compound that, when administered via intraperitoneal injection into mice was well tolerated and gave excellent brain exposure. Such compounds will be tested for efficacy in animal models of neurodegeneration.

REFERENCES

Abraham, C. R., Chen, C., Cuny, G. D., Glicksman, M. A., & Zeldich, E. (2012). Small-molecule Klotho enhancers as novel treatment of neurodegeneration. *Future Medicinal Chemistry*, *4*(13), 1671–1679. http://dx.doi.org/10.4155/fmc.12.134.

Anamizu, Y., Kawaguchi, H., Seichi, A., Yamaguchi, S., Kawakami, E., Kanda, N., … Oyanagi, K. (2005). Klotho insufficiency causes decrease of ribosomal RNA gene transcription activity, cytoplasmic RNA and rough ER in the spinal anterior horn cells. *Acta Neuropathologica*, *109*(5), 457–466. http://dx.doi.org/10.1007/s00401-004-0971-7.

Arancio, O., Kiebler, M., Lee, C. J., Lev-Ram, V., Tsien, R. Y., Kandel, E. R., & Hawkins, R. D. (1996). Nitric oxide acts directly in the presynaptic neuron to produce long-term potentiation in cultured hippocampal neurons. *Cell*, *87*, 1025–1035.

Arking, D. E., Krebsova, A., Macek, M., Sr., Macek, M., Jr., Arking, A., Mian, I. S., … Dietz, H. C. (2002). Association of human aging with a functional variant of klotho. *Proceedings of the National Academy of Science of the United States of America*, *99*(2), 856–861.

Barber, S. C., Mead, R. J., & Shaw, P. J. (2006). Oxidative stress in ALS: A mechanism of neurodegeneration and a therapeutic target. *Biochimica et Biophysica Acta*, *1762*(11–12), 1051–1067. http://dx.doi.org/10.1016/j.bbadis.2006.03.008.

Barria, A., & Malinow, R. (2005). NMDA receptor subunit composition controls synaptic plasticity by regulating binding to CaMKII. *Neuron*, *48*(2), 289–301. http://dx.doi.org/10.1016/j.neuron.2005.08.034.

Bliss, T. V., & Collingridge, G. L. (1993). A synaptic model of memory: Long-term potentiation in the hippocampus. *Nature*, *361*(6407), 31–39.

Bloch, L., Sineshchekova, O., Reichenbach, D., Reiss, K., Saftig, P., Kuro-o, M., & Kaether, C. (2009). Klotho is a substrate for alpha-, beta- and gamma-secretase. *FEBS Letters*, *583*(19), 3221–3224. http://dx.doi.org/10.1016/j.febslet.2009.09.009.

Brim, B. L., Haskell, R., Awedikian, R., Ellinwood, N. M., Jin, L., Kumar, A., … Magnusson, K. R. (2013). Memory in aged mice is rescued by enhanced expression of the GluN2B subunit of the NMDA receptor. *Behavioural Brain Research*, *238*, 211–226. http://dx.doi.org/10.1016/j.bbr.2012.10.026.

Brobey, R. K., German, D., Sonsalla, P. K., Gurnani, P., Pastor, J., Hsieh, C. C., … Rosenblatt, K. P. (2015). Klotho protects dopaminergic neuron oxidant-induced degeneration by modulating ASK1 and p38 MAPK signaling pathways. *PLoS One*, *10*(10), e0139914. http://dx.doi.org/10.1371/journal.pone.0139914.

Cha, S. K., Hu, M. C., Kurosu, H., Kuro-o, M., Moe, O., & Huang, C. L. (2009). Regulation of renal outer medullary potassium channel and renal K(+) excretion by Klotho. *Molecular Pharmacology*, *76*(1), 38–46. http://dx.doi.org/10.1124/mol.109.055780. mol.109.055780 [pii].

Cha, S. K., Ortega, B., Kurosu, H., Rosenblatt, K. P., Kuro, O. M., & Huang, C. L. (2008). Removal of sialic acid involving Klotho causes cell-surface retention of TRPV5 channel via binding to galectin-1. *Proceedings of the National Academy of Sciences of the United States of America*, *105*(28), 9805–9810. http://dx.doi.org/10.1073/pnas.0803223105. 0803223105 [pii].

Chang, Q., Hoefs, S., van der Kemp, A. W., Topala, C. N., Bindels, R. J., & Hoenderop, J. G. (2005). The beta-glucuronidase klotho hydrolyzes and activates the TRPV5 channel. *Science*, *310*(5747), 490–493. http://dx.doi.org/10.1126/science.1114245. 310/5747/490 [pii].

Chen, C. D., Li, H., Liang, J., Hixson, K., Zeldich, E., & Abraham, C. R. (2015). The anti-aging and tumor suppressor protein Klotho enhances differentiation of a human oligodendrocytic hybrid cell line. *Journal of Molecular Neuroscience*, *55*(1), 76–90. http://dx.doi.org/10.1007/s12031-014-0336-1.

Chen, C. D., Podvin, S., Gillespie, E., Leeman, S. E., & Abraham, C. R. (2007). Insulin stimulates the cleavage and release of the extracellular domain of Klotho by ADAM10 and ADAM17. *Proceedings of the National Academy of Sciences of the United States of America*, *104*(50), 19796–19801. http://dx.doi.org/10.1073/pnas.0709805104. 0709805104 [pii].

Chen, C. D., Sloane, J. A., Li, H., Aytan, N., Giannaris, E. L., Zeldich, E., … Abraham, C. R. (2013). The antiaging protein Klotho enhances oligodendrocyte maturation and myelination of the CNS. *The Journal of Neuroscience*, *33*(5), 1927–1939. http://dx.doi.org/10.1523/JNEUROSCI.2080-12.2013. 33/5/1927 [pii].

Chen, C. D., Tung, T. Y., Liang, J., Zeldich, E., Tucker Zhou, T. B., Turk, B. E., & Abraham, C. R. (2014). Identification of cleavage sites leading to the shed form of the anti-aging protein klotho. *Biochemistry*, *53*(34), 5579–5587. http://dx.doi.org/10.1021/bi500409n.

Clinton, S. M., Glover, M. E., Maltare, A., Laszczyk, A. M., Mehi, S. J., Simmons, R. K., & King, G. D. (2013). Expression of klotho mRNA and protein in rat brain parenchyma from early postnatal development into adulthood. *Brain Research*, *1527*, 1–14. http://dx.doi.org/10.1016/j.brainres.2013.06.044.

Daniel, J. M., Hulst, J. L., & Berbling, J. L. (2006). Estradiol replacement enhances working memory in middle-aged rats when initiated immediately after ovariectomy but not after a long-term period of ovarian hormone deprivation. *Endocrinology*, *147*(1), 607–614. http://dx.doi.org/10.1210/en.2005-0998.

Deary, I. J., Harris, S. E., Fox, H. C., Hayward, C., Wright, A. F., Starr, J. M., & Whalley, L. J. (2005). KLOTHO genotype and cognitive ability in childhood and old age in the same individuals. *Neuroscience Letters*, *378*(1), 22–27. http://dx.doi.org/10.1016/j.neulet.2004.12.005.

Degaspari, S., Tzanno-Martins, C. B., Fujihara, C. K., Zatz, R., Branco-Martins, J. P., Viel, T. A., … Kawamoto, E. M. (2015). Altered KLOTHO and NF-kappaB-TNF-alpha signaling are correlated with nephrectomy-induced cognitive impairment in rats. *PLoS One*, *10*(5), e0125271. http://dx.doi.org/10.1371/journal.pone. 0125271.

Dubal, D. B., Yokoyama, J. S., Zhu, L., Broestl, L., Worden, K., Wang, D., … Mucke, L. (2014). Life extension factor klotho enhances cognition. *Cell Reports*, 7(4), 1065–1076. http://dx.doi.org/10.1016/j.celrep.2014.03.076.

Dubal, D. B., Zhu, L., Sanchez, P. E., Worden, K., Broestl, L., Johnson, E., … Mucke, L. (2015). Life extension factor klotho prevents mortality and enhances cognition in hAPP transgenic mice. *The Journal of Neuroscience*, *35*(6), 2358–2371. http://dx.doi.org/10.1523/JNEUROSCI.5791-12.2015.

Duce, J. A., Hollander, W., Jaffe, R., & Abraham, C. R. (2006). Activation of early components of complement targets myelin and oligodendrocytes in the aged rhesus monkey brain. *Neurobiology of Aging*, *27*(4), 633–644. http://dx.doi.org/10.1016/j.neurobiolaging.2005.03.027.

Duce, J. A., Podvin, S., Hollander, W., Kipling, D., Rosene, D. L., & Abraham, C. R. (2008). Gene profile analysis implicates Klotho as an important contributor to aging changes in brain white matter of the rhesus monkey. *Glia*, *56*(1), 106–117. http://dx.doi.org/10.1002/glia.20593.

Emami Aleagha, M. S., Siroos, B., Ahmadi, M., Balood, M., Palangi, A., Haghighi, A. N., & Harirchian, M. H. (2015). Decreased concentration of Klotho in the cerebrospinal fluid of patients with relapsing-remitting multiple sclerosis. *Journal of Neuroimmunology*, *281*, 5–8. http://dx.doi.org/10.1016/j.jneuroim.2015.02.004.

Farinelli, P., Arango-Gonzalez, B., Volkl, J., Alesutan, I., Lang, F., Zrenner, E., … Ekstrom, P. A. (2013). Retinitis pigmentosa: Over-expression of anti-ageing protein Klotho in degenerating photoreceptors. *Journal of Neurochemistry*, *127*(6), 868–879. http://dx.doi.org/10.1111/jnc.12353.

Foran, E., & Trotti, D. (2009). Glutamate transporters and the excitotoxic path to motor neuron degeneration in amyotrophic lateral sclerosis. *Antioxidants & Redox Signaling*, *11*(7), 1587–1602. http://dx.doi.org/10.1089/ars.2009.2444.

Foy, M. R., Xu, J., Xie, X., Brinton, R. D., Thompson, R. F., & Berger, T. W. (1999). 17beta-estradiol enhances NMDA receptor-mediated EPSPs and long-term potentiation. *Journal of Neurophysiology*, *81*(2), 925–929.

Franklin, R. J. (2002). Why does remyelination fail in multiple sclerosis? *Nature Reviews. Neuroscience*, *3*(9), 705–714. http://dx.doi.org/10.1038/nrn917.

Galantucci, S., Agosta, F., Gagliardi, D., Caso, F., Stankovic, I., Petrovic, I., … Filippi, M. (2015). White matter microstructural damage and cognitive impairment in Parkinson's disease. *Neurology*, *84*(14), Supplement S40.005 (Abstract). Published online before print April 8, 2015.

German, D. C., Khobahy, I., Pastor, J., Kuro, O. M., & Liu, X. (2012). Nuclear localization of Klotho in brain: An anti-aging protein. *Neurobiology of Aging*, *33*(7), 1483.e1425–1483.e1430. http://dx.doi.org/10.1016/j.neurobiolaging.2011.12.018. S0197-4580(11)00550-1 [pii].

Giannaris, E. L., & Rosene, D. L. (2012). A stereological study of the numbers of neurons and glia in the primary visual cortex across the lifespan of male and female rhesus monkeys. *The Journal of Comparative Neurology*, *520*(15), 3492–3508. http://dx.doi.org/10.1002/cne.23101.

Heckman, K. L., DeCoteau, W., Estevez, A., Reed, K. J., Costanzo, W., Sanford, D., … Erlichman, J. S. (2013). Custom cerium oxide nanoparticles protect against a free radical mediated autoimmune degenerative disease in the brain. *ACS Nano*, 7(12), 10582–10596. http://dx.doi.org/10.1021/nn403743b.

Hensel, N., Schön, A., Konen, T., Lübben, V., Förthmann, B., Baron, O., … Haffner, D. (2016 Feb 20). Fibroblast growth factor 23 signaling in hippocampal cells: Impact on neuronal morphology and synaptic density. *Journal of Neurochemistry*. http://dx.doi.org/10.1111/jnc.13585. [Epub ahead of print]. PMID:26896818.

Hinman, J. D., & Abraham, C. R. (2007). What's behind the decline? The role of white matter in brain aging. *Neurochemical Research*, *32*(12), 2023–2031. http://dx.doi.org/10.1007/s11064-007-9341-x.

Huang, Y. Y., Nguyen, P. V., Abel, T., & Kandel, E. R. (1996). Long-lasting forms of synaptic potentiation in the mammalian hippocampus. *Learning & Memory*, *3*(2–3), 74–85.

Huerta, P. T., Sun, L. D., Wilson, M. A., & Tonegawa, S. (2000). Formation of temporal memory requires NMDA receptors within CA1 pyramidal neurons. *Neuron*, *25*(2), 473–480.

Imura, A., Iwano, A., Tohyama, O., Tsuji, Y., Nozaki, K., Hashimoto, N., ... Nabeshima, Y. (2004). Secreted Klotho protein in sera and CSF: Implication for post-translational cleavage in release of Klotho protein from cell membrane. *FEBS Letters*, *565*(1–3), 143–147. http://dx.doi.org/10.1016/j.febslet.2004.03.090. S0014579304003990 [pii].

King, G. D., Chen, C., Huang, M. M., Zeldich, E., Brazee, P. L., Schuman, E. R., ... Abraham, C. R. (2012). Identification of novel small molecules that elevate Klotho expression. *The Biochemical Journal*, *441*(1), 453–461. http://dx.doi.org/10.1042/BJ20101909. BJ20101909 [pii].

King, G. D., Rosene, D. L., & Abraham, C. R. (2012). Promoter methylation and age-related downregulation of Klotho in rhesus monkey. *Age*, *34*(6), 1405–1419. http://dx.doi.org/10.1007/s11357-011-9315-4.

Kohler, S., Tschopp, O., Sze, L., Neidert, M., Bernays, R. L., Spanaus, K. S., ... Schmid, C. (2013). Monitoring for potential residual disease activity by serum insulin-like growth factor 1 and soluble Klotho in patients with acromegaly after pituitary surgery: Is there an impact of the genomic deletion of exon 3 in the growth hormone receptor (d3-GHR) gene on "safe" GH cut-off values? *General and Comparative Endocrinology*, *188*, 282–287. PMID:23648743.

Kokkinaki, M., Abu-Asab, M., Gunawardena, N., Ahern, G., Javidnia, M., Young, J., & Golestaneh, N. (2013). Klotho regulates retinal pigment epithelial functions and protects against oxidative stress. *The Journal of Neuroscience*, *33*(41), 16346–16359. http://dx.doi.org/10.1523/JNEUROSCI.0402-13.2013.

Kuro-o, M. (2012). Klotho in health and disease. *Current Opinion in Nephrology and Hypertension*, *21*(4), 362–368. http://dx.doi.org/10.1097/MNH.0b013e32835422ad.

Kuro-o, M., Matsumura, Y., Aizawa, H., Kawaguchi, H., Suga, T., Utsugi, T., ... Nabeshima, Y. I. (1997). Mutation of the mouse klotho gene leads to a syndrome resembling ageing. *Nature*, *390*(6655), 45–51. http://dx.doi.org/10.1038/36285.

Kurosu, H., Ogawa, Y., Miyoshi, M., Yamamoto, M., Nandi, A., Rosenblatt, K. P., ... Kuro-o, M. (2006). Regulation of fibroblast growth factor-23 signaling by klotho. *The Journal of Biological Chemistry*, *281*(10), 6120 6123. http://dx.doi.org/10.1074/jbc.C500457200.

Kurosu, H., Yamamoto, M., Clark, J. D., Pastor, J. V., Nandi, A., Gurnani, P., ... Kuro-o, M. (2005). Suppression of aging in mice by the hormone Klotho. *Science*, *309*(5742), 1829–1833. http://dx.doi.org/10.1126/science.1112766. C500457200 [pii].

Lau, A., & Tymianski, M. (2010). Glutamate receptors, neurotoxicity and neurodegeneration. *Pflügers Archiv*, *460*(2), 525–542. http://dx.doi.org/10.1007/s00424-010-0809-1.

Li, S. A., Watanabe, M., Yamada, H., Nagai, A., Kinuta, M., & Takei, K. (2004). Immunohistochemical localization of Klotho protein in brain, kidney, and reproductive organs of mice. *Cell Structure and Function*, *29*(4), 91–99. JST.JSTAGE/csf/29.91 [pii].

Madero, M., Gul, A., & Sarnak, M. J. (2008). Cognitive function in chronic kidney disease. *Seminars in Dialysis*, *21*(1), 29–37. http://dx.doi.org/10.1111/j.1525-139X.2007. 00384.x.

Makris, N., Papadimitriou, G. M., van der Kouwe, A., Kennedy, D. N., Hodge, S. M., Dale, A. M., ... Rosene, D. L. (2007). Frontal connections and cognitive changes in normal aging rhesus monkeys: A DTI study. *Neurobiology of Aging*, *28*(10), 1556–1567. http://dx.doi.org/10.1016/j.neurobiolaging.2006.07.005.

Malenka, R. C., & Nicoll, R. A. (1993). NMDA-receptor-dependent synaptic plasticity: Multiple forms and mechanisms. *Trends in Neurosciences*, *16*(12), 521–527.

Massó, A., Sánchez, A., Gimenez-Llort, L., Lizcano, J. M., Cañete, M., García, B., ... Chillon, M. (2015). Secreted and transmembrane αklotho isoforms have different spatio-temporal profiles in the brain during aging and Alzheimer's disease progression. *PLoS One*, *10*(11), e0143623. http://dx.doi.org/10.1371/journal.pone.0143623. eCollection 2015. PMID:26599613.

Moreno, J. A., Izquierdo, M. C., Sanchez-Nino, M. D., Suarez-Alvarez, B., Lopez-Larrea, C., Jakubowski, A., … Sanz, A. B. (2011). The inflammatory cytokines TWEAK and TNFα reduce renal klotho expression through NFκB. *Journal of the American Society of Nephrology, 22*(7), 1315–1325. http://dx.doi.org/10.1681/ASN.2010101073.

Morris, R. G., Anderson, E., Lynch, G. S., & Baudry, M. (1986). Selective impairment of learning and blockade of long-term potentiation by an N-methyl-D-aspartate receptor antagonist, AP5. *Nature, 319*(6056), 774–776.

Nagai, T., Yamada, K., Kim, H. C., Kim, Y. S., Noda, Y., Imura, A., … Nabeshima, T. (2003). Cognition impairment in the genetic model of aging klotho gene mutant mice: A role of oxidative stress. *FASEB Journal, 17*(1), 50–52. http://dx.doi.org/10.1096/fj.02-0448fje.

Park, S. J., Shin, E. J., Min, S. S., An, J., Li, Z., Hee Chung, Y., … Kim, H. C. (2013). Inactivation of JAK2/STAT3 signaling axis and downregulation of M1 mAChR cause cognitive impairment in klotho mutant mice, a genetic model of aging. *Neuropsychopharmacology, 38*(8), 1426–1437. http://dx.doi.org/10.1038/npp.2013.39.

Peters, A. (1996). Age-related changes in oligodendrocytes in monkey cerebral cortex. *The Journal of Comparative Neurology, 371*(1), 153–163. http://dx.doi.org/10.1002/(SICI)1096-9861(19960715)371:1<153::AID-CNE9>3.0.CO;2-2.

Peters, A., Leahu, D., Moss, M. B., & McNally, K. J. (1994). The effects of aging on area 46 of the frontal cortex of the rhesus monkey. *Cerebral Cortex, 4*(6), 621–635.

Peters, A., Nigro, N. J., & McNally, K. J. (1997). A further evaluation of the effect of age on striate cortex of the rhesus monkey. *Neurobiology of Aging, 18*(1), 29–36.

Popa-Wagner, A., Mitran, S., Sivanesan, S., Chang, E., & Buga, A. M. (2013). ROS and brain diseases: The good, the bad, and the ugly. *Oxidative Medicine and Cellular Longevity, 2013*, 963520. http://dx.doi.org/10.1155/2013/963520.

Reish, N. J., Maltare, A., McKeown, A. S., Laszczyk, A. M., Kraft, T. W., Gross, A. K., & King, G. D. (2013). The age-regulating protein klotho is vital to sustain retinal function. *Investigative Ophthalmology & Visual Science, 54*(10), 6675–6685. http://dx.doi.org/10.1167/iovs.13-12550.

Rizzo, B., Maltese, G., Paraskevi, M. P., Hrelia, S., Mann, G., & Siow, R. (2014). Induction of antioxidant genes by sulforaphane and klotho in human aortic smooth muscle cells. *Free Radical Biology & Medicine, 75*(Suppl. 1), S14–S15. http://dx.doi.org/10.1016/j.freeradbiomed.2014.10.588.

Rodrigues, S. M., LeDoux, J. E., & Sapolsky, R. M. (2009). The influence of stress hormones on fear circuitry. *Annual Review of Neuroscience, 32*, 289–313. http://dx.doi.org/10.1146/annurev.neuro.051508.135620.

Rondi-Reig, L., Libbey, M., Eichenbaum, H., & Tonegawa, S. (2001). CA1-specific N-methyl-D-aspartate receptor knockout mice are deficient in solving a nonspatial transverse patterning task. *Proceedings of the National Academy of Sciences of the United States of America, 98*(6), 3543–3548.

Sarvari, M., Kallo, I., Hrabovszky, E., Solymosi, N., Rodolosse, A., Vastagh, C., … Liposits, Z. (2015). Hippocampal gene expression is highly responsive to estradiol replacement in middle-aged female rats. *Endocrinology, 156*(7), 2632–2645. http://dx.doi.org/10.1210/en.2015-1109.

Scoville, W. B., & Milner, B. (2000). Loss of recent memory after bilateral hippocampal lesions. 1957. *The Journal of Neuropsychiatry and Clinical Neurosciences, 12*, 103–113.

Semba, R. D., Moghekar, A. R., Hu, J., Sun, K., Turner, R., Ferrucci, L., & O'Brien, R. (2014). Klotho in the cerebrospinal fluid of adults with and without Alzheimer's disease. *Neuroscience Letters, 558*, 37–40. http://dx.doi.org/10.1016/j.neulet.2013.10.058.

Shahmoon, S., Rubinfeld, H., Wolf, I., Cohen, Z. R., Hadani, M., Shimon, I., & Rubinek, T. (2014). The aging suppressor klotho: a potential regulator of growth hormone secretion. *American Journal of Physiology. Endocrinology and Metabolism, 307*(3), E326–E334. PMID:24939736.

Shiozaki, M., Yoshimura, K., Shibata, M., Koike, M., Matsuura, N., Uchiyama, Y., & Gotow, T. (2008). Morphological and biochemical signs of age-related neurodegenerative changes in klotho mutant mice. *Neuroscience, 152*(4), 924–941. http://dx.doi.org/10.1016/j.neuroscience.2008.01.032. S0306-4522(08)00145-0 [pii].

Slemmer, J. E., De Zeeuw, C. I., & Weber, J. T. (2005). Don't get too excited: Mechanisms of glutamate-mediated Purkinje cell death. *Progress in Brain Research, 148*, 367–390. http://dx.doi.org/10.1016/S0079-6123(04)48029-7.

Sloane, J. A., Hollander, W., Moss, M. B., Rosene, D. L., & Abraham, C. R. (1999). Increased microglial activation and protein nitration in white matter of the aging monkey. *Neurobiology of Aging, 20*(4), 395–405.

Sloane, J. A., Hollander, W., Rosene, D. L., Moss, M. B., Kemper, T., & Abraham, C. R. (2000). Astrocytic hypertrophy and altered GFAP degradation with age in subcortical white matter of the rhesus monkey. *Brain Research, 862*(1–2), 1–10.

Smith, C. C., & McMahon, L. L. (2006). Estradiol-induced increase in the magnitude of long-term potentiation is prevented by blocking NR2B-containing receptors. *The Journal of Neuroscience, 26*(33), 8517–8522. http://dx.doi.org/10.1523/JNEUROSCI.5279-05.2006.

Smith, C. C., Smith, L. A., Bredemann, T. M., & McMahon, L. L. (2015). 17beta estradiol recruits GluN2B-containing NMDARs and ERK during induction of long-term potentiation at temporoammonic-CA1 synapses. *Hippocampus, 26*(1), 110–117. http://dx.doi.org/10.1002/hipo.22495.

Sweatt, J. D. (2004). Mitogen-activated protein kinases in synaptic plasticity and memory. *Current Opinion in Neurobiology, 14*(3), 311–317. http://dx.doi.org/10.1016/j.conb.2004.04.001.

Talboom, J. S., Williams, B. J., Baxley, E. R., West, S. G., & Bimonte-Nelson, H. A. (2008). Higher levels of estradiol replacement correlate with better spatial memory in surgically menopausal young and middle-aged rats. *Neurobiology of Learning and Memory, 90*(1), 155–163. http://dx.doi.org/10.1016/j.nlm.2008.04.002.

Teocchi, M. A., Ferreira, A. E., da Luz de Oliveira, E. P., Tedeschi, H., & D'Souza-Li, L. (2013). Hippocampal gene expression dysregulation of Klotho, nuclear factor kappa B and tumor necrosis factor in temporal lobe epilepsy patients. *Journal of Neuroinflammation, 10*, 53. http://dx.doi.org/10.1186/1742-2094-10-53.

Thurston, R. D., Larmonier, C. B., Majewski, P. M., Ramalingam, R., Midura-Kiela, M., Laubitz, D., ... Ghishan, F. K. (2010). Tumor necrosis factor and interferon-gamma down-regulate Klotho in mice with colitis. *Gastroenterology, 138*(4), 1384–1394. http://dx.doi.org/10.1053/j.gastro.2009.12.002. 1394 e1381–1382.

Tohyama, O., Imura, A., Iwano, A., Freund, J. N., Henrissat, B., Fujimori, T., & Nabeshima, Y. (2004). Klotho is a novel beta-glucuronidase capable of hydrolyzing steroid beta-glucuronides. *The Journal of Biological Chemistry, 279*(11), 9777–9784. http://dx.doi.org/10.1074/jbc.M312392200.

Tu, P. H., Robinson, K. A., de Snoo, F., Eyer, J., Peterson, A., Lee, V. M., & Trojanowski, J. Q. (1997). Selective degeneration fo Purkinje cells with Lewy body-like inclusions in aged NFHLACZ transgenic mice. *The Journal of Neuroscience, 17*(3), 1064–1074.

Tucker Zhou, T. B., King, G. D., Chen, C., & Abraham, C. R. (2013). Biochemical and functional characterization of the Klotho-VS polymorphism implicated in aging and disease risk. *The Journal of Biological Chemistry, 288*(51), 36302–36311. http://dx.doi.org/10.1074/jbc.M113.490052.

Urakawa, I., Yamazaki, Y., Shimada, T., Iijima, K., Hasegawa, H., Okawa, K., ... Yamashita, T. (2006). Klotho converts canonical FGF receptor into a specific receptor for FGF23. *Nature*, *444*(7120), 770–774. http://dx.doi.org/10.1038/nature05315. nature05315 [pii].

West, M. J. (1993). Regionally specific loss of neurons in the aging human hippocampus. *Neurobiology of Aging*, *14*(4), 287–293.

Woolley, C. S., Weiland, N. G., McEwen, B. S., & Schwartzkroin, P. A. (1997). Estradiol increases the sensitivity of hippocampal CA1 pyramidal cells to NMDA receptor-mediated synaptic input: Correlation with dendritic spine density. *The Journal of Neuroscience*, *17*(5), 1848–1859.

Yamamoto, M., Clark, J. D., Pastor, J. V., Gurnani, P., Nandi, A., Kurosu, H., … Kuro-o, M. (2005). Regulation of oxidative stress by the anti-aging hormone klotho. *The Journal of Biological Chemistry*, *280*(45), 38029–38034. http://dx.doi.org/10.1074/jbc.M509039200. M509039200 [pii].

Yun, S. H., Park, K. A., Kwon, S., Woolley, C. S., Sullivan, P. M., Pasternak, J. F., & Trommer, B. L. (2007). Estradiol enhances long term potentiation in hippocampal slices from aged apoE4-TR mice. *Hippocampus*, *17*(12), 1153–1157. http://dx.doi.org/10.1002/hipo.20357.

Zeldich, E., Chen, C. D., Avila, R., Medicetty, S., & Abraham, C. R. (2015). The anti-aging protein Klotho enhances remyelination following cuprizone-induced demyelination. *Journal of Molecular Neuroscience*, *57*(2), 185–196. http://dx.doi.org/10.1007/s12031-015-0598-2.

Zeldich, E., Chen, C. D., Colvin, T. A., Bove-Fenderson, E. A., Liang, J., Tucker Zhou, T. B., … Abraham, C. R. (2014). The neuroprotective effect of Klotho is mediated via regulation of members of the redox system. *The Journal of Biological Chemistry*, *289*, 24700–24715. http://dx.doi.org/10.1074/jbc.M114.567321.

CHAPTER TEN

Function and Change with Aging of α-Klotho in the Kidney

K. Akasaka-Manya, H. Manya, T. Endo[1]
Molecular Glycobiology, Research Team for Mechanism of Aging, Tokyo Metropolitan Geriatric Hospital and Institute of Gerontology, Tokyo, Japan
[1]Corresponding author: e-mail address: endo@tmig.or.jp

Contents

Abstract

The α-Klotho mouse is an animal model that prematurely shows phenotypes resembling human aging, such as osteoporosis, arteriosclerosis, pulmonary emphysema, and kidney damage. Interestingly, these abnormalities are triggered by a deficiency of a single protein, α-Klotho. The kidney is an organ that highly expresses α-Klotho, suggesting that α-Klotho is important for kidney function. Recent studies suggest that α-Klotho is associated with phosphate, vitamin D, and calcium homeostasis. The calcium imbalance in α-Klotho mice may induce calpain overactivation, leading to cell death and tissue destruction. α-Klotho is predicted to have glycosidase activity, capable of modifying the N-glycans of channels and transporters and regulating transmembrane movement of several ions, including calcium. Interestingly, N-glycan changes are observed in the kidney of α-Klotho mice and normal aged mice in association with decreased α-Klotho levels. These results imply that glycobiology and α-Klotho function are interesting targets for future studies.

Vitamins and Hormones, Volume 101
ISSN 0083-6729
http://dx.doi.org/10.1016/bs.vh.2016.02.006

1. INTRODUCTION

α-Klotho is a type I single-pass membrane protein with a short intracellular domain (Kuro-o et al., 1997). In its long extracellular region, the α-Klotho protein contains two internal repeats that are homologous to the amino acid sequence of β-glucosidase, although each domain lacks the glutamic acid residues necessary for enzyme activity (Kuro-o et al., 1997). α-Klotho is predominantly expressed in the renal tubules of the kidney, epithelial cells of the choroid plexus in the brain, and parathyroid glands. However, the abnormalities of α-Klotho mice are observed not only in these organs but also in other organs, such as the lung, heart, bone, gastric wall, and skin. The α-Klotho mouse is an animal model that prematurely shows phenotypes resembling human aging triggered by a deficit of α-Klotho, which is caused by a mutation—the insertion of ectopic DNA into the 5′ promoter region of the *α-Klotho* gene. α-Klotho mice appear normal at birth but show multiple age-associated abnormalities, including the following: arteriosclerosis; osteoporosis; muscle and skin atrophy; hyperphosphatemia; pulmonary emphysema; ectopic vascular and soft tissue calcification in the lung, heart, and skin; depletion of Purkinje cells in the cerebellum; and infertility after 3 and 4 weeks. α-Klotho mice die at approximately 2 months old, whereas wild-type mice live for approximately 3 years. The overexpression of α-Klotho extends the life span in mice by 20–30% (Kurosu et al., 2005). These facts suggest that α-Klotho functions in the suppression of aging.

A number of changes in the kidney are observed during the aging process. Even without suffering from diseases such as hypertension or diabetes mellitus, renal morphology and function are affected by aging. In the aged kidney, there are glomerulosclerosis, interstitial fibrosis, arteriosclerosis, loss of renal mass, reduction in glomerular filtration rate, increased vascular resistance, degeneration in urine concentration mechanism, and acid–base imbalance (Percy, Power, & Gobe, 2008; Yang & Fogo, 2010; Zhou et al., 2008). These symptoms exacerbate the development of chronic kidney disease (CKD). Here, we describe the recent progress of the understanding of α-Klotho function in the kidney and its change with aging.

2. α-KLOTHO EXPRESSION IN THE KIDNEY

α-Klotho is expressed in the highest abundance in the kidney compared with other organs (Kuro-o et al., 1997). Immunohistological analysis showed that α-Klotho localizes to the renal tubular epithelia in distal convoluted and connecting tubules in the kidney, which are nephron segments that are responsible for transepithelial calcium transport (Kato et al., 2000; Li et al., 2004). The α-Klotho protein is observed as fine puncta in the cytoplasm of the tubular epithelia (Li et al., 2004). Previously, α-Klotho was thought to be exclusively expressed in the distal tubules, but two groups recently reported that α-Klotho also localizes to the proximal tubules (Andrukhova et al., 2012; Hu, Shi, Zhang, Pastor, et al., 2010).

Deficiency of α-Klotho causes disorders not only in organs where α-Klotho is expressed but also in other organs, such as the lung and skin, where α-Klotho is not expressed, indicating that both a membrane-bound α-Klotho and the circulating form of α-Klotho protein should be present. In fact, soluble α-Klotho is present in the blood, urine, and cerebrospinal fluid of rodents and humans (Hu, Shi, Zhang, Pastor, et al., 2010; Imura et al., 2004; Li et al., 2004). Membrane-anchored α-Klotho is cleaved by a disintegrin and metalloproteinase (ADAM) family, ADAM10 and ADAM17 (Chen, Podvin, Gillespie, Leeman, & Abraham, 2007), and β-secretase BACE-1 (Bloch et al., 2009). Thus, α-Klotho exists in both a membranous form and a soluble form. Recently, it has been found that the kidney is the principal organ in the circulation of α-Klotho production and the clearing of α-Klotho from the blood into the urine (Hu et al., 2015). It is of interest that clinical and experimental animal studies show reduced kidney α-Klotho expression in both acute kidney injury and CKD (Asai et al., 2012; Hu, Shi, Zhang, Quinones, et al., 2010; Koh et al., 2001); in the future, it will be important to elucidate the mechanism of this reduction. Additionally, it is of interest to determine whether the upregulation of α-Klotho production via external sources improves the kidney function when renal production fails. The decrease in α-Klotho protein expression with age may be associated with the deterioration of kidney function during aging (Akasaka-Manya, Manya, Kizuka, Oka, & Endo, 2014; Manya et al., 2002).

3. α-KLOTHO, PHOSPHATE, AND VITAMIN D

3.1 α-Klotho and FGF23

The FGF (fibroblast growth factor) superfamily is a group of polypeptide growth factors with multiple functions, such as cell proliferation and differentiation (Bottcher & Niehrs, 2005; Ornitz & Itoh, 2001; Ornitz & Marie, 2002). FGF23 is the newest member of the FGF superfamily. FGF23 maintains homeostasis in phosphate and vitamin D_3 metabolism in the kidney by regulating the sodium phosphate cotransporter (NaPi-IIa), vitamin D_3-metabolizing enzymes, 25-hydroxyvitamin D-1α-hydrolase (CYP27B1), the vitamin D-activating enzyme, and 24-hydroxylase (CYP24A1), which inactivates vitamin D (ADHRConsortium, 2000; Baum, Schiavi, Dwarakanath, & Quigley, 2005; Shimada et al., 2001; Shimada et al., 2005; White et al., 2001). Additionally, FGF23 abnormality causes autosomal dominant hypophosphatemic rickets (ADHR; ADHRConsortium, 2000). In FGF23 transgenic mice, a high level of FGF23 causes hypophosphatemia and a reduction in serum $1,25(OH)_2D_3$ (Larsson et al., 2004). These phenomena are induced by the binding of FGF23 to its specific receptor (FGFR). Only four FGFR genes are known, in contrast to many types of FGFs (Zhang et al., 2006). Two groups previously reported that the α-Klotho protein induces a specific interaction between FGF23 and FGFRs, although the results of the two studies were slightly different (Kurosu et al., 2006; Urakawa et al., 2006). One study showed that α-Klotho binds to FGFR1c, FGFR3c, and FGFR4 (Kurosu et al., 2006), and the other study showed that α-Klotho binds to only FGFR1(IIIc) (Urakawa et al., 2006). Additional studies are required to understand how and which FGFR is definitively and preferentially used in FGF23 signaling. Although the difference is unclear, α-Klotho protein seems to help FGF23 transmit signals into cells by converting canonical FGFR into a specific receptor for FGF23 (Fig. 1A).

The serum level of $1,25(OH)_2D_3$ in α-Klotho mice is higher than in wild-type mice (Kuro-o et al., 1997; Yoshida, Fujimori, & Nabeshima, 2002). CYP27B1 is suppressed via a negative feedback mechanism when the serum $1,25(OH)_2D_3$ level is high. However, in α-Klotho mice, CYP27B1 is not suppressed, even if the serum $1,25(OH)_2D_3$ level is high (Yoshida et al., 2002), suggesting that α-Klotho is a physiological regulator of $1,25(OH)_2D_3$ homeostasis. It is of interest that both α-Klotho mice and FGF23-deficient mice show multiple phenotypes resembling premature human aging (Kuro-o et al., 1997; Sitara et al., 2004; Yoshida et al., 2002). Both mice also show aberrant $1,25(OH)_2D_3$ synthesis, leading to the elevation of serum concentrations of phosphate and $1,25(OH)_2D_3$. FGF23 upregulates *CYP27B1* transcription and inhibits *CYP24A1* in the kidney (Chanakul et al., 2013). Because α-Klotho induces a specific interaction between FGF23 and FGFR (Fig. 1A), the activation of the membranous α-Klotho–FGFR–FGF23 complex may be crucial for the function of FGF23 and α-Klotho. Therefore, both α-Klotho mice and FGF23-deficient

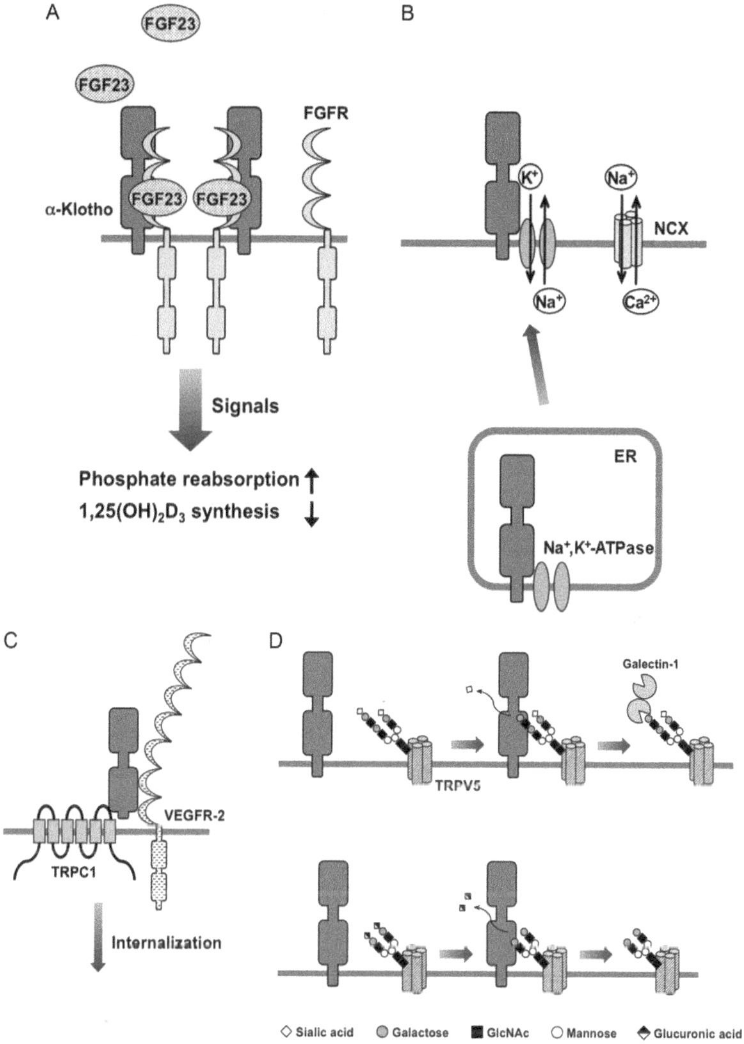

Fig. 1 Proposed α-Klotho functions in the kidney. (A) α-Klotho protein binds to FGFR and converts it into an FGF23-specific receptor. The FGF23–α-Klotho–FGFR complex is necessary for the signaling of phosphate and vitamin D homeostasis. (B) α-Klotho binds to the Na^+,K^+-ATPase complex in the endoplasmic reticulum (ER) and recruits it to the plasma membrane. The accumulation of Na^+,K^+-ATPase in the plasma membrane creates the sodium gradient that upregulates calcium reabsorption in cooperation with NCX-1 (sodium calcium exchanger-1). (C) α-Klotho forms a complex with TRPC-1 (transient receptor potential canonical-1) and VEGFR-2 (vascular endothelial growth factor receptor-2) in the vascular endothelium. α-Klotho promotes the internalization of the complex in response to VEGF stimulation. (D) α-Klotho maintains the calcium channels in the plasma membrane by hydrolyzing N-glycans on transient receptor potential vanilloid 5 (TRPV5). α-Klotho is assumed to modify N-glycans by sialidase (upper flow) and/or β-glucuronidase (lower flow) activity. Galectin-1 binds to the newly appeared LacNAc structure on TRPV5 and then efficiently retains it in the membrane, resulting in the increase in calcium transport. *Tandem shadow box* in the figure indicates α-Klotho protein. *GlcNAc*, *N*-acetylglucosamine.

mice show common phenotypes. The following observations also support the cooperation of α-Klotho and FGF23 in signaling. The serum concentration of FGF23 in α-Klotho mice is 150–2000-fold higher than that in wild-type mice (Segawa et al., 2007; Urakawa et al., 2006). Despite the high concentration of FGF23 in α-Klotho mice, FGF23 does not function properly in α-Klotho mice. FGF23 is a regulator of phosphate homeostasis (Shimada et al., 2004), but the serum phosphate concentration in α-Klotho mice is markedly high. Taken together, the common phenomena that are observed in α-Klotho mice and FGF23-deficient mice may be attributed to the disruption of the FGF23 signaling pathway caused by the loss of FGF23 or α-Klotho.

As previously described, α-Klotho mice exhibit hyperphosphatemia (Kuro-o et al., 1997). Phosphate homeostasis is mainly regulated by reabsorption in the kidney. Renal phosphate reabsorption is conducted by type IIa and type IIc sodium-dependent Pi cotransporters (NaPi-IIa and NaPi-IIc). Seventy to eighty percent of phosphate is reabsorbed in the proximal tubules, whereas the remainder occurs in the distal tubules in the kidney (Miyamoto, Segawa, Ito, & Kuwahata, 2004). NaPi-IIa and NaPi-IIc localize to the apical membrane of proximal tubular cells (Murer, Hernando, Forster, & Biber, 2003), and NaPi-IIa expression is regulated by PTH (parathyroid hormone), $1,25(OH)_2D_3$, and FGF23, whereas NaPi-IIc expression is regulated only by $1,25(OH)_2D_3$ (Friedlaender et al., 2001; Pfister et al., 1998; Segawa et al., 2003). However, the expression of NaPi-IIa and NaPi-IIc in the proximal tubules is decreased in FGF23 transgenic mice (Larsson et al., 2004; Saito et al., 2003). Furthermore, the translocation of NaPi-IIa to the apical plasma membrane is aberrant in α-Klotho mice (Li et al., 2004). In wild-type mice, NaPi-IIa is present both in the apical plasma membrane and inside the cells. However, most of the NaPi-IIa inside cells is absent in α-Klotho mice. These results suggest that FGF23 signaling through the α-Klotho protein regulates NaPi cotransporter activity. In addition, Gattineni et al. reported that FGF23 utilizes different FGFRs: FGF23 suppresses $1,25(OH)_2D_3$ synthesis via FGFR3 and FGFR4, whereas FGF23 inhibits phosphate transport through NaPi-IIa and NaPi-IIc via FGFR1 (Gattineni et al., 2009; Gattineni & Baum, 2010). However, this hypothesis seems unlikely because of the location of target molecules of FGF23; for example, NaPi-IIa and α-Klotho localize differently. The former localizes to the proximal tubules, whereas the latter localizes mainly to the distal tubules. Therefore, two possibilities should be considered: one possibility is that a small quantity of α-Klotho in the proximal tubule functions as an FGF23 coreceptor, and another is that secreted α-Klotho (not membranous α-Klotho) from the distal tubule acts as an FGF23 coreceptor.

FGF23 is believed to act as a circulating metabolic regulator and is a physiological regulator of serum phosphate and $1,25(OH)_2D_3$. These important regulatory functions of FGF23 seem to be associated with α-Klotho. As described earlier, the α-Klotho mice and FGF23 knockout mice show elevated serum concentrations of phosphate and $1,25(OH)_2D_3$. The reduction in serum concentrations of $1,25(OH)_2D_3$ with vitamin D-deficient diets rescued the abnormal phenotypes in α-Klotho mice (Tsujikawa, Kurotaki, Fujimori, Fukuda, & Nabeshima, 2003). In addition, CYP27B1 deficiency in both mice dispersed serum $1,25(OH)_2D_3$ and ameliorated age-related features (Ohnishi, Nakatani, Lanske, & Razzaque, 2009; Razzaque, Sitara, Taguchi, St-Arnaud, & Lanske, 2006). NaPi-IIa and α-Klotho double-knockout mice demonstrated an improvement in the symptoms observed in α-Klotho mice (Ohnishi & Razzaque, 2010). Finally, dietary phosphate restriction rescued premature aging in α-Klotho mice (Morishita et al., 2001). Taken together, α-Klotho interacts with FGFR and functions as a coreceptor for FGF23, and this complex formation is essential for signaling to exhibit the biological functions of FGF23, vitamin D, and phosphate homeostasis.

3.2 α-Klotho and Na^+,K^+-ATPase

PTH is an important hormone in calcium homeostasis. Calcium reabsorption in the kidney is regulated by PTH in the distal convoluted tubules. PTH is secreted when the serum calcium concentration is decreased, after which PTH accelerates the calcium absorption in the renal tubules. The serum level of calcium in α-Klotho mice is higher than that in wild-type mice, and the PTH concentration is lower (Kuro-o et al., 1997; Yoshida et al., 2002). In α-Klotho mice, PTH-induced calcium absorption from the lumen is impaired (Tsuruoka et al., 2006), and the response of PTH secretion to low serum calcium is decreased (Imura et al., 2007). These data indicate that α-Klotho plays important roles in the calcium homeostasis regulatory network.

Na^+,K^+-ATPase is expressed in the distal convoluted tubules, and the association of α-Klotho with Na^+,K^+-ATPase regulates calcium homeostasis (Imura et al., 2007). The α-Klotho protein binds to the α1 subunit of Na^+,K^+-ATPase in the ER, Golgi apparatus, and early endosome in the kidney and, after binding α-Klotho, transports Na^+,K^+-ATPase from these intracellular organelles to the cell surface depending on the extracellular calcium concentration (Fig. 1B). Thus, cell-surface Na^+,K^+-ATPase accumulates rapidly in response to a low concentration of calcium. However, the rapid recruitment of Na^+,K^+-ATPase to the cell surface in response to low calcium is not observed in α-Klotho mice, and the amount of cell-surface Na^+,K^+-ATPase in α-Klotho mice is lower than that in wild-type mice. An increase in Na^+,K^+-ATPase creates the sodium gradient that drives calcium reabsorption in cooperation with NCX-1, sodium calcium exchanger-1, which is highly expressed in the apical membrane of the epithelium. Hence, impaired calcium reabsorption is caused by the reduction of surface Na^+,K^+-ATPase in α-Klotho mice. When α-Klotho is required for the rapid recruitment of Na^+,K^+-ATPase to the cell surface in response to the calcium concentration, the increased Na^+ gradient that is created by Na^+,K^+-ATPase activity may drive calcium transport in cooperation with ion channels and transporters in the kidney. Thus, α-Klotho regulates calcium homeostasis.

An interaction between Na^+,K^+-ATPase and α-Klotho is also observed in the parathyroid glands. In wild-type mice, decreased serum calcium causes increased PTH secretion but decreased PTH in α-Klotho mice, suggesting that the reduced response of serum PTH in α-Klotho mice is largely caused by an impairment in the Na^+,K^+-ATPase-dependent secretion of PTH. Thus, Na^+,K^+-ATPase and α-Klotho are involved in PTH secretion.

4. α-KLOTHO AND CALPAIN

4.1 Calpain Overactivation in the Kidney

We previously reported that the level of αII-spectrin, a cell cytoskeletal protein, is decreased in the kidneys of 4-week-old α-Klotho mice compared with wild-type mice (Manya et al., 2002). αII-spectrin is cleaved at a specific site by calpain, and the breakdown products are detected only in α-Klotho mice (Fig. 2A). These results indicate that αII-spectrin is degraded by calpain in α-Klotho mice. Calpain is a calcium-dependent protease that localizes to the cytosol and is involved in various proteolytic physiological events (Vanderklish & Bahr, 2000). There are two types of calpain isozymes: m-calpain and μ-calpain. The former is activated by a millimolar concentration of calcium and the latter by micromolar calcium. Interestingly, μ-calpain but not m-calpain is specifically activated in α-Klotho mice (Manya et al., 2002), suggesting that the elevated calcium concentration

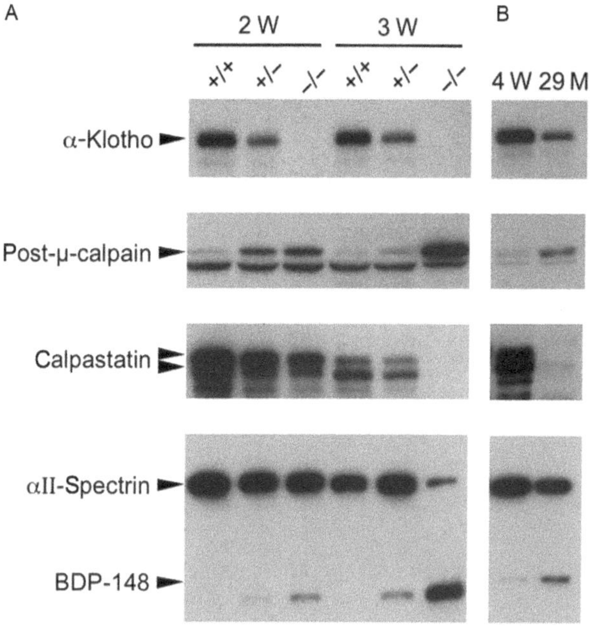

Fig. 2 Alterations in the expression of calpain, calpastatin, and αII-spectrin in 2- and 3-week-old α-Klotho mice (A) and normal young (*4 W*, 4 weeks) and aged mice (*29 M*, 29 months) (B). The active μ-calpain and breakdown product of αII-spectrin increase in α-Klotho mice (A). Similar phenomena are observed in normal aged mice (B). *Post-μ-calpain*, active μ-calpain; *BDP-148*, breakdown product of αII-spectrin; *2 W*, 2 weeks; *3 W*, 3 weeks; +/+, wild-type mice; +/−, heterozygous mice; −/−, α-Klotho mice. *Modified from Manya, H., Inomata, M., Fujimori, T., Dohmae, N., Sato, Y., Takio, K., et al. (2002). Klotho protein deficiency leads to overactivation of mu-calpain.* The Journal of Biological Chemistry, 277, *35503–35508. Copyright (2002) The American Society for Biochemistry and Molecular Biology, USA.*

inside cell may be micromolar rather than millimolar. Calpastatin, which is an endogenous inhibitor of calpain (Sorimachi, Ishiura, & Suzuki, 1997), is barely detected α-Klotho mouse kidneys. In α-Klotho mice, the amount of activated μ-calpain (postcalpain) and cleaved αII-spectrin increases gradually from 2 to 4 weeks after birth. In 2-week-old α Klotho mice, αII-spectrin expression is largely unaffected, although μ-calpain is already activated and the calpastatin level is low. By contrast, the amounts of calpastatin and inactivated μ-calpain (precalpain) are decreased and undetectable in 3-week-old mice (Fig. 2A). In addition, calcification is observed in the kidney of 4-week-old α-Klotho mice, but morphological changes are not.

μ-Calpain is gradually activated, and calpastatin is gradually decreased during pathological change; finally, the intensive cleavage of αII-spectrin is observed after the complete disappearance of calpastatin. No calcification is observed in 3-week-old α-Klotho mice, indicating that the degradation of αII-spectrin in the kidney precedes any tissue damage. It is noteworthy that the upregulation of activated μ-calpain is also observed in normal aged mice (29 months; Fig. 2B). α-Klotho expression is decreased in association with aging. Therefore, kidney impairment along with aging is attributed to aberrant μ-calpain activation that is caused by the decreased expression of α-Klotho.

4.2 Calpain as a Therapeutic Target for Aging-Related Disorders

Recently, it has been reported that the inhibition of overactivated μ-calpain ameliorates aging-related phenotypes in mice deficient in the α-*Klotho* gene (Nabeshima et al., 2014). The mice are treated with BDA-410, which inhibits μ-calpain rather than m-calpain. The administration of BDA-410 starts at 2 weeks of age, before the appearance of abnormalities in various tissues. All of the phenotypes, such as pulmonary emphysema, skin aging, decrease in bone mineral density, ectopic calcification in aorta, and atrophic changes in organs, are improved in BDA-410-treated mice. In the kidney, vascular calcification becomes undetectable, although tiny calcium precipitates are still detected in the cortical layer in nephron segments. Furthermore, it is of interest that the administration of BDA-410 beginning at 4 weeks of age is effective even after the appearance of characteristic abnormalities. The administration of BDA-410 shows no effects on the serum levels of $1,25(OH)_2D_3$, phosphate, or calcium in α-Klotho mice, and these levels are still higher than those in wild-type mice, indicating that μ-calpain is likely activated downstream of $1,25(OH)_2D_3$ and hyperphosphatemia (Nabeshima et al., 2014). This study strengthens our hypothesis that the hyperactivation of μ-calpain is responsible for the age-associated tissue damages and that μ-calpain would be a novel therapeutic target for aging-related disorders.

4.3 Calpain Overactivation via Aberrant Calcium Transport

μ-Calpain is responsible for tissue damage in α-Klotho mice, but the question is how μ-calpain is activated in their cells (Manya et al., 2002). In other words, how is the calcium concentration inside the cell elevated? As described earlier, calcium transport change may be one reason that the decrease in α-Klotho induces the deterioration of calcium homeostasis.

Another reason is that α-Klotho forms a complex with transient receptor potential canonical-1 (TRPC-1) and vascular endothelial growth factor receptor-2 (VEGF receptor-2) and promotes the internalization of the complex in response to VEGF stimulation vascular endothelium (Fig. 1C; Kusaba et al., 2010). Thus, α-Klotho downregulates TRPC-1 expression on the cell surface and regulates calcium influx. In α-Klotho-deficient endothelial cells, the VEGF-mediated internalization of the VEGFR-2/TRPC-1 complex is impaired, and the increase of surface TRPC-1 causes the elevation of calcium influx. μ-Calpain and caspase-3 are activated by increased calcium influx, which leads to the apoptosis of endothelial cells and then to vascular hyperpermeability. Similar sequential changes in calcium influx may occur in the aged kidney, and μ-calpain may be overactivated inside the cell.

5. α-KLOTHO AND NONSULFATED HNK-1 GLYCAN IN THE AGING KIDNEY

Although the α-Klotho protein has homologous amino acid sequences to β-glucosidase, α-Klotho was thought originally not to have any glycosidase activity because it lacks the glutamic acid residues that are necessary for enzyme activity (Kuro-o et al., 1997). However, α-Klotho has weak β-glucuronidase activity and hydrolyzes steroid glucuronides (Tohyama et al., 2004). Several studies have examined whether α-Klotho has any glycosidase activity. α-Klotho hydrolyzes the extracellular N-glycans on the calcium channel, transient receptor potential vanilloid 5 (TRPV5), to entrap and activate TRPV5 at the plasma membrane in the kidney (Fig. 1D; Cha et al., 2008; Chang et al., 2005). However, β-glucuronidase activity is extremely low, and it is not clear whether α-Klotho itself actually hydrolyzes the β-glucuronic acid linkage on N-glycan of TRPV5. However, another study showed that α-Klotho cleaves terminal α2,6-linked sialic acids of the N-glycans of TRPV5, and the cleavage of sialic acids exposes galactose β1–4-*N*-acetylglucosamine (LacNAc), a ligand for galectin-1 (Cha et al., 2008). Galectin-1 binds to TRPV5 through the newly exposed LacNAc structure and accumulates TRPV5 in the plasma membrane by preventing endocytosis. The accumulation of TRPV5 thus enhances calcium transport activity (Fig. 1D). It is of interest that a single protein α-Klotho has different activities of glycosidase, β-glucuronidase, and sialidase. By contrast, Leunissen et al. (2013) compared the effect of α-Klotho and sialidase on TRPV5 stabilization in the plasma membrane and found that sialidase stabilized TRPV5 by the inhibition of lipid raft-mediated internalization, whereas α-Klotho stabilized TRPV5 depending on its N-glycan. That is, TRPV5 stabilization in the plasma membrane by α-Klotho and sialidase may be regulated in different manners. Collectively, these studies are interesting but controversial. Furthermore, Hu et al. reported that α-Klotho deglycosylates the NaPi-IIa protein (Hu, Shi, Zhang, Pastor, et al., 2010). These authors showed that the addition of α-Klotho or β-glucuronidase to the brush border membrane vesicle of the kidney decreases sodium-dependent phosphate uptake activity and that β-glucuronidase inhibitor abrogates the inhibition of transport activity by both α-Klotho and β-glucuronidase, suggesting that α-Klotho may be a β-glucuronidase. However, it is unknown whether NaPi-IIa itself is a substrate of β-glucuronidase or α-Klotho. Additional studies are necessary to determine whether α-Klotho carries the enzymatic activity or indirectly activates an endogenous uncharacterized glycosidase(s).

Several proteins carrying the nonsulfated human natural killer-1 (HNK-1) glyco-epitope (GlcAβ1–3Galβ1–4GlcNAc) are present in the kidney (Tagawa et al., 2005; Fig. 3A). Nonsulfated HNK-1 is a kidney-specific glyco-epitope and has glucuronic acid on its nonreducing terminus. Therefore, we examined whether the nonsulfated HNK-1 glyco-epitope is changed in α-Klotho mice, and an increase in nonsulfated HNK-1 glyco-epitope was found in α-Klotho mice (Fig. 3B; Akasaka-Manya et al., 2014). It is interesting that this phenomenon is also observed in normal aged mice and is correlated with decreased α-Klotho expression (Fig. 3C). The distribution of increased nonsulfated HNK-1 glyco-epitope coincides with α-Klotho protein expression in the renal cortex. To clarify the cause of the α-Klotho-associated increase in the nonsulfated HNK-1 glyco-epitope, the expression of glucuronyltransferase S (GlcAT-S) and the activity of β-glucuronidase were also examined. GlcAT-S is the key enzyme involved in the formation of nonsulfated HNK-1 carbohydrate in the kidney (Tagawa et al., 2005); however, its expression level is not increased in α-Klotho mice compared with wild-type mice. Therefore, it does not appear that GlcAT-S induces the accumulation of the nonsulfated HNK-1 glyco-epitope. Moreover, β-glucuronidase activity is lower in α-Klotho mice than in wild-type mice. These results suggest that increased nonsulfated HNK-1 epitope levels in α-Klotho mice may be due to decreased β-glucuronidase activity. However, whether this activity is related directly to α-Klotho protein is not clear. The relationship between renal function and increased levels of nonsulfated HNK-1 glyco-epitope during aging is important in the future.

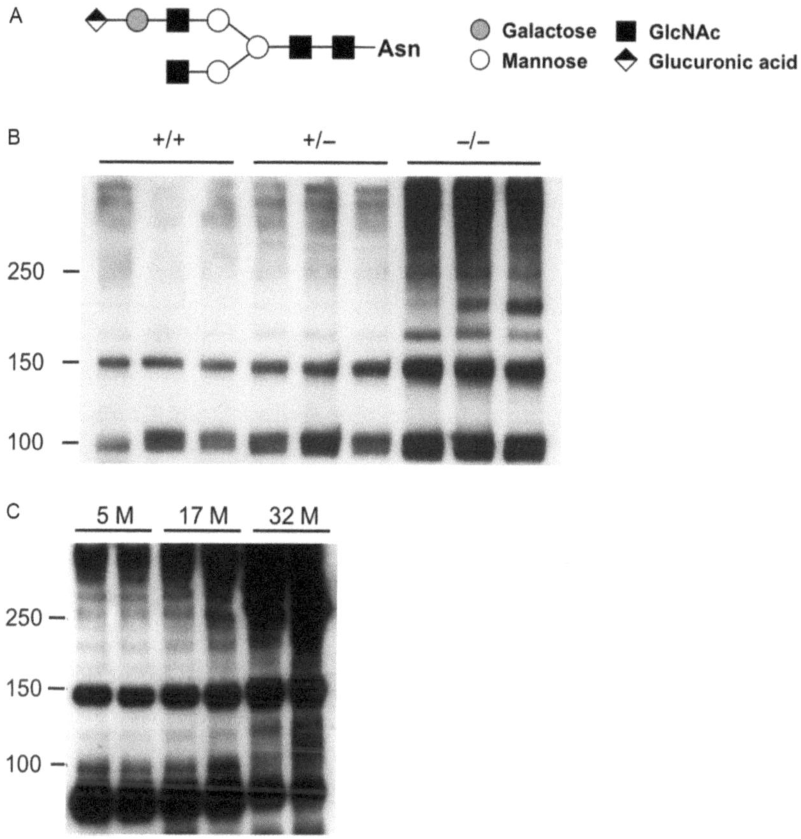

Fig. 3 Typical structure of N-glycan containing nonsulfated HNK-1 (A), the expression of the nonsulfated HNK-1 glyco-epitope in α-Klotho mice (B), and wild-type mice at different ages (*5 M*, 5 months; *17 M*, 17 months; *32 M*, 32 months) (C). Nonsulfated HNK-1 glyco-epitope expression is increased in α-Klotho mice and in wild-type mice during aging. +/+, wild-type mice; +/−, heterozygous; −/−, α-Klotho mice; *GlcNAc*, *N*-acetylglucosamine. *Modified from Akasaka-Manya, K., Manya, H., Kizuka, Y., Oka, S., & Endo, T. (2014). alpha-Klotho mice demonstrate increased expression of the nonsulfated N-glycan form of the HNK-1 glyco-epitope in kidney tissue.* The Journal of Biochemistry, *156, 107–113. Copyright (2014) The Japanese Biochemical Society, Japan.*

6. CONCLUSION

The kidney is an important organ that functions in the control of the acid–base balance, the regulation of blood pressure, and the excretion of excess organic molecules as waste products. α-Klotho is expressed most abundantly in the kidney and decreases with age. Furthermore, α-Klotho disruption causes kidney damage similar to that found in the aged kidney.

Many studies have reported that α-Klotho is indispensable to the maintenance of kidney function and have shown that α-Klotho functions in calcium and phosphate homeostasis and vitamin D synthesis. One way to express its function is that α-Klotho may modulate the functional molecule, receptors, and channels after forming a complex (Fig. 1A–C). Another possibility is that α-Klotho may modify the N-glycan structures of receptors and channels as glycosidase does (Fig. 1D). However, the details are unclear yet. Although further studies are required to determine whether both mechanisms are correct, the proposed relationship between α-Klotho deficiency and renal damage is shown in Fig. 4. Because μ-calpain inhibitor ameliorates aging-related symptoms, μ-calpain overactivation seems to be the final stage of kidney tissue dysfunction. Interestingly, although similar phenomena occur in the aging kidney, it is unclear whether α-Klotho is the sole molecule responsible for kidney deterioration during aging. Further investigations are necessary to elucidate the functions of α-Klotho during the normal aging process in the kidney.

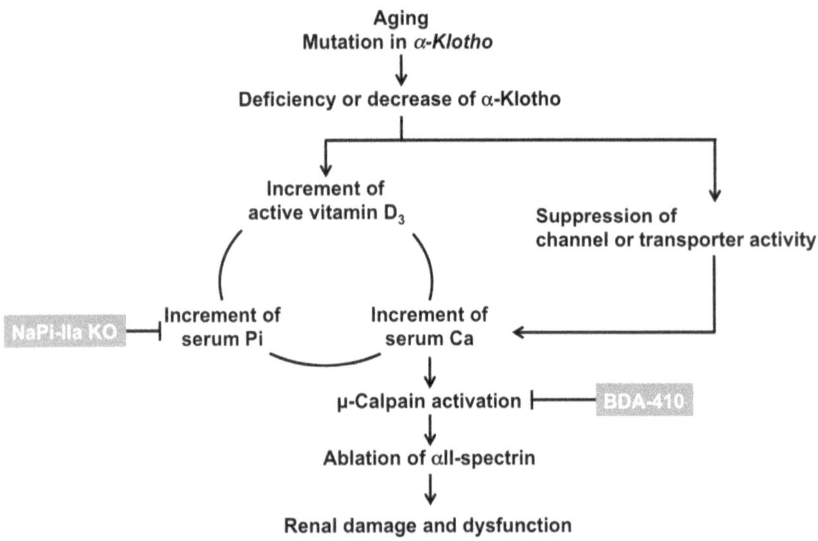

Fig. 4 Proposed pathway of deficiency in α-Klotho causing renal damage. At first, α-Klotho deficiency increases serum phosphate (Pi) concentration or suppresses the glycan modification of channels and transporter recruitment to the plasma membrane. Both induce the increment of calcium (Ca) flow to inside the cell. Thus, increased calcium leads to μ-calpain overactivation and finally results in renal damage. NaPi-IIa gene knockout and BDA-410 treatment ameliorate aging-related symptoms.

REFERENCES

ADHRConsortium (2000). Autosomal dominant hypophosphataemic rickets is associated with mutations in FGF23. *Nature Genetics, 26*, 345–348.

Akasaka-Manya, K., Manya, H., Kizuka, Y., Oka, S., & Endo, T. (2014). alpha-Klotho mice demonstrate increased expression of the non-sulfated N-glycan form of the HNK-1 glyco-epitope in kidney tissue. *Journal of Biochemistry, 156*, 107–113.

Andrukhova, O., Zeitz, U., Goetz, R., Mohammadi, M., Lanske, B., & Erben, R. G. (2012). FGF23 acts directly on renal proximal tubules to induce phosphaturia through activation of the ERK1/2-SGK1 signaling pathway. *Bone, 51*, 621–628.

Asai, O., Nakatani, K., Tanaka, T., Sakan, H., Imura, A., Yoshimoto, S., et al. (2012). Decreased renal alpha-Klotho expression in early diabetic nephropathy in humans and mice and its possible role in urinary calcium excretion. *Kidney International, 81*, 539–547.

Baum, M., Schiavi, S., Dwarakanath, V., & Quigley, R. (2005). Effect of fibroblast growth factor-23 on phosphate transport in proximal tubules. *Kidney International, 68*, 1148–1153.

Bloch, L., Sineshchekova, O., Reichenbach, D., Reiss, K., Saftig, P., Kuro-o, M., et al. (2009). Klotho is a substrate for alpha-, beta- and gamma-secretase. *FEBS Letters, 583*, 3221–3224.

Bottcher, R. T., & Niehrs, C. (2005). Fibroblast growth factor signaling during early vertebrate development. *Endocrine Reviews, 26*, 63–77.

Cha, S. K., Ortega, B., Kurosu, H., Rosenblatt, K. P., Kuro, O. M., & Huang, C. L. (2008). Removal of sialic acid involving Klotho causes cell-surface retention of TRPV5 channel via binding to galectin-1. *Proceedings of the National Academy of Sciences of the United States of America, 105*, 9805–9810.

Chanakul, A., Zhang, M. Y., Louw, A., Armbrecht, H. J., Miller, W. L., Portale, A. A., et al. (2013). FGF-23 regulates CYP27B1 transcription in the kidney and in extra-renal tissues. *PLoS One, 8*, e72816.

Chang, Q., Hoefs, S., van der Kemp, A. W., Topala, C. N., Bindels, R. J., & Hoenderop, J. G. (2005). The beta-glucuronidase klotho hydrolyzes and activates the TRPV5 channel. *Science, 310*, 490–493.

Chen, C. D., Podvin, S., Gillespie, E., Leeman, S. E., & Abraham, C. R. (2007). Insulin stimulates the cleavage and release of the extracellular domain of Klotho by ADAM10 and ADAM17. *Proceedings of the National Academy of Sciences of the United States of America, 104*, 19796–19801.

Friedlaender, M. M., Wald, H., Dranitzki-Elhalel, M., Zajicek, H. K., Levi, M., & Popovtzer, M. M. (2001). Vitamin D reduces renal NaPi-2 in PTH-infused rats: Complexity of vitamin D action on renal P(i) handling. *American Journal of Physiology, 281*, F428–F433.

Gattineni, J., Bates, C., Twombley, K., Dwarakanath, V., Robinson, M. L., Goetz, R., et al. (2009). FGF23 decreases renal NaPi-2a and NaPi-2c expression and induces hypophosphatemia in vivo predominantly via FGF receptor 1. *American Journal of Physiology, 297*, F282–F291.

Gattineni, J., & Baum, M. (2010). Regulation of phosphate transport by fibroblast growth factor 23 (FGF23): Implications for disorders of phosphate metabolism. *Pediatric Nephrology, 25*, 591–601.

Hu, M. C., Shi, M., Zhang, J., Addo, T., Cho, H. J., Barker, S. L., et al. (2015). Renal production, uptake, and handling of circulating alphaKlotho. *Journal of the American Society of Nephrology, 27*, 79–90.

Hu, M. C., Shi, M., Zhang, J., Pastor, J., Nakatani, T., Lanske, B., et al. (2010). Klotho: A novel phosphaturic substance acting as an autocrine enzyme in the renal proximal tubule. *The FASEB Journal, 24*, 3438–3450.

Hu, M. C., Shi, M., Zhang, J., Quinones, H., Kuro-o, M., & Moe, O. W. (2010). Klotho deficiency is an early biomarker of renal ischemia-reperfusion injury and its replacement is protective. *Kidney International, 78*, 1240–1251.

Imura, A., Iwano, A., Tohyama, O., Tsuji, Y., Nozaki, K., Hashimoto, N., et al. (2004). Secreted Klotho protein in sera and CSF: Implication for post-translational cleavage in release of Klotho protein from cell membrane. *FEBS Letters, 565*, 143–147.

Imura, A., Tsuji, Y., Murata, M., Maeda, R., Kubota, K., Iwano, A., et al. (2007). alpha-Klotho as a regulator of calcium homeostasis. *Science, 316*, 1615–1618.

Kato, Y., Arakawa, E., Kinoshita, S., Shirai, A., Furuya, A., Yamano, K., et al. (2000). Establishment of the anti-Klotho monoclonal antibodies and detection of Klotho protein in kidneys. *Biochemical and Biophysical Research Communications, 267*, 597–602.

Koh, N., Fujimori, T., Nishiguchi, S., Tamori, A., Shiomi, S., Nakatani, T., et al. (2001). Severely reduced production of klotho in human chronic renal failure kidney. *Biochemical and Biophysical Research Communications, 280*, 1015–1020.

Kuro-o, M., Matsumura, Y., Aizawa, H., Kawaguchi, H., Suga, T., Utsugi, T., et al. (1997). Mutation of the mouse klotho gene leads to a syndrome resembling ageing. *Nature, 390*, 45–51.

Kurosu, H., Ogawa, Y., Miyoshi, M., Yamamoto, M., Nandi, A., Rosenblatt, K. P., et al. (2006). Regulation of fibroblast growth factor-23 signaling by klotho. *The Journal of Biological Chemistry, 281*, 6120–6123.

Kurosu, H., Yamamoto, M., Clark, J. D., Pastor, J. V., Nandi, A., Gurnani, P., et al. (2005). Suppression of aging in mice by the hormone Klotho. *Science, 309*, 1829–1833.

Kusaba, T., Okigaki, M., Matui, A., Murakami, M., Ishikawa, K., Kimura, T., et al. (2010). Klotho is associated with VEGF receptor-2 and the transient receptor potential canonical-1 Ca^{2+} channel to maintain endothelial integrity. *Proceedings of the National Academy of Sciences of the United States of America, 107*, 19308–19313.

Larsson, T., Marsell, R., Schipani, E., Ohlsson, C., Ljunggren, O., Tenenhouse, H. S., et al. (2004). Transgenic mice expressing fibroblast growth factor 23 under the control of the alpha1(I) collagen promoter exhibit growth retardation, osteomalacia, and disturbed phosphate homeostasis. *Endocrinology, 145*, 3087–3094.

Leunissen, E. H., Nair, A. V., Bull, C., Lefeber, D. J., van Delft, F. L., Bindels, R. J., et al. (2013). The epithelial calcium channel TRPV5 is regulated differentially by klotho and sialidase. *The Journal of Biological Chemistry, 288*, 29238–29246.

Li, S. A., Watanabe, M., Yamada, H., Nagai, A., Kinuta, M., & Takei, K. (2004). Immunohistochemical localization of Klotho protein in brain, kidney, and reproductive organs of mice. *Cell Structure and Function, 29*, 91–99.

Manya, H., Inomata, M., Fujimori, T., Dohmae, N., Sato, Y., Takio, K., et al. (2002). Klotho protein deficiency leads to overactivation of mu-calpain. *The Journal of Biological Chemistry, 277*, 35503–35508.

Miyamoto, K., Segawa, H., Ito, M., & Kuwahata, M. (2004). Physiological regulation of renal sodium-dependent phosphate cotransporters. *The Japanese Journal of Physiology, 54*, 93–102.

Morishita, K., Shirai, A., Kubota, M., Katakura, Y., Nabeshima, Y., Takeshige, K., et al. (2001). The progression of aging in klotho mutant mice can be modified by dietary phosphorus and zinc. *Journal of Nutrition, 131*, 3182–3188.

Murer, H., Hernando, N., Forster, I., & Biber, J. (2003). Regulation of Na/Pi transporter in the proximal tubule. *Annual Review of Physiology, 65*, 531–542.

Nabeshima, Y., Washida, M., Tamura, M., Maeno, A., Ohnishi, M., Shiroishi, T., et al. (2014). Calpain 1 inhibitor BDA-410 ameliorates alpha-klotho-deficiency phenotypes resembling human aging-related syndromes. *Scientific Reports, 4*, 5847.

Ohnishi, M., Nakatani, T., Lanske, B., & Razzaque, M. S. (2009). Reversal of mineral ion homeostasis and soft-tissue calcification of klotho knockout mice by deletion of vitamin D 1alpha-hydroxylase. *Kidney International, 75*, 1166–1172.

Ohnishi, M., & Razzaque, M. S. (2010). Dietary and genetic evidence for phosphate toxicity accelerating mammalian aging. *The FASEB Journal, 24*, 3562–3571.

Ornitz, D. M., & Itoh, N. (2001). Fibroblast growth factors. *Genome Biology, 2*. REVIEWS3005.1–REVIEWS3005.12.

Ornitz, D. M., & Marie, P. J. (2002). FGF signaling pathways in endochondral and intramembranous bone development and human genetic disease. *Genes & Development, 16*, 1446–1465.

Percy, C. J., Power, D., & Gobe, G. C. (2008). Renal ageing: Changes in the cellular mechanism of energy metabolism and oxidant handling. *Nephrology (Carlton, Vic.), 13*, 147–152.

Pfister, M. F., Ruf, I., Stange, G., Ziegler, U., Lederer, E., Biber, J., et al. (1998). Parathyroid hormone leads to the lysosomal degradation of the renal type II Na/Pi cotransporter. *Proceedings of the National Academy of Sciences of the United States of America, 95*, 1909–1914.

Razzaque, M. S., Sitara, D., Taguchi, T., St-Arnaud, R., & Lanske, B. (2006). Premature aging-like phenotype in fibroblast growth factor 23 null mice is a vitamin D-mediated process. *The FASEB Journal, 20*, 720–722.

Saito, H., Kusano, K., Kinosaki, M., Ito, H., Hirata, M., Segawa, H., et al. (2003). Human fibroblast growth factor-23 mutants suppress Na^+-dependent phosphate co-transport activity and 1alpha,25-dihydroxyvitamin D3 production. *The Journal of Biological Chemistry, 278*, 2206–2211.

Segawa, H., Kawakami, E., Kaneko, I., Kuwahata, M., Ito, M., Kusano, K., et al. (2003). Effect of hydrolysis-resistant FGF23-R179Q on dietary phosphate regulation of the renal type-II Na/Pi transporter. *Pflügers Archiv: European Journal of Physiology, 446*, 585–592.

Segawa, H., Yamanaka, S., Onitsuka, A., Tomoe, Y., Kuwahata, M., Ito, M., et al. (2007). Parathyroid hormone-dependent endocytosis of renal type IIc Na-Pi cotransporter. *American Journal of Physiology, 292*, F395–F403.

Shimada, T., Kakitani, M., Yamazaki, Y., Hasegawa, H., Takeuchi, Y., Fujita, T., et al. (2004). Targeted ablation of Fgf23 demonstrates an essential physiological role of FGF23 in phosphate and vitamin D metabolism. *The Journal of Clinical Investigation, 113*, 561–568.

Shimada, T., Mizutani, S., Muto, T., Yoneya, T., Hino, R., Takeda, S., et al. (2001). Cloning and characterization of FGF23 as a causative factor of tumor-induced osteomalacia. *Proceedings of the National Academy of Sciences of the United States of America, 98*, 6500–6505.

Shimada, T., Yamazaki, Y., Takahashi, M., Hasegawa, H., Urakawa, I., Oshima, T., et al. (2005). Vitamin D receptor-independent FGF23 actions in regulating phosphate and vitamin D metabolism. *American Journal of Physiology, 289*, F1088–F1095.

Sitara, D., Razzaque, M. S., Hesse, M., Yoganathan, S., Taguchi, T., Erben, R. G., et al. (2004). Homozygous ablation of fibroblast growth factor-23 results in hyperphosphatemia and impaired skeletogenesis, and reverses hypophosphatemia in Phex-deficient mice. *Matrix Biology, 23*, 421–432.

Sorimachi, H., Ishiura, S., & Suzuki, K. (1997). Structure and physiological function of calpains. *Biochemical Journal, 328*(Pt. 3), 721–732.

Tagawa, H., Kizuka, Y., Ikeda, T., Itoh, S., Kawasaki, N., Kurihara, H., et al. (2005). A non-sulfated form of the HNK-1 carbohydrate is expressed in mouse kidney. *The Journal of Biological Chemistry, 280*, 23876–23883.

Tohyama, O., Imura, A., Iwano, A., Freund, J. N., Henrissat, B., Fujimori, T., et al. (2004). Klotho is a novel beta-glucuronidase capable of hydrolyzing steroid beta-glucuronides. *The Journal of Biological Chemistry, 279*, 9777–9784.

Tsujikawa, H., Kurotaki, Y., Fujimori, T., Fukuda, K., & Nabeshima, Y. (2003). Klotho, a gene related to a syndrome resembling human premature aging, functions in a negative regulatory circuit of vitamin D endocrine system. *Molecular Endocrinology, 17*, 2393–2403.

Tsuruoka, S., Nishiki, K., Ioka, T., Ando, H., Saito, Y., Kurabayashi, M., et al. (2006). Defect in parathyroid-hormone-induced luminal calcium absorption in connecting tubules of Klotho mice. *Nephrology, Dialysis, Transplantation, 21*, 2762–2767.

Urakawa, I., Yamazaki, Y., Shimada, T., Iijima, K., Hasegawa, H., Okawa, K., et al. (2006). Klotho converts canonical FGF receptor into a specific receptor for FGF23. *Nature, 444*, 770–774.

Vanderklish, P. W., & Bahr, B. A. (2000). The pathogenic activation of calpain: A marker and mediator of cellular toxicity and disease states. *International Journal of Experimental Pathology, 81*, 323–339.

White, K. E., Carn, G., Lorenz-Depiereux, B., Benet-Pages, A., Strom, T. M., & Econs, M. J. (2001). Autosomal-dominant hypophosphatemic rickets (ADHR) mutations stabilize FGF-23. *Kidney International, 60*, 2079–2086.

Yang, H., & Fogo, A. B. (2010). Cell senescence in the aging kidney. *Journal of the American Society of Nephrology, 21*, 1436–1439.

Yoshida, T., Fujimori, T., & Nabeshima, Y. (2002). Mediation of unusually high concentrations of 1,25-dihydroxyvitamin D in homozygous klotho mutant mice by increased expression of renal 1alpha-hydroxylase gene. *Endocrinology, 143*, 683–689.

Zhang, X., Ibrahimi, O. A., Olsen, S. K., Umemori, H., Mohammadi, M., & Ornitz, D. M. (2006). Receptor specificity of the fibroblast growth factor family. The complete mammalian FGF family. *The Journal of Biological Chemistry, 281*, 15694–15700.

Zhou, X. J., Rakheja, D., Yu, X., Saxena, R., Vaziri, N. D., & Silva, F. G. (2008). The aging kidney. *Kidney International, 74*, 710–720.

CHAPTER ELEVEN

αKlotho and Chronic Kidney Disease

J.A. Neyra*,†, M.C. Hu*,†,1

*University of Texas Southwestern Medical Center, Dallas, TX, United States

†Charles and Jane Pak Center for Mineral Metabolism and Clinical Research, University of Texas Southwestern Medical Center, Dallas, TX, United States

[1]Corresponding author: e-mail address: ming-chang.hu@utsouthwestern.edu

Contents

Vitamins and Hormones, Volume 101
ISSN 0083-6729
http://dx.doi.org/10.1016/bs.vh.2016.02.007

Abstract

Alpha-Klotho (αKlotho) protein is encoded by the gene, *Klotho*, and functions as a coreceptor for endocrine fibroblast growth factor-23. The extracellular domain of αKlotho is cleaved by secretases and released into the circulation where it is called soluble αKlotho. Soluble αKlotho in the circulation starts to decline in chronic kidney disease (CKD) stage 2 and urinary αKlotho in even earlier CKD stage 1. Therefore soluble αKlotho is an early and sensitive marker of decline in kidney function. Preclinical data from numerous animal experiments support αKlotho deficiency as a pathogenic factor for CKD progression and extrarenal CKD complications including cardiac and vascular disease, hyperparathyroidism, and disturbed mineral metabolism. αKlotho deficiency induces cell senescence and renders cells susceptible to apoptosis induced by a variety of cellular insults including oxidative stress. αKlotho deficiency also leads to defective autophagy and angiogenesis and promotes fibrosis in the kidney and heart. Most importantly, prevention of αKlotho decline, upregulation of endogenous αKlotho production, or direct supplementation of soluble αKlotho are all associated with attenuation of renal fibrosis, retardation of CKD progression, improvement of mineral metabolism, amelioration of cardiac function and morphometry, and alleviation of vascular calcification in CKD. Therefore in rodents, αKlotho is not only a diagnostic and prognostic marker for CKD but the enhancement of endogenous or supplement of exogenous αKlotho are promising therapeutic strategies to prevent, retard, and decrease the comorbidity burden of CKD.

1. INTRODUCTION

The *Klotho* gene was discovered in 1997 when mice with silencing of this gene developed multiple organ dysfunction and failure with shortened life span resembling human premature aging (Kuro-o et al., 1997). The overexpression of the *Klotho* transgene with a ubiquitous promoter (Kurosu et al., 2005), viral-based transfer (Masuda, Chikuda, Suga, Kawaguchi, & Kuro-o, 2005), or direct parenteral administration (Chen, Kuro, et al., 2013) can extend mouse life span compared to normal mouse and rescue the most phenotypes observed in *Klotho*-deficient mouse (Kurosu et al., 2005). Two other paralogs βKlotho (Ito et al., 2000) and γKlotho (Ito, Fujimori, Hayashizaki, & Nabeshima, 2002) were identified, then *Klotho* gene was designated *αKlotho* to distinguish from the other two paralogs (Hu, Shiizaki, Kuro-o, & Moe, 2013).

αKlotho is highly expressed in the kidney, brain, and in lesser extent in other organs (Kato et al., 2000; Kuro-o et al., 1997). Human αKlotho is a single transmembrane 1012 amino acid 130 kDa protein encoded by human *Klotho* gene, while rodent αKlotho is a 1014 amino acid protein (Kuro-o et al., 1997; Matsumura et al., 1998; Shiraki-Iida et al., 1998; Tohyama et al., 2004). The extracellular domain of membrane αKlotho consisting of two repeat sequences (kl1 and kl2) can be shed by secretases and released into the circulation (Bloch et al., 2009; Chen, Podvin, Gillespie, Leeman, & Abraham, 2007; Chen, Tung, et al., 2014; Hu, Shi, Zhang, et al., 2015). This released extracellular domain of membrane αKlotho is referred as soluble or cleaved αKlotho. It is a main functional form in the circulation (Hu, Shi, Zhang, et al., 2015; Hu, Shi, Zhang, Pastor, et al., 2010; Hu, Shi, Zhang, Quinones, et al., 2010; Imura et al., 2004; Kurosu et al., 2005). Soluble αKlotho protein is also present in cerebrospinal fluid (Chen et al., 2015; Degaspari et al., 2015; Emami Aleagha et al., 2015; Imura et al., 2004; Semba et al., 2014) and urine of mammals (Akimoto et al., 2012; Hu, Shi, Zhang, Pastor, et al., 2010; Hu et al., 2011; Lau et al., 2012). Soluble αKlotho functions as a circulating substance exerting multiple systemic biological actions on distant organs and directly protects cells against a variety of insults including hypoxia, hyperoxia, oxidative stress, and cytotoxic medication and suppresses apoptosis (Cheng et al., 2015; Hu, Shi, Cho, et al., 2013; Panesso et al., 2014; Ravikumar et al., 2014; Sun et al., 2015; Wang et al., 2013).

Chronic kidney disease (CKD) is characterized by progressive deterioration of renal function with high risk of end-stage renal disease (ESRD) regardless of whether initial kidney insults have regressed or are continuously present (D'Hoore et al., 2015; Ferenbach & Bonventre, 2015; Rimes-Stigare et al., 2015; Venkatachalam, Weinberg, Kriz, & Bidani, 2015). As expected, CKD risk increases with age, and about half of the CKD stage ≥3 cases occurs in subjects >70 years old. CKD can be viewed as a state of accelerating aging (Kooman et al., 2013; Stenvinkel & Larsson, 2013). The relative risk for cardiovascular (CV) mortality of a 25- to 34-year-old dialysis patient is similar to a non-CKD patient of >75 years of age (Foley, Parfrey, & Sarnak, 1998). The similar phenotypes between αKlotho-deficient mice and CKD subjects also suggest a potential pathogenic role of αKlotho deficiency in CKD development and progression (Hu, Kuro-o, & Moe, 2012, 2013a, 2013b; Hu et al., 2011; Hu, Shiizaki, Kuro-o, et al., 2013; Shi et al., 2015).

In this chapter, we aim to summarize the current state of knowledge on αKlotho biology and pathophysiology in CKD, and provide a possible novel perspective on potential clinical applications of αKlotho in CKD.

2. CKD IS A STATE OF KLOTHO DEFICIENCY

2.1 The Kidney Is the Main Origin for Systemic αKlotho

Compared to the wide distribution of *αKlotho* mRNA in many organs and tissues, αKlotho protein expression is restricted to only a few tissues including the kidney, brain, heart, parathyroid gland, and testis (Kuro-o et al., 1997; Takeshita et al., 2004). αKlotho protein was also found in vascular endothelial cells and smooth muscle cells in humans and rodents (Fang et al., 2014; Jimbo et al., 2014; Lim et al., 2012; Ritter, Zhang, Delmez, Finch, & Slatopolsky, 2015), but this is still in debate because there is equally convincing evidence that do not support the presence of αKlotho protein in the vasculature (Hu, 2016; Lau et al., 2012; Lindberg et al., 2013; Mencke et al., 2015; Scialla et al., 2013). Therefore, whether αKlotho is expressed in the vasculature remains to be resolved. Among tissues expressing αKlotho protein, the kidney has the highest level. In mammalian kidney, αKlotho is prominently expressed in distal convoluted tubules (DCTs; Kato et al., 2000; Kuro-o et al., 1997), but is also unequivocally found in the proximal convoluted tubules although at lower levels compared to DCT (Hu, Shi, Zhang, Pastor, et al., 2010; Lim et al., 2015).

Despite the fact that the kidney is the organ expressing the highest levels of αKlotho protein, the confirmation that circulating αKlotho in serum mainly derived from the kidney under physiological conditions was demonstrated by Lindberg et al. (2014). The strongest evidence comes from mouse with renal tubule-specific partial deletion of αKlotho. This mouse line has reduced serum αKlotho levels and systemic features resembling the phenotype of global αKlotho deletion or the αKlotho hypomorphic mice, indicating that the kidney may be the principal organ mediating the systemic αKlotho effects (Lindberg et al., 2014). More direct evidence to support this notion is that αKlotho protein in the suprarenal vein is higher than that in infrarenal vein in rodents and humans (Hu, Shi, Zhang, et al., 2015) and circulating αKlotho levels were dramatically and quickly dropped in rodents that underwent bilateral nephrectomy (Hu, Shi, Zhang, et al., 2015) which strongly suggest that the kidney is the main source of αKlotho in the circulation under physiological conditions (Fig. 1). In living human kidney donors (Akimoto et al., 2013; Ponte et al., 2014), αKlotho was shown to drop after nephrectomy but it is difficult to distinguish this from the effect of surgery. However, under pathological renal conditions such as ESRD, circulating αKlotho is low rather absent, suggesting the possibility that

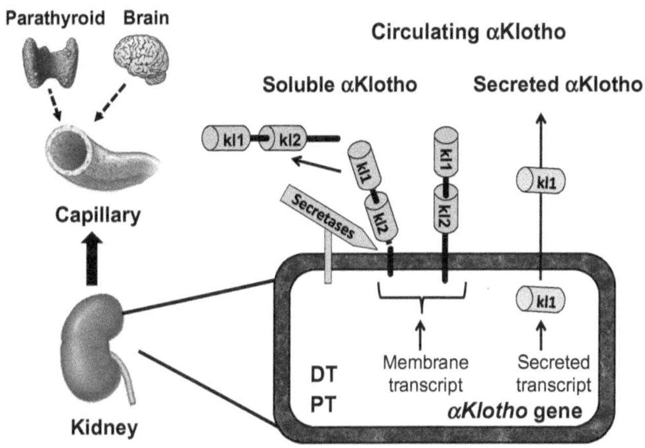

Fig. 1 Source of circulatory αKlotho. αKlotho protein is expressed in a few organs, but the kidney is a main resource of circulating αKlotho under physiological conditions. The contribution of parathyroid gland and brain is not clear. Both renal proximal (PT) and distal tubules (DT) express membrane αKlotho protein and may also produce a secreted αKlotho protein which only contains kl1 domain and is directly secreted into the blood circulation. Extracellular domain of membrane αKlotho-containing kl1 and kl2 repeats is shed and cleaved by α and β-secretases, and released into the blood circulation. (See the color plate.)

αKlotho may come from extrarenal source(s), although its origin is not clear to date (Lau et al., 2012). Establishing extrarenal sources of αKlotho and characterizing how this can be upregulated when renal production fails are of paramount importance.

2.2 Renal αKlotho Deficiency in CKD

With the renal source of circulating αKlotho established (Fig. 1), the next step is to understand why and how αKlotho is drastically reduced in kidney disease. As a general principle, if the organ of origin of an endocrine substance is diseased, it is logical to suspect that endocrine deficiency of that substance ensues. Therefore, it is not surprising to witness the reduction of αKlotho protein in the diseased kidney. However, whether the reduction of αKlotho is due to destruction of kidney, loss ability to produce/secrete αKlotho, or a maladaptive response, remains to be explored.

It has been shown that there is a significant reduction of renal αKlotho transcript and protein in the diseased kidney resulting from a wide variety of etiologies from glomerular and tubulointerstitial diseases, obstructive nephropathy, diabetic nephropathy, ischemic injury, subtotal nephrectomy, oxidative stress, chronic allograft rejection, and exposure to cisplatin, angiotensin II (Ang II), and calcineurin inhibitors in both humans and rodents (Hu et al., 2012; Hu et al., 2013a, 2013b; Hu, Shiizaki, Kuro-o, et al., 2013; Panesso et al., 2014; Sastre et al., 2013; Shi et al., 2015; Zhou, Li, Zhou, Tan, & Liu, 2013). The mechanisms underlying the relationship between renal αKlotho downregulation and kidney diseases therefore need to be further illustrated.

It is proposed that renal αKlotho deficiency in early stages of CKD may be attributed mainly to suppression of αKlotho expression rather than loss of viable renal tubules. Several intermediates are shown to be involved in the reduction of αKlotho expression: high serum phosphate (Hu, Kuro-o, & Moe, 2014), hypermethylation (Azuma et al., 2012; King, Rosene, & Abraham, 2012; Lee, Jeong, et al., 2010; Sun, Chang, & Wu, 2012; Young & Wu, 2012), and hyperdeacetylation (Moreno et al., 2011) in αKlotho gene promoter induced by inflammatory cytokines and the uremic toxin, indoxyl sulfate (Fig. 2). If these observations

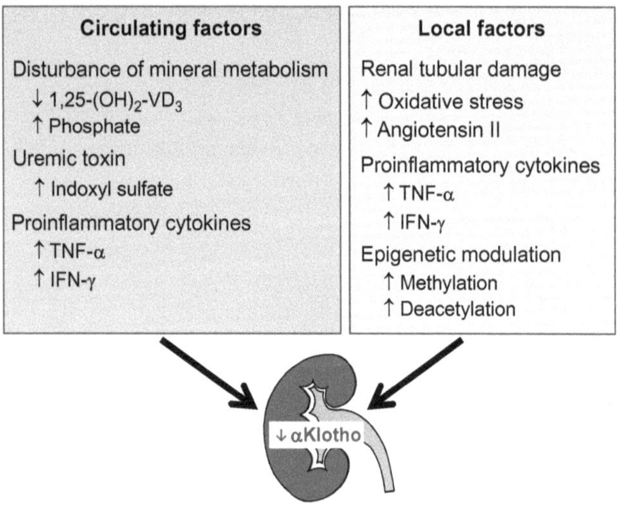

Fig. 2 Circulating and local renal factors involved in the reduction of αKlotho expression in the kidney. In acute and chronic kidney disease, a variety of circulating factors including disturbed mineral metabolism, and accumulation of indoxyl sulfate and proinflammatory cytokines (*left panel*), can downregulate renal αKlotho expression. On the other hand, the elevation of reactive oxygen species, Ang II, and inflammatory cytokines in the diseased kidney can also downregulate renal αKlotho expression (*right panel*). Epigenetic modulation of αKlotho promoter via hypermethylation and deacetylation can reduce αKlotho expression and contribute to αKlotho deficiency in chronic kidney disease. (See the color plate.)

are correct, they provide an opportunity to reactivate αKlotho expression by modulation of these factors and thereby correcting circulating and renal αKlotho deficiency in early stages of CKD.

2.3 Circulating αKlotho Deficiency in CKD

αKlotho transcript and protein expression in diseased kidneys from humans and animals is clearly decreased. However, the relationship between renal αKlotho expression and serum and/or urinary αKlotho levels remains to be confirmed.

In a rodent model of CKD from uninephrectomy plus contralateral ischemia reperfusion, serum αKlotho concentration was remarkably decreased, and the degree of its reduction was similar in magnitude to that of decreased αKlotho protein in the kidneys and in the urine (Hu et al., 2011). Thus in rodents, CKD is a state of endocrine (circulating and urinary) αKlotho deficiency in addition to renal αKlotho deficiency (Hu et al., 2011).

In CKD patients, urinary αKlotho levels are significantly decreased at very early stages (stage 1) and sustainably reduced with progression of CKD (Hu et al., 2011), while the reduction of serum αKlotho starts later at stage 2 CKD (Barker et al., 2015). Moreover, human urinary αKlotho excretion is significantly decreased and the amount of urinary αKlotho decrease is directly correlated with estimated glomerular filtration rate (eGFR) decline (Yamazaki et al., 2010). These observations suggest that urinary soluble αKlotho may be a good biomarker for early CKD detection. A growing body of evidence has shown a reduction of circulating αKlotho in CKD and ESRD patients (Devaraj, Syed, Chien, & Jialal, 2012; Fliser, Seiler, Heine, & Ketteler, 2012; Pavik et al., 2013; Scholze et al., 2014; Shimamura et al., 2012; Zhou et al., 2013), therefore identifying the plausible mechanisms and clinical implications of serum and/or urinary αKlotho reduction in CKD/ESRD patients is of paramount importance.

3. αKlotho DEFICIENCY CONTRIBUTES TO CKD DEVELOPMENT AND PROGRESSION

αKlotho deficiency is not only an early biomarker of CKD but also a pathogenic intermediate for CKD development and progression, and extrarenal complications. Compared to wild-type mice, αKlotho-deficient mice have more severe kidney damage and faster progression to CKD with more fibrosis, and αKlotho overexpressors have milder kidney dysfunction and less fibrosis after exposure to renal insults including ischemic injury (Hu et al., 2011; Hu, Shi, Zhang, Quinones, et al., 2010; Shi et al., 2015), cisplatin (Panesso et al., 2014), Adriamycin (Zhou et al., 2013), and ureteric ligation (Sugiura et al., 2012; Zhou et al., 2013). αKlotho is a multifaceted protein. Different forms of αKlotho may be involved in different biological functions. Membrane αKlotho is confirmed to participate in maintenance of mineral homeostasis, while soluble αKlotho protein plays a more important and systemic role in cytoprotection, antifibrosis, and angiogenesis.

3.1 Increased Cell Senescence and Reduced Ability of Regeneration

Stem cells in most mammalian tissues participate in maintenance of tissue homeostasis and are involved in tissue repair or regeneration (Li & Clevers, 2010; Weissman, 2000). The dysfunction and depletion of stem cells and progenitor cells contribute to aging and aging-associated diseases including kidney disease. CKD can be a consequence of incomplete or failed tubule recovery after AKI (D'Hoore et al., 2015; Ferenbach & Bonventre, 2015; Kramann, Tanaka, & Humphreys, 2014; Polichnowski et al., 2014; Venkatachalam et al., 2015; Zhang et al., 2013). The repeated administration of bone marrow-derived mesenchymal stem cells improved renal function and histology, reduced blood pressure, and attenuated the infiltration of inflammatory cells on a remnant rat kidney (Lee, Lee, et al., 2010). More recently, evidence has shown human-induced pluripotent stem cells derived from any human somatic cell type after the introduction of reprogramming transcription factors contributing to kidney regeneration and improvement in kidney function (Schmitt, Susnik, & Melk, 2015).

The decrease in stem cell number is associated with an increase in progenitor cell senescence, a complicated process present not only in normal aging but also in pathophysiological states (Dmitrieva & Burg, 2007; Haruna et al., 2007; Jennings et al., 2007; Kailong et al., 2007; Nakano-Kurimoto et al., 2009; Yang et al., 2009; Yang & Fogo, 2010). Excessive senescence and subsequent stem cell deletion may decrease the ability of the kidney to defend against renal insults and impair regeneration (Schmitt et al., 2015).

αKlotho deficiency is associated with stem cell dysfunction and depletion which is part of normal aging (Bian, Neyra, Zhan, & Hu, 2015; Liu et al., 2007). αKlotho deficiency in CKD could enhance renal tubular and vascular cell senescence induced by oxidative stress, uremic toxin, and high phosphate (Carracedo et al., 2013; Clements, Chaber, Ledbetter, & Zuk, 2013; de Oliveira, 2006; Niwa & Shimizu, 2012; Small et al., 2012; Tsirpanlis, 2008; Verbeke, Van Biesen, & Vanholder, 2011; Yamada et al., 2015). Wnt signaling activity is significantly increased in tissues from *kl/kl* mice, which can be rescued by genetic αKlotho overexpression (Liu et al., 2007). Administration of exogenous Wnt stimulates Wnt signal transduction, and triggers or accelerates cell senescence both in vitro and in vivo. αKlotho appears to be a secreted Wnt antagonist and may utilize this mechanism to retard mammalian aging. Suppression of cell senescence may be one of many novel strategies for promotion of kidney regeneration after AKI and retardation of CKD progression (Camilli et al., 2011; Liu et al., 2007; Satoh et al., 2012; Zhou et al., 2013).

3.2 Defective Endothelial Function and Impaired Vasculogenesis

In CKD patients, there is endothelial dysfunction and impaired bone marrow-derived endothelial progenitor cells-mediated vascular regeneration and kidney repair (Jie et al., 2010; Mohandas & Segal, 2010), both of which can contribute to progression of CKD and aging of the kidney (Chade et al., 2006; Mu et al., 2009; Reinders, Rabelink, & Briscoe, 2006; Taniyama & Morishita, 2010; Westerweel et al., 2007). Recent studies further indicated a potential causal link between vascular rarefaction and CKD progression. *kl/kl* mice do not only have abnormal vasodilatation due to abnormal endothelial function (Nakamura et al., 2002) and low blood flow after hind limb ischemia (Fukino et al., 2002) but also impaired angiogenesis and vasculogenesis (Shimada et al., 2004). The HMG coenzyme A reductase inhibitor, cerivastatin, increases αKlotho levels in cultured kidney cell lines (Narumiya et al., 2004) and mice (Yoon et al., 2012) and also restores impaired neovascularization in *kl/kl* mice (Shimada et al., 2004), but the causal relation between increased Klotho and restoration of vasculogenesis remains to be confirmed. The impaired vasculogenesis and angiogenesis might be attributable to downregulation of vascular endothelial growth factor (VEGF) in the aorta (Nakamura et al., 2002). Recent in vivo and in vitro studies showed that αKlotho is associated with VEGF receptor-2 (VEGFR-2) and the transient receptor potential canonical-1 (TRPC-1) Ca^{2+} channel to maintain endothelial integrity because in αKlotho-deficient endothelial cells, VEGF-mediated internalization of the VEGFR-2/TRPC-1 complex is impaired, and surface TRPC-1 expression increases which can be reversed by αKlotho protein (Kusaba et al., 2010). In addition, αKlotho mitigates the increased cell senescence and apoptosis triggered by oxidative stress in endothelial cells (Ikushima et al., 2006); and αKlotho also suppresses TNF-β-induced expression of intracellular adhesion molecule-1 and vascular cell adhesion molecule-1, attenuates NF-kappaB activation, and reverses the inhibition of eNOS phosphorylation by TNF-α. Thus αKlotho protein also protects vascular endothelium by inhibition of endothelial inflammation (Maekawa et al., 2009).

3.3 Promotion of Renal Fibrosis

Renal fibrosis is a histological hallmark of CKD and also believed to be a pathogenic intermediate for CKD progression (Ardura, Rayego-Mateos, Ramila, Ruiz-Ortega, & Esbrit, 2010; Iwano et al., 2002; Kalluri & Neilson, 2003; Liu, 2010; Zeisberg & Duffield, 2010; Zeisberg et al., 2003). *kl/kl* mice have more renal tubulointerstitial fibrosis (Sugiura et al., 2012) which is associated with upregulation of TGF-β in the kidneys. The renal fibrosis induced by unilateral ureteral obstruction (UUO) is accompanied by upregulation of TGF-β and fibronectin, and down-regulation of αKlotho mRNA and protein. These alterations are exaggerated in αKlotho-deficient UUO mice compared to *WT* UUO mice. Along the same line, more renal fibrosis was found in αKlotho-deficient mice injected with Adriamycin (Zhou et al., 2013). Soluble αKlotho alleviates renal fibrosis induced by UUO and suppresses expression of fibrosis markers and TGF-β1 target genes (eg, Snail, Twist), but does not reduce TGF-β1 expression in UUO kidney (Doi et al., 2011), suggesting that αKlotho suppresses renal fibrosis primarily through inhibiting TGF-β1 downstream signaling (Doi et al., 2011).

As discussed earlier, αKlotho is an antagonist for Wnt signaling and Wnt is associated with renal fibrosis. αKlotho's suppression of renal fibrosis is conceivably attributable to inhibition of Wnt signal transduction. Mice with UUO and Adriamycin injection have high Wnt levels and β-catenin activity as well as myofibroblast activation which can be suppressed by administration of expression vector encoding the extracellular domain of αKlotho (Zhou et al., 2013).

When renal fibrosis and Wnt signaling are compared between αKlotho overexpression and αKlotho deficiency in UUO, Wnt signaling and tubulointerstitial fibrosis were attenuated in αKlotho-overexpressing compared to *WT* mice. In contrast, Wnt signaling and tubulointerstitial fibrosis were dramatically augmented in αKlotho heterozygous-deficient (*kl/+*)

mice after UUO compared with *WT* mice. Interestingly, after transferring plasmid overexpressing αKlotho into skeletal muscle, *kl/+* mutant mice had much lower Wnt signaling and extracellular matrix deposition. Therefore, αKlotho is a critical negative regulator of Wnt signaling and a suppressor of renal fibrosis in the obstructed kidney model (Satoh et al., 2012). In addition, αKlotho also promotes clearance of collagen I through upregulation of autophagy (Shi et al., 2015). Therefore, αKlotho suppresses fibrosis and enhances removal of collagen. Exogenous αKlotho administration may be a novel therapeutic agent for renal fibrosis.

4. αKlotho DEFICIENCY EXACERBATES DISORDERS OF MINERAL METABOLISM IN CKD

The fact that mice with complete αKlotho deficiency share similar features such as hyperphosphatemia, hyper-FGF23-temia, and high morbidity and mortality from CV disease than CKD subjects suggests that αKlotho deficiency may participate in CKD development (Hu et al., 2013a, 2013b; Hu, Shiizaki, Kuro-o, et al., 2013). Furthermore, αKlotho-deficient mice with AKI progress to CKD more rapidly and exhibit more severe vascular lesions (Hu et al., 2011; Shi et al., 2015) and uremic cardiac remodeling (Hu, Shi, Cho, et al., 2015), supporting the concept that αKlotho deficiency might be a pathogenic intermediate in CKD (Fig. 3). Given that disturbed mineral metabolism contributes to the high morbidity and mortality of CV disease in CKD (Davidovich, Davidovits, Peretz, Shapira, & Aframian, 2009; Fernandez-Martin et al., 2015; Kaisar, Isbel, & Johnson, 2007; Kestenbaum & Belozeroff, 2007; London, Marchais, Guerin, & Metivier, 2005; Obi, Hamano, & Isaka, 2015; Siomou & Stefanidis, 2012; van Ballegooijen, Rhee, Elmariah, de Boer, & Kestenbaum, 2016; Wesseling-Perry, 2015), a better understanding of the molecular mechanisms of how αKlotho deficiency dysregulates mineral metabolism will aid in the exploration of novel therapeutic strategies in CKD.

4.1 Hyperphosphatemia

The role of αKlotho in phosphate homeostasis was recognized as soon as αKlotho was discovered because αKlotho-deficient mice have severe hyperphosphatemia (Hu, Shi, Zhang, Pastor, et al., 2010; Kuro-o et al., 1997). This was further confirmed by the fact that there is low serum phosphate in αKlotho-overexpressing mice (Kurosu et al., 2005). A patient with homozygous missense mutation (H193R) in the *αKLOTHO* gene had

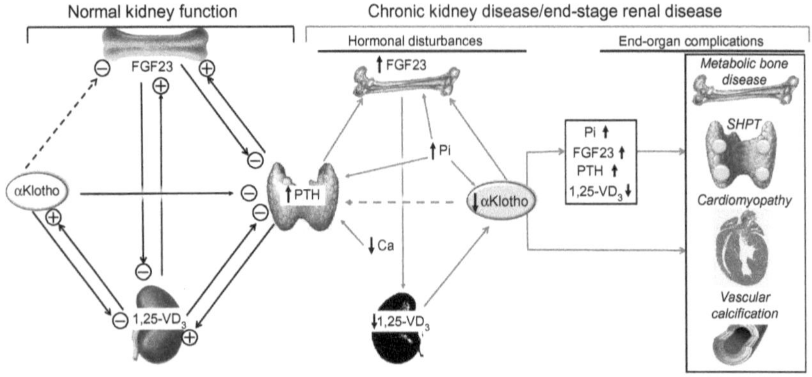

Fig. 3 Proposed physiological role of αKlotho in mineral metabolism and pathophysiological consequences of αKlotho deficiency in CKD. In the setting of normal kidney function with normal αKlotho levels (*left panel*), αKlotho may suppress FGF23 production and release from the bone. But there is no direct evidence to prove it. αKlotho functions as coreceptor of FGFR to allow FGF23 to suppress PTH production and release from parathyroid gland. PTH stimulates and increases plasma levels of FGF23 and 1,25-$(OH)_2$-vitamin D_3. Increased 1,25-$(OH)_2$-vitamin D_3 further stimulates FGF23, and directly and indirectly suppresses PTH levels. Increased 1,25-$(OH)_2$-vitamin D_3 also stimulates αKlotho production in the kidney. Taken together, through several negative- or positive-feedback loops, αKlotho functions as both a phosphate and calcium regulatory hormone to directly or indirectly suppress PTH, 1,25-$(OH)_2$-vitamin D_3, and FGF23 production and release. αKlotho's action on the kidney is to prevent renal Pi retention and to prevent renal Ca loss. In CKD and ESRD (*right panel*), the network is deranged (*red arrows*). Renal αKlotho is decreased followed by decrease in plasma αKlotho. The downregulation of αKlotho increases FGF23 production via unknown mechanism, which in turn suppresses 1,25-$(OH)_2$-vitamin D_3 production in the kidney. Whether low plasma αKlotho renders parathyroid gland resistant to the suppressive effect of FGF23 on PTH production is not proven. However, decreased FGFR1/3 and αKlotho expression in the uremic parathyroid gland could make the gland resistant to FGF23, and triggers and/or promotes secondary hyperparathyroidism (SHPT). Low plasma Ca also participates in SHPT development. Hyperphosphatemia amplifies the high FGF23 and PTH levels, and low αKlotho levels in the blood. The high plasma PTH, Pi, and FGF23, and low plasma 1,25-$(OH)_2$-vitamin D_3 and αKlotho in concert contribute to the development of complications such as metabolic bone disease, SHPT, cardiomyopathy, and vascular calcification. *Dash line*: unproven putative roles of αKlotho. *Ca*, ion calcium; *CKD*, chronic kidney disease; *ESRD*, end-stage renal disease; *FGFR*, FGF receptor; *Pi*, phosphate; *PTH*, parathyroid hormone; *SHPT*, secondary hyperparathyroidism; *1,25-VD3*, 1,25-$(OH)_2$-vitamin D_3. (See the color plate.)

severe calcinosis, dural and carotid artery calcifications, severe hyperphosphatemia, hypercalcemia, and high serum 1,25-$(OH)_2$-vitamin D_3 and fibroblast growth factor (FGF23) (Ichikawa et al., 2007). This mutation conceivably destabilizes kl1 domain of αKlotho, thereby attenuating production of membrane-bound and soluble αKlotho protein (Ichikawa et al., 2007). Therefore, in one human, the manifestations are similar to those observed in αKlotho-deficient mice. Phosphate overload suppresses αKlotho expression in the kidney (Hu, Shi, Cho, et al., 2015; Shi et al., 2015; Fig. 2). Normal mice fed a high Pi diet have dramatically decreased αKlotho protein and mRNA in the kidney, while αKlotho hypomorphic mice fed low Pi diet can regain part of their Klotho expression (Hu, Shi, Cho, et al., 2015; Morishita et al., 2001; Shi et al., 2015).

Accumulating evidence showed that kidney disease is a status of αKlotho deficiency. Although the mechanism of reduced renal and circulating αKlotho is not understood, it is conceivable that αKlotho deficiency might be involved in the development of hyperphosphatemia, one of components of CKD–metabolic bone disease (CKD-MBD). αKlotho deficiency impairs phosphaturia (Hu, Shi, Zhang, Pastor, et al., 2010) and consequently accelerates Pi accumulation in CKD. The higher the level of serum Pi, the greater the degree of soft-tissue calcification, and the greater the risk of mortality (Kestenbaum et al., 2005). Higher serum Pi levels have been shown to be associated with high mortality in incident ESRD patients (Gutiérrez et al., 2008) and control of serum Pi may help decrease vascular calcification and suppress proliferation of parathyroid glands (Cannata-Andia & Rodriguez-Garcia, 2002; Isakova et al., 2009; Kestenbaum et al., 2005). αKlotho administration could be a novel strategy for the correction of hyperphosphatemia in CKD patients.

4.2 Increased FGF23 Levels

FGF23, a phosphatonin, is thought to be implicated in the systemic balance of phosphate maintained by the interaction of intestine, bone, and kidneys (Hu, Shiizaki, Kuro-o, et al., 2013) through interplay with αKlotho, parathyroid hormone (PTH), and 1,25-$(OH)_2$-vitamin D_3 (Bian, Xing, & Hu, 2014). One principal stimulus for FGF23 secretion is currently believed to be high serum phosphate caused by dietary phosphate load (Nishida et al., 2006). In CKD, there is an increase in FGF23 levels in parallel with the deterioration of renal function (Fliser et al., 2007) and the increase of serum phosphate and PTH (Ben-Dov et al., 2007; Nagano et al., 2006; Silver & Naveh-Many, 2010). High serum FGF23 in CKD may not only serve as a diagnostic biomarker of early CKD and predictor of CV disease and mortality in CKD/ESRD patients, but recently, it is proposed to be the necessary and sufficient contributor to uremic cardiomyopathy

(Faul et al., 2011; Grabner et al., 2015) through activation of FGFR4 and independently from αKlotho. In contrast, recent data also showed that the relationship between FGF23 and cardiac remodeling depends on αKlotho, and the association of FGF23 with cardiac hypertrophy and fibrosis is only evident in the presence of αKlotho deficiency (Hu, Shi, Cho, et al., 2015).

High serum FGF23 levels antedate high serum levels of phosphate, suggesting a disrupted feedback loop resulting in very high levels of serum FGF23. αKlotho-deficient mice have very high serum levels of FGF23 further supporting that αKlotho might be a negative regulator of FGF23 regardless of the unknown precise mechanisms of how it suppresses FGF23 synthesis in bone (Fig. 3). Currently, there are no experimental data to show direct suppression of FGF23 by αKlotho in osteoblast or osteocytes in vitro. But other animal experiments have shown that extremely high circulating αKlotho with viral delivery can induce severe hypophosphatemia and increase blood FGF23 through stimulation of FGF23 production in the bone although the molecular mechanisms remain to be clarified (Smith et al., 2012). Because phosphate is a potent stimulus for FGF23 production in the bone, therefore, whether αKlotho directly or indirectly increases FGF23 remains to be explored. On the other hand, αKlotho-deficient mice have high levels of serum phosphate and conceivably high FGF23 results from high phosphate due to defective phosphate excretion induced by αKlotho deficiency (Hu, Shi, Zhang, Pastor, et al., 2010; Fig. 3).

4.3 Hypovitaminosis D

Low 1,25-$(OH)_2$-vitamin D_3 is a major component of disorders of mineral metabolism, and is conventionally attributed to cause bone disease and secondary hyperparathyroidism in CKD (Lips, 2001). However, hypervitaminosis D is present in αKlotho deficiency (Kuro-o et al., 1997) and removal of key components involved in vitamin D metabolism or function can rescue these phenotypes in αKlotho-deficient mice (Lanske & Razzaque, 2007; Ohnishi, Nakatani, Lanske, & Razzaque, 2009; Razzaque, 2012) suggesting that αKlotho is a suppressor of vitamin D signaling. If CKD is a state of αKlotho deficiency, why αKlotho deficiency does not raise the serum levels of 1,25-$(OH)_2$-vitamin D_3 in CKD? The increase in plasma FGF23 in CKD is thought to suppress 1α-hydroxylase in the kidney and initiate or accelerate vitamin D deficiency (Gutiérrez, 2010; Liu et al., 2006). Because 1,25-$(OH)_2$-vitamin D_3 induces αKlotho expression in the kidney (Tsujikawa, Kurotaki, Fujimori, Fukuda, & Nabeshima, 2003), it is plausible that low vitamin D levels in CKD may exacerbate renal αKlotho deficiency (Fig. 3).

4.4 Secondary Hyperparathyroidism

Secondary hyperparathyroidism is a common complication of CKD/ESRD and is induced by retention of phosphate as a result of reduced glomerular filtration. The "trade off" hypothesis formulated by Slatopolsky and Bricker has been used for several decades to explain the role of hyperphosphatemia in secondary hyperparathyroidism (Slatopolsky & Bricker, 1973). It was proposed that in the early stages of CKD, an increase in serum phosphate concentrations can be overcome by an increased rate of PTH release, which may be also a result of hypocalcemia. However, as CKD progresses to advanced stages or ESRD, hyperphosphatemia becomes sustained and PTH chronically elevated, suppressing the synthesis of 1α,25(OH)D_3 in the kidney. The reduction in the synthesis of 1α,25(OH)D_3 also results from hyperphosphatemia and reduced nephron mass. 1α,25(OH)D_3 deficiency also contributes to an increase in PTH synthesis (Delmez & Slatopolsky, 1992).

Currently, secondary hyperparathyroidism is considered as part of the syndrome of CKD-metabolic bone disease (MBD; Galitzer, Ben-Dov, Silver, & Naveh-Many, 2010; Khan, 2007). Even mild increments in PTH levels are associated with an increased CV risk, regardless of the serum levels of calcium and phosphorus and whether vitamin D therapy is given, suggesting that decreasing PTH levels may improve mineral metabolism disorders (Floege et al., 2011; Panichi et al., 2010; Patel et al., 2011; Pontoriero, Cozzolino, Locatelli, & Brancaccio, 2010).

CKD patients have high blood FGF23 levels (Shimada et al., 2010) and low αKlotho and FGFR(s) in parathyroid gland (Canalejo et al., 2010; Krajisnik et al., 2010). In subjects with normal kidney function, FGF23 plays a crucial role, both as a phosphaturic factor (Gattineni & Baum, 2010; Goetz et al., 2010; Weber, Liu, Indridason, & Quarles, 2003) and as a calciotropic hormone to suppress 1,25-$(OH)_2$-vitamin D production in the kidney (Liu et al., 2006) and PTH production in the parathyroid gland (Ben-Dov et al., 2007), and to increase renal calcium reabsorption through modulation of TRPV5 channel (Andrukhova et al., 2014). In contrast, in a CKD setting, FGF23 fails to inhibit PTH production probably due to downregulation of αKlotho and FGFR(s) in parathyroid gland (Fig. 3; Canalejo et al., 2010; Krajisnik et al., 2010). The mechanisms of downregulation of αKlotho and FGFR(s) in uremic parathyroid gland remain to be explored.

Renal and circulating αKlotho deficiency is associated with development and progression of CKD-MBD (Fahrleitner-Pammer et al., 2008; Hruska, Saab, Mathew, & Lund, 2007; Kalantar-Zadeh et al., 2010;

Patel et al., 2011). Although whether αKlotho is present in the vasculature is still under debate, it has been shown that even early CKD-MBD may cause a reduction of vascular αKlotho (Fang et al., 2014), stimulate vascular osteoblastic transition, increase osteocytic secreted proteins, and consequently induce vascular calcification. Correction of αKlotho and maintenance of mineral homeostasis do not only benefit bone and mineral metabolism but also may attenuate CV disease and improve the quality of life of CKD/ESRD patients (Fernandez-Martin et al., 2015).

5. αKlotho DEFICIENCY IN CV DISEASE IN CKD

CKD confers significant CV morbidity and mortality (Go, Chertow, Fan, McCulloch, & Hsu, 2004; Gross & Ritz, 2008; Taddei, Nami, Bruno, Quatrini, & Nuti, 2011). A large number of CKD patients die from CV disease even before initiation of dialysis. The main clinical features of CV disease in CKD include uremic cardiomyopathy and vascular calcification.

5.1 Pathological Uremic Cardiomyopathy

Uremic cardiomyopathy or cardiomyopathy of advanced CKD, characterized by cardiac hypertrophy and fibrosis, is a major cause of CV disease, by causing congestive heart failure, cardiac dysrhythmias, and sudden cardiac death (Glassock, Pecoits-Filho, & Barberato, 2009; Go et al., 2004; Gross & Ritz, 2008; Taddei et al., 2011). There are traditional risk factors such as hypertension, coronary disease, atherosclerosis, anemia, and volume overload (Glassock et al., 2009; Gross & Ritz, 2008), and also CKD-specific factors such as hyperphosphatemia (Block, Hulbert-Shearon, Levin, & Port, 1998; Glassock et al., 2009; Gross & Ritz, 2008; Kestenbaum et al., 2005). Recent data have shown an association between elevated FGF23 levels and uremic cardiac remodeling (Faul et al., 2011; Gutiérrez et al., 2009). Soluble αKlotho deficiency may also be an intermediate mediator of the pathological cardiac remodeling observed in CKD (Hu, Shi, Cho, et al., 2015). Furthermore, αKlotho may protect the heart against stress-induced cardiac hypertrophy by inhibiting TRPC-6 channel-mediated abnormal Ca^{2+} signaling in the heart (Xie et al., 2012; Xie, Yoon, An, Kuro-o, & Huang, 2015) or against uremic solute indoxyl sulfate-induced myocardial hypertrophy probably by suppressing NADPH oxidase Nox2/Nox4-derived reactive oxygen species (ROS) production and its downstream signaling (Yang et al., 2015; Fig. 4).

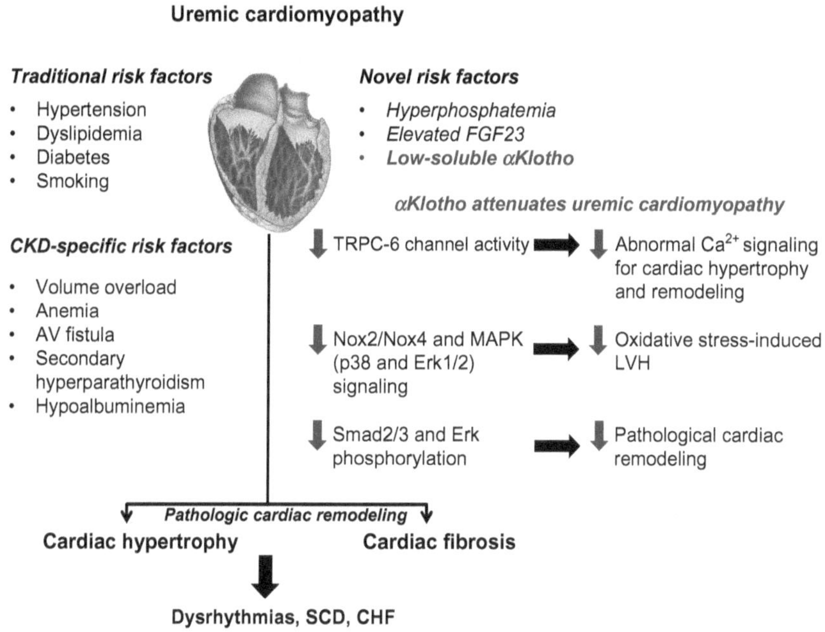

Fig. 4 Risk factors for uremic cardiomyopathy and proposed mechanisms of attenuation of pathological cardiac remodeling by αKlotho. αKlotho deficiency is a novel risk factor for uremic cardiomyopathy. Soluble αKlotho deficiency is an intermediate mediator of the pathological cardiac remodeling observed in CKD. Experimental αKlotho overexpression (*Tg-Kl*) suppressed phosphorylation of Smad2/3 and Erk, which are known to be involved in uremic cardiac fibrosis. Furthermore, αKlotho may protect the heart against stress-induced cardiac hypertrophy by inhibiting TRPC-6 channel-mediated abnormal Ca^{2+} signaling in the heart or against uremic solute indoxyl sulfate-induced myocardial hypertrophy probably by suppressing NADPH oxidase Nox2/Nox4-derived reactive oxygen species production and its downstream signaling. *CHF*, congestive heart failure; *Erk*, extracellular signal-regulated kinase; *LVH*, left ventricular hypertrophy; *MAPK*, mitogen-activated protein kinases; *Nox*, NADPH oxidase; *SCD*, sudden cardiac death; *TRPC-6*, transient receptor potential canonical-6. (See the color plate.)

5.1.1 αKlotho as a Modulator of Pathological Cardiac Remodeling

Uremic cardiomyopathy is a state of pathological cardiac remodeling characterized by left ventricular hypertrophy (LVH) and extensive fibrosis (Foley et al., 1995; Tyralla & Amann, 2003). Both primary genetic αKlotho deficiency (heterozygous αKlotho-deficient, *kl*/+ mice) and secondary αKlotho deficiency (from phosphate loading, aging, and CKD) triggered cardiac hypertrophy and fibrosis in mice, such that higher plasma phosphate and lower plasma αKlotho levels were associated with more severe cardiac hypertrophy and fibrosis (Hu, Shi, Cho, et al., 2015). Furthermore, higher plasma FGF23 levels were associated with more severe cardiac hypertrophy and fibrosis but only in the presence of moderate or low plasma αKlotho levels (Hu, Shi, Cho, et al., 2015). This suggests that FGF23 may not be cardiotoxic unless there is simultaneous αKlotho deficiency.

CKD models of secondary αKlotho deficiency included: (1) unilateral nephrectomy and contralateral ischemic–reperfusion injury followed by high-phosphate diet (2% phosphate) and (2) 5/6th nephrectomy. Both CKD models showed cardiac hypertrophy and left ventricular fibrosis. αKlotho levels in kidney tissue, plasma, and urine were decreased by high-phosphate diet starting at 6 months of age. When high-phosphate diet was given to older mice (12 months of age), additional reductions in plasma and kidney αKlotho were observed (Hu, Shi, Cho, et al., 2015). Cardiac hypertrophy and fibrosis were exaggerated in *kl*/+ mice and lessened in transgenic αKlotho-overexpressing mice (*Tg-Kl*) compared to WT mice and changes were more severe at age 15 months compared with 9 months. Aging exacerbated phosphate or αKlotho deficiency-induced pathological cardiac remodeling. Notably, αKlotho suppressed cardiac fibrosis triggered by high dietary phosphate. αKlotho overexpression (*Tg-Kl*) suppressed phosphorylation of Smad2/3 and extracellular signal-regulated kinase (Erk; Hu, Shi, Cho, et al., 2015), which are known to be involved in uremic cardiac fibrosis (Olson, Naugle, Zhang, Bomser, & Meszaros, 2005).

In vitro, αKlotho blocked TGF-β1- and angiotensin II (Ang II)-induced hypertrophy in cardiomyocytes (primary culture of neonatal rat) by inhibiting Smad2/3 phosphorylation. αKlotho also attenuated TGF-β1-, Ang II-, and high phosphate-induced upregulation of fibrosis markers in cultured cardiac fibroblasts by inhibiting Erk phosphorylation (Hu, Shi, Cho, et al., 2015).

Cardiac hypertrophy and fibrosis scores correlated negatively with plasma αKlotho levels and positively with plasma phosphate levels. In multivariable analysis, adjusting for plasma creatinine, FGF23, PTH, and 1,25-$(OH)_2$-vitamin D_3 levels, only plasma αKlotho and phosphorus levels were independent factors associated with pathological cardiac remodeling (Hu, Shi, Cho, et al., 2015).

5.1.2 αKlotho Protection Against Stress-Induced Cardiac Hypertrophy

Xie and coworkers reported that cardioprotection by αKlotho in normal mice is mediated by downregulation of TRPC-6 channels in the heart (Xie et al., 2012). In their experiments, deletion of TRPC-6 prevented stress-induced exaggerated cardiac remodeling in αKlotho-deficient mice (*kl*/+). In contrast, mice with heart-specific overexpression of TRPC-6 developed spontaneous cardiac hypertrophy and remodeling. Furthermore, αKlotho overexpression (*Tg-Kl* mice) ameliorated pathological cardiac remodeling and improved long-term survival (Xie et al., 2012). In addition, they proposed that soluble αKlotho inhibits TRPC-6 currents in cardiomyocytes by blocking phosphoinositide-3-kinase-dependent exocytosis of TRPC-6 channels (Xie et al., 2012).

Subsequently, the same investigators inferred that the decrease in soluble αKlotho in CKD is not only an important cause of uremic cardiomyopathy but independent of FGF23 and phosphotoxicity (Xie et al., 2015). They reported that αKlotho levels in αKlotho-deficient mice (*kl*/+) were about one half of those of WT mice, and they further decreased in *kl*/+ CKD mice to barely detectable levels (Xie et al., 2015). Heart weight-to-body weight ratio (a measure of cardiac hypertrophy) was significantly increased in both *WT* and *kl*/+ CKD mice, but the increase in *kl*/+ CKD mice was significantly more prominent than that in WT CKD mice. Similarly, the degree of fibrosis in *kl*/+ CKD was much more severe than that in *WT* CKD heart (Xie et al., 2015). *WT* CKD mice had ventricular hypertrophy, normal chamber size, and preserved contractility (diastolic dysfunction) in contrast to systolic dysfunction with dilated cardiomyopathy and impaired contractility in *kl*/+ CKD mice. Dietary phosphate restriction was successfully utilized to normalize serum phosphate and FGF23 levels in CKD mice (serum phosphate and FGF23 levels were similar in *WT* and *kl*/+ CKD mice compared with sham controls). Notably, dietary phosphate restriction did not significantly alter the pattern of cardiac hypertrophy in *WT* or *kl*/+ CKD mice (Xie et al., 2015). Moreover, viral-based deliver of αKlotho transgene significantly ameliorated cardiac hypertrophy and fibrosis in *kl*/+ CKD mice when compared with empty vector-injected mice. Functional TRPC-6-mediated currents were increased in cardiac myocytes isolated from CKD mice (vs sham), and the increase was more pronounced in *kl*/+ CKD vs WT CKD mice. Extracellular application of soluble αKlotho decreased these currents, confirming that αKlotho directly affects TRPC-6 functionality (Xie et al., 2015). The mechanisms of how the increase in TRPC-6 induces cardiomyopathy are not currently known.

5.1.3 αKlotho Protection Against Indoxyl Sulfate-Induced Myocardial Hypertrophy

Yang and colleges studied 86 patients with CKD and showed higher levels of indoxyl sulfate (Yang et al., 2015), a uremic solute derived from dietary protein and excreted by the kidney. Indoxyl sulfate accumulates with progressive loss of kidney function and can induce vascular endothelial cell dysfunction by enhancing oxidative stress (Tumur & Niwa, 2009; Tumur, Shimizu, Enomoto, Miyazaki, & Niwa, 2010). They showed a negative correlation between serum levels of indoxyl sulfate and αKlotho ($r = -0.59$, $p < 0.001$). Importantly, serum levels of indoxyl sulfate and αKlotho were independently associated with LVH (Yang et al., 2015). This was further confirmed by experiments in normal mice in which intraperitoneal injection of indoxyl sulfate for 8 weeks induced LVH, accompanied by substantial renal αKlotho downregulation. Notably, indoxyl sulfate-induced LVH was more severe in heterozygous αKlotho-deficient (*kl*/+) mice relative to *WT* mice, indicating that αKlotho deficiency may exacerbate indoxyl sulfate-mediated LVH (Yang et al., 2015) and αKlotho supplementation may be a strategy to counteract indoxyl sulfate-mediated LVH.

In vitro experiments showed that indoxyl sulfate induces cardiomyocyte hypertrophy through activation of Nox/ROS/MAPK (p38 and Erk1/2) signaling pathways and this activation can be attenuated by pretreatment with αKlotho protein, possibly through inhibition of ROS signaling. Indoxyl sulfate-induced cardiomyocyte hypertrophy is not mediated through TRPC-6 signaling pathway (Yang et al., 2015). Interestingly, the in vivo administration of exogenous αKlotho protein significantly alleviated the development of LVH in a mouse model of CKD-associated LVH characterized by high serum indoxyl sulfate levels, which confirmed in vitro findings (Yang et al., 2015). Moreover, indoxyl sulfate has been shown to suppress αKlotho deficiency in mice (Adijiang, Shimizu, Higuchi, Nishijima, & Niwa, 2011) and downregulate αKlotho expression in cultured cells (Shimizu et al., 2011; Sun et al., 2012). Therefore, indoxyl sulfate has a dual effect: induction of cardiomyocyte hypertrophy and induction of αKlotho deficiency to enhance cardiomyocyte hypertrophy.

5.2 Vascular Medial Calcification

Apart from traditional risk factors, the high CV morbidity and mortality in CKD have been linked to CKD-specific mechanisms of vascular calcification through modulation of the endothelium–vascular smooth muscle network (Hu et al., 2014; Vervloet, Adema, Larsson, & Massy, 2014). Calcium and phosphate play an important role in the initiation of osteochondrogenic changes of cellular elements in the arterial wall, and also in the final common pathway of alleged ectopic bone formation (Vervloet et al., 2014). Although the expression of αKlotho in the vasculature is highly controversial, there are data associating changes in circulating αKlotho levels with uremic vasculopathy (Hu et al., 2014; Vervloet et al., 2014). Blood vessels are composed of endothelial cells, mural cells (smooth muscle cells and pericytes), their shared basement membrane, and extracellular matrix. Vascular smooth muscle cells (VSMCs) and endothelial cells work synergistically for maintenance of the integrity of the vasculature (Heydarkhan-Hagvall et al., 2003; Fig. 5).

5.2.1 αKlotho and Endothelium

Endothelial dysfunction is associated with CV morbidity and mortality in CKD (Ravani et al., 2005). Endothelial cells damaged by high phosphate or uremic solutes show increase in apoptosis, ROS, proinflammatory cytokines, profibrotic and proangiogenic growth factors, and impaired nitric oxide production (Carracedo et al., 2013; Di Marco et al., 2008). Vascular endothelium can be a source of osteoprogenitor cells in vascular calcification (Di Marco et al., 2008). Moreover, the endothelium could also stimulate VSMCs to initiate or participate in vascular calcification (Yao et al., 2013).

A functional vascular tone and low levels of oxidative stress are maintained by releasing nitric oxide, prostacyclin, and endothelin-1, and by controlling local angiotensin II activity. In addition, the endothelium also regulates vascular permeability, platelet and leukocyte adhesion and aggregation, and thrombosis (Sitia et al., 2010). Elevated asymmetric dimethylarginine, a known inhibitor of nitric oxide synthase, and consequent reduced nitric oxide production and reduced flow-mediated dilatation (FMD) of the vessels have been characterized in CKD (Schwedhelm & Boger, 2011; Yilmaz et al., 2006). High phosphate impairs FMD in experimental CKD and FMD is inversely related to serum phosphate level in humans (Shuto et al., 2009; Van et al., 2012). Similarly, oxidative stress and inflammation are implicated in the development of endothelial dysfunction in CKD (Recio-Mayoral, Banerjee, Streather, & Kaski, 2011).

Interestingly, circulating αKlotho regulates vasodilation through modulation of nitric oxide production in vascular endothelium (Nagai et al., 2000; Saito et al., 1998; Six et al., 2014; Yamagishi et al., 2001). In addition, treatment of cultured endothelial cells with αKlotho alleviates tumor necrosis factor α-mediated ROS activity, cell apoptosis, and induction of adhesion molecules (Carracedo et al., 2012; Maekawa et al., 2009; Yang et al., 2012).

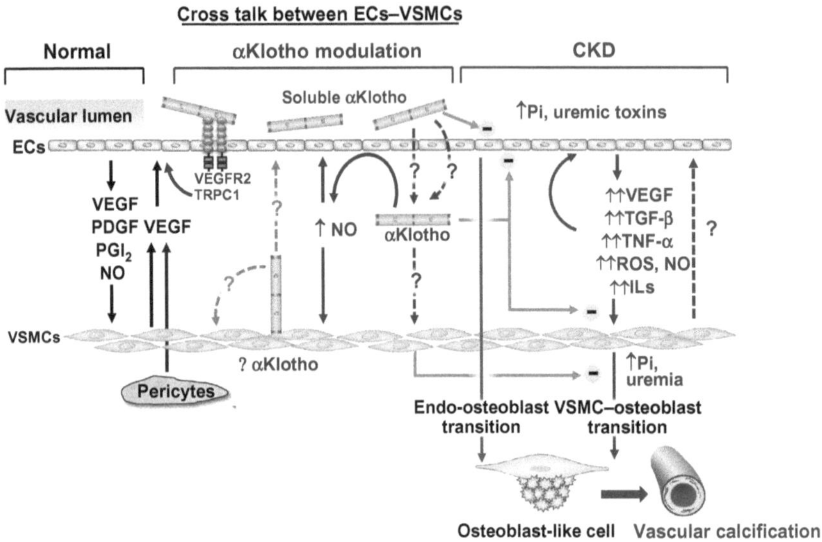

Fig. 5 Proposed model of αKlotho as intermediate of endothelial cells (ECs)–vascular smooth muscle cells (VSMCs) cross talk. *Left panel*: Normal ECs–VSMCs cross talk. Normal ECs modulate VSMCs growth via release of growth factors (*black line* in *left panel*). *NO*, nitric oxide; *PDGF*, platelet-derived growth factor; *PGI_2*, prostaglandin I_2; *VEGF*, vascular endothelial growth factor. VEGF released from pericytes and/or VSMCs could also regulate endothelial cell. *Right panel*: In CKD, uremic toxins including high plasma Pi, damage ECs and induce release of growth factors, proinflammatory cytokines, and profibrotic factors, which exacerbate ECs injury and also induce VSMCs transition to osteoblast and promote vascular calcification in medial layer (*red solid lines* in *right panel*). Impaired ECs also directly contributes to vascular calcification through endo-osteoblast transition (*green line* in *right panel*). Whether dedifferentiated or damaged VSMCs could further modulate function of endothelial cells is speculative (*red dash line* in *right panel*). *Central panel*: αKlotho is a vascular protective protein. Whether resident aortic αKlotho protein in VSMCs functions in autocrine and/or paracrine mode to modulate VSMCs and/or ECs remains to be clarified (*brown dash line* in *central panel*). The mechanisms of how αKlotho is able to reach the VSMCs from the circulation and function as an endocrine factor remain to be defined (*blue dash line* in *central panel*). αKlotho, regardless of source, could increase NO production from ECs and NO consequently modulates VSMCs and ECs function in an autocrine mode (*blue solid line* in *middle panel*). αKlotho could protect EC from high phosphate and other uremic toxins and also attenuate oxidative stress and proinflammatory cytokines-induced cell senescence and apoptosis in VSMC (*orange line* in *central panel*). αKlotho also directly inhibits osteoblast transition induced by hyperphosphatemia and uremic milieu (*orange line* in *central panel*). Current experimental and clinical observations suggest that both ECs and VSMCs endothelium may be potential targets of soluble αKlotho to protect the vasculature from vascular calcification in CKD. *ECs*, endothelial cells; *Pi*, phosphorus; *VSMCs*, vascular smooth muscle cells. (See the color plate.)

Importantly, αKlotho-deficient mice have increased VEGF-mediated calcium influx, downregulation of cadherin surface expression, increased apoptosis, and increased permeability (Kusaba et al., 2010). It is suggested that the *Kl2* domain of αKlotho protein binds directly to VEGFR-2 and endothelial TRPC-1 Ca^{2+} channel and promotes their cointernalization and consequent reduction of cellular Ca^{2+} influx limiting the activity of Ca^{2+}-dependent proteases that disrupt endothelial integrity (Kusaba et al., 2010). αKlotho protein is also capable of attenuating indoxyl sulfate-induced endothelial dysfunction, partly through inhibition of ROS/p38 mitogen-activated protein kinase and downstream nuclear factor-κB signaling pathways (Yang et al., 2012).

It is possible that αKlotho may act on the endothelium and induce a secondary effect via endothelial-VSMC crosstalk, or that VSMC-resident αKlotho regulates VSMC function in an autocrine mode or even endothelium in a paracrine mode (Fig. 5). The role of αKlotho protein in the disturbed endothelium–vascular smooth muscle network in CKD requires further investigation.

5.2.2 αKlotho and Vascular Smooth Muscle

Measurement of aortic αKlotho mRNA expression has not been consistent (Hu et al., 2014; Lim et al., 2012; Mencke et al., 2015; Navarro-Gonzalez et al., 2014; Ritter et al., 2015; Six et al., 2014). However, the association between αKlotho deficiency and medial vascular calcification has been well documented in hypomorphic αKlotho mice (Kuro-o et al., 1997), a phenotype rescued by transgenic overexpression, viral delivery of αKlotho, or recombinant αKlotho protein (Chen, Kuro, et al., 2013; Masuda et al., 2005; Shiraki-Iida et al., 2000). Furthermore, transgenic mice overexpressing αKlotho had significantly less vascular calcification after CKD induction in comparison to αKlotho-haploinsufficient mice with CKD that exhibited more severe vascular calcification (Hu et al., 2011).

The potential mechanisms underpinning the association between high serum phosphate and vascular calcification have been described in experimental models of CKD (Jono et al., 2000; Lomashvili, Cobbs, Hennigar, Hardcastle, & O'Neill, 2004; Mathew et al., 2008). The beneficial effect of αKlotho on vascular calcification in CKD is thought to be a result of more than its effect on amelioration of renal dysfunction and hyperphosphatemia. In vitro experiments have shown that αKlotho suppresses type III Na^{+}-dependent uptake of phosphate (Pit-1 and Pit-2 cotransporters) and mineralization induced by high phosphate in VSMCs

(Hu et al., 2011). Runt-related transcription factor-2 (Runx2) expression, an early marker of ectopic osteogenesis, was decreased in the aortas of overexpressing αKlotho mice (Hu et al., 2011). The contractile phenotype of VSMCs was lost with exposure to high phosphate or resident αKlotho knockdown (Lim et al., 2012). Resident αKlotho knockdown in VSMCs accelerated the development of vascular calcification through Runx2 and myocardin-serum response factor-dependent pathway (Lim et al., 2012). Therefore, αKlotho may regulate VSMCs differentiation under CKD procalcific stressors (Hu et al., 2011; Lim et al., 2012). However, other experiments have resulted in conflicting evidence. One study did not show any effect of αKlotho protein on FGF23 and high phosphate-mediated vascular calcification in human or mouse VSMCs (Scialla et al., 2013). Another study performed in uremic rats showed that FGF23 augmented phosphate-induced aortic calcification in αKlotho-overexpressing but not naive VSMCs through Erk1/2 phosphorylation pathway (Jimbo et al., 2014). The conflicting in vitro data may be, besides plausibility, a reflection of differences in VSMCs or αKlotho protein preparations utilized.

Vitamin D receptor agonists (eg, calcitriol or paracalcitol) were shown to increase serum and urine αKlotho levels and abate aortic calcification in CKD mice likely through modulation of osteopontin, an anticalcification factor in VSMCs (Lau et al., 2012). Importantly, no αKlotho mRNA expression was found in the aorta in these in vivo experiments (Lau et al., 2012). One independent study further confirmed that there is no membrane αKlotho expression in either healthy or uremic vessels in humans (Mencke et al., 2015). In contrast, Lim and colleagues showed that calcitriol effectively restored mRNA αKlotho expression in VSMCs (Lim et al., 2012). In a different experiment, αKlotho mRNA was detected in mouse aorta but specific deletion of αKlotho in mouse VSMCs did not induce vascular calcification (Lindberg et al., 2013) challenging the principal role of αKlotho in VSMCs. Moreover, αKlotho protein was found to be increased in atherosclerotic arteries (Donate-Correa et al., 2013) contradicting the suggested protective role of resident αKlotho in vasculature. The existence and role of resident αKlotho protein in the vasculature need to be clarified and further investigated (Table 1).

In contrast to the inconclusive evidence of vascular αKlotho expression, the administration of exogenous recombinant αKlotho, αKlotho gene delivery, and increased endogenous circulating αKlotho significantly reduced vascular calcification and improved endothelial function, suggesting that soluble αKlotho may play a pivotal role in the protection of vasculature

Table 1 The Determination of Aortic αKlotho mRNA Expression Has Been Inconsistently Reproduced in Different Studies

Aortic αKlotho mRNA Expression = Yes	Aortic αKlotho mRNA Expression = No
Klotho mRNA and protein is present in aorta and vascular smooth muscle cells Resident αKlotho knockdown induced vascular calcification through *Runx2* and myocardin-serum response factor-dependent pathway Calcitriol effectively restored mRNA αKlotho expression in VSMCs (Lim et al., 2015, 2012)	No αKlotho mRNA expression was found in the aorta in in vivo experiments of aortic calcification in CKD (Lau et al., 2012)
αKlotho mRNA was detected in mouse aorta but specific deletion of αKlotho in mouse VSMCs did not induce vascular dysfunction and vascular calcification (Lindberg et al., 2013)	No membrane αKlotho expression in either healthy or uremic human vessels was found (Mencke et al., 2015)
αKlotho protein was found to be increased in atherosclerotic arteries (Donate-Correa et al., 2013)	
Human umbilical vein endothelial cells (HUVECs) express αKlotho. The decline in αKlotho preceded the manifestations of cell aging induced by repeated passage and those of senescence by TNF-α. The exogenous αKlotho administration prevented TNF-α-mediated senescence (Carracedo et al., 2012)	

integrity as an endocrine factor (Chen, Kuro, et al., 2013; Lau et al., 2012; Masuda et al., 2005; Saito et al., 2000; Utsugi et al., 2000; Fig. 5).

It was shown that αKlotho protein attenuates endothelial cell damage from high phosphate and oxidative stress, and inhibits osteogenic transformation of VSMCs induced by high phosphate. The administration of exogenous αKlotho or modulators of αKlotho expression may represent novel therapies for the management of vascular calcification in CKD patients. Studies of endothelial cells or VSMCs in isolation may not fully represent the vascular system to dissect out the role of the individual players. Coculture of endothelial cells and VSMCs may be a viable intermittent system to further elucidate the role of αKlotho in the vasculature under normal and pathological conditions (Hu et al., 2014).

6. αKlotho DEFICIENCY AS A BIOMARKER OF CKD

CKD is a global public health problem that affects over 20 million people in the United States (Snyder, Foley, & Collins, 2009). The major complications in this population are progression to ESRD and CV morbidity and mortality (Eckardt et al., 2013; Hemmelgarn et al., 2010; Levey et al., 2011).

There has been intense search for highly sensitive (diagnostic value) or highly specific (treatment effect value) biomarkers of CKD onset and/or prognosis of progression. An ideal prognostic CKD biomarker should be able to predict CKD onset and progression, characterize the severity of CKD stage, display similar reliability across multiple species (particularly humans), and be accessible in readily available body fluids or tissues.

In the following sections, novel functional (detecting primarily loss of kidney function) and injury biomarkers (with or without loss of kidney function) of CKD and the potential role of FGF23 and αKlotho as early diagnostic and prognostic biomarkers of CKD will be discussed.

6.1 Functional Biomarkers in Human CKD

New filtration markers such as β-trace protein (BTP), β2 microglobulin (β2M), and cystatin C were associated with mortality risk in a representative sample of 6445 US adults from the Third National Health and Nutrition Examination Survey (Foster et al., 2013). The highest quintile for cystatin C, BTP, and β2M were associated with increased all-cause mortality risk, whereas the association was weaker for serum creatinine-based eGFR (Foster et al., 2013). A >50% decline in serum creatinine-based eGFR is an established surrogate marker for ESRD in clinical trials but a >30% decline in kidney function assessed using novel filtration markers (cystatin C and β2M) has been strongly associated with ESRD (Rebholz, Grams, Matsushita, Selvin, & Coresh, 2015).

6.2 Injury Biomarkers in Human CKD

The assessment of the plasma proteome through mass spectrometry analysis identified three fragments of high-molecular-weight kininogen associated with early progressive renal function decline in microalbuminuric patients with type 1 diabetes (Merchant et al., 2013). The performance of urine neutrophil gelatinase-associated lipocalin (NGAL) was analyzed in a cohort of

3386 patients with CKD in the Chronic Renal Insufficiency Cohort (CRIC) study. Urine NGAL was independently associated with 50% decreased eGFR or incident ESRD development over a mean follow-up of 3.2 years. However, it did not improve prediction models for CKD progression (Liu et al., 2013). More recently, urine NGAL has been independently associated with ischemic atherosclerotic events but not heart failure events or death (Fufaa et al., 2015). In a cohort of 124 patients with type 1 diabetes and proteinuria, serum kidney injury molecule-1 (KIM-1) levels at baseline strongly predicted rate of eGFR loss and risk of ESRD during 5–15 years of follow-up, after adjustment for baseline urinary albumin-to-creatinine ratio, eGFR, and hemoglobin A1C (Sabbisetti et al., 2014). Similarly, urinary NGAL and liver fatty acid-binding protein (L-FABP) were independently associated with incident ESRD and mortality but did not meaningfully improved clinical prediction models of CKD progression in a cohort of 260 Pima Indians with median follow-up of 14 years (Fufaa et al., 2015).

6.3 FGF23 in Human CKD

Current evidence favors a direct pathogenic role for dysregulated FGF23-α Klotho in CKD-related adverse outcomes, in particular CV disease. The relationship between baseline serum intact FGF23 and incident ESRD was evaluated in 13,448 Atherosclerosis Risk in Communities (ARIC) study participants during a median follow-up of 19 years. After adjustment for demographics, baseline eGFR, and traditional CKD risk factors, the highest FGF23 quintile (>54.6 pg/mL) compared with the lowest quintile (<32.0 pg/mL) was associated with risk of developing ESRD (Rebholz, Grams, Coresh, et al., 2015). Similarly, elevated FGF23 has been proposed as an early indicator of kidney injury or CKD progression (Fliser et al., 2007; Wolf, 2012). Elevated serum FGF23 has been independently associated with LVH and a causal inference through Klotho-independent activation of FGF receptor-dependent activation of the calcineurin–NFAT signaling pathway in rat cardiomyocytes has been proposed (Faul et al., 2011). Increased FGF23 was independently associated with mortality among incident hemodialysis patients (Gutiérrez et al., 2008) and has been linked to mortality and higher risk of ESRD or CV disease in patients with CKD during a median follow-up of 3.5 years (Faul et al., 2011; Ix et al., 2012). In contrast, FGF23 was not associated with arterial calcification in 1501 patients from the CRIC study (Scialla et al., 2013).

6.4 The Role of αKlotho as a Marker of Adverse Outcomes in Human CKD

CKD is a state of αKlotho deficiency in multiple tissues. αKlotho mRNA levels in parathyroid gland declined in parallel with decreasing eGFR over CKD stages (Krajisnik et al., 2010). Similarly, αKlotho mRNA expression in kidney tissue was greatly reduced and positively and significantly correlated with eGFR in CKD patients (Asai et al., 2012,; Koh et al., 2001). In a larger sample of 236 CKD patients with available kidney biopsies, αKlotho mRNA levels were significantly and positively correlated with eGFR ($p < 0.001$) in multiple regression analysis including CKD-MBD parameters (Sakan et al., 2014). Most importantly, αKlotho mRNA in the kidney was the only independent contributing factor to serum αKlotho across all strata of CKD patients. Renal αKlotho was significantly correlated with serum calcium, serum phosphorus, 1,25-$(OH)_2$-vitamin D_3, FGF23, and intact PTH (Sakan et al., 2014). Correspondingly, circulating serum αKlotho levels were progressively lower with each CKD stage when compared to healthy controls (Pavik et al., 2013). This was also observed in kidney transplant recipients vs healthy controls (Sawires, Essam, Morgan, & Mahmoud, 2015). Furthermore, adjusted mean serum αKlotho decrease was 3.2 pg/mL for each 1 mL/min eGFR decrease in adult CKD patients (Pavik et al., 2013). A positive correlation between αKlotho levels (serum and urine) and eGFR has been further characterized in adult CKD patients (Akimoto et al., 2012; Hu et al., 2011; Kim et al., 2013; Kitagawa et al., 2013; Ozeki et al., 2014), although only 24-h urine αKlotho but not serum αKlotho has been shown to be independently associated with eGFR change (Akimoto et al., 2012). However, questions about the stability of αKlotho in the urine have emerged (Adema, Vervloet, Blankenstein, & Heijboer, 2015). Therefore, a standardized urine protocol is required. A similar positive correlation between plasma αKlotho levels and eGFR was shown in CKD children without kidney transplant (Wan et al., 2013). Moreover, serum αKlotho levels in children on chronic peritoneal dialysis were significantly lower (~41%) than healthy controls (Cano et al., 2014). The progressive decline of serum αKlotho in adults with early stages of CKD (eg, stage 2) has been subsequently demonstrated, suggesting that the decrease in serum αKlotho may antecede high FGF23, PTH, and hyperphosphatemia (Barker et al., 2015; Kim et al., 2013; Rotondi et al., 2015; Shimamura et al., 2012; Table 2).

Table 2 Clinical Studies Demonstrating the Implication of Kidney and Circulating αKlotho in Different Strata of Chronic Kidney Disease and Complications of Chronic Kidney Disease

Study	Design	Sample	Exclusion	Measurement	Observations	Comments
Koh et al. (2001)	Cross sectional	10 CKD and 15 controls	None	Kidney tissue *kl* mRNA expression	αKlotho expression was greatly reduced in CKD vs controls	Controls included healthy parts of resected kidneys with renal cell carcinoma or angiomyolipoma
Asai et al. (2012)	Cross sectional	31 pts with DN, 31 IgAN, and 7 MCD	None	Kidney tissue *kl* mRNA expression	Kidney *kl* mRNA expression levels were positively and significantly correlated with eGFR ($r=0.35$, $p=0.003$)	Different slopes depending on CKD cause were observed, particularly in patients with eGFR >60
Pavik et al. (2013)	Cross sectional	87 CKD pts (stages 1–5) and 21 controls	PKD, kidney transplantation	Serum αKlotho	Adjusted mean αKlotho decrease was 3.2 pg/mL for each 1 mL/min eGFR decrease	Age and eGFR were independently associated with αKlotho
Wan et al. (2013)	Cohort	154 children with CKD (1–5, 28 on dialysis, 44 posttransplant)	None	*Predictor*: Plasma αKlotho *Outcome*: eGFR Follow-up: 1 year	αKlotho levels decreased with decreasing eGFR in CKD children without kidney transplant	No independent association between plasma αKlotho and eGFR. Decreased αKlotho was associated with increased FGF23 and PTH levels in CKD children without kidney transplant

Continued

Table 2 Clinical Studies Demonstrating the Implication of Kidney and Circulating αKlotho in Different Strata of Chronic Kidney Disease and Complications of Chronic Kidney Disease—cont'd

Study	Design	Sample	Exclusion	Measurement	Observations	Comments
Kitagawa et al. (2013)	Cross sectional	114 CKD pts	Patients with established atherosclerotic complications (CAD, CHF, PVOD) or those treated with vitamin D or phosphate binders	Serum αKlotho	αKlotho was a significant determinant of arterial stiffness (adjusted OR per 100 pg/mL increase 0.60, 95% CI 0.39–0.98, $p=0.008$)	Positive correlation between serum αKlotho and eGFR. Arterial stiffness was determined by ankle–brachial pulse wave velocity
Kim et al. (2013)[a]	Cohort	243 pts with CKD (stages 1–5), post hoc analysis	RRT, organ transplantation, heart failure, cirrhosis, malignancy, pregnancy, acute coronary syndrome, or ischemic stroke within 3 months prior to the study, progressive CKD within 3 months prior to the study	*Predictor*: Serum αKlotho *Outcome*: composite of doubling SCr, ESRD, or death Follow-up: median 29.7 months	αKlotho level independently predicted the composite outcome after multivariate adjustment (adjusted HR per 10 pg/mL increase 0.96, 95% CI 0.94–0.98, $p<0.001$). If serum αKlotho was $\leq$396.3 pg/mL, 35.2% reached the composite outcome vs 15.7% if >396.3 pg/mL ($p=0.03$)	αKlotho levels were lower at more advanced CKD stages (p for trend <0.001) and correlated positively with eGFR and negatively with FGF23 and phosphate level. In multivariate linear regression analysis, αKlotho was independently associated with eGFR ($p<0.001$)

Akimoto et al. (2012)	Cross sectional	131 CKD pts (stages 1–5)	RRT	Serum and urine αKlotho	Positive correlation between serum and urine αKlotho and eGFR	24-h urine αKlotho was independently associated with eGFR
Sakan et al. (2014)	Cross sectional	236 CKD pts (stages 1–5), 43 on dialysis	None	Kidney tissue *kl* mRNA expression and serum αKlotho	Kidney *kl* mRNA levels were significantly and positively correlated with eGFR ($p < 0.001$) in multiple regression analysis including CKD-MBD parameters	Kidney *kl* mRNA was the only independent contributing factor to serum αKlotho across all strata of CKD patients ($p < 0.001$)
Seiler et al. (2014)	Cohort	444 CKD pts (stages 2–4)	None	*Predictor*: plasma αKlotho *Outcome*: (1) composite of atherosclerotic event or death or (2) time until ADHF admission or death Follow-up: median of 2.6 years	Plasma αKlotho levels (highest vs lowest tertile) did not predict atherosclerotic events/death (HR 0.75, 95% CI 0.43–1.30, $p = 0.30$) or ADHF/death	Events were adjudicated by two independent nephrologists

Continued

Table 2 Clinical Studies Demonstrating the Implication of Kidney and Circulating αKlotho in Different Strata of Chronic Kidney Disease and Complications of Chronic Kidney Disease—cont'd

Study	Design	Sample	Exclusion	Measurement	Observations	Comments
Ozeki et al. (2014)	Cross sectional	185 CKD pts (stages 1–5)	RRT	Serum αKlotho	eGFR was an independent predictor of serum αKlotho levels in multivariate linear regression analysis, specifically when eGFR was <60	Serum αKlotho was not significantly correlated with serum calcium or phosphate
Cano et al. (2014)	Cohort	31 children on chronic PD and 45 healthy controls	None	*Predictor*: CKD-MBD parameters *Outcome*: serum αKlotho Follow-up: 1 year	Baseline αKlotho levels were lower than controls (132 ± 58 vs 320 ± 119 pg/mL, $p < 0.001$) and remained virtually unchanged throughout the observation period	αKlotho levels did not correlated with FGF23 and phosphorus levels
Park et al. (2015)	Cross sectional	24 HTN pts (12 with RVH and 12 EH) and 12 HV controls	eGFR < 30, uncontrolled BP, diabetes, recent CV event (within 6 months), pregnancy, kidney transplant	Serum αKlotho	Serum αKlotho was significantly reduced in HTN pts vs HV controls after adjustment by eGFR	eGFR correlated directly with serum αKlotho levels

Sawires et al. (2015)	Cross sectional	40 CKD pts (stages 2–5), 44 pts with ESRD on HD, 40 kidney transplant recipients, and 40 healthy controls	Marked hypocalcemia, dysfunctional HD access, combined organ transplantation, surgical parathyroidectomy, sarcoidosis	Serum αKlotho	Using multivariate regression analysis, only serum calcium was an independent predictor of serum αKlotho in all groups with kidney disease	There was an inverse significant correlation between serum calcium and αKlotho
Shroff et al. (2016)	Cohort	ESCAPE cohort post hoc analysis, 167 children with CKD on ACEI	None	*Predictor*: ACEI therapy *Outcome*: serum αKlotho Follow-up: 8 months	ACEI therapy significantly increased serum αKlotho levels without any associated changes in serum calcium or phosphate	αKlotho levels did not correlate with eGFR at baseline ($r=0.02$, $p=0.82$) or at 8-month follow-up ($r=0.06$, $p=0.45$)

[a]Highlighted as an important observational study.

6.4.1 *α*Klotho and CV Disease in CKD

A cross-sectional study of 114 CKD patients (mean eGFR 48 ± 29 mL/min/1.73 m^2) revealed that serum αKlotho was a significant marker of arterial stiffness measured by ankle–brachial pulse waive velocity (Kitagawa et al., 2013). In contrast, a large cohort study of 444 patients with CKD stages 2–4 showed that plasma αKlotho levels (highest vs lowest tertile) did not predict atherosclerotic events or death at 2.6 years follow-up. Serum αKlotho was also significantly reduced in hypertensive (essential and renovascular) patients with mild CKD when compared to healthy controls, even after adjustment by eGFR (Park et al., 2015). The proposed cross talk between the renin–angiotensin–aldosterone system and the vitamin D–FGF23–α Klotho pathways supports the concept that modulation of one system can have positive effects on the other (de Borst, Vervloet, ter Wee, & Navis, 2011). In this context, a post hoc analysis of the ESCAPE trial in children with CKD (all received fixed dose of ramipril 6 mg/m^2 per day) showed that 25(OH)D $\geq$50 nmol/L was associated with greater preservation of renal function. Interestingly, ACEI therapy significantly increased serum αKlotho levels without any associated changes in serum calcium or phosphate (Shroff et al., 2016; Table 2).

6.4.2 *α*Klotho and Progressive CKD

The most conclusive evidence thus far about the role of αKlotho as a predictor of adverse outcomes in CKD is based on a post hoc cohort study of 243 adult patients with CKD. In this study, serum αKlotho levels independently predicted the composite outcome of doubling S_{Cr}, ESRD, or death after multivariable adjustment. If serum αKlotho was $\leq$396.3 pg/mL, 35.2% reached the composite outcome vs 15.7% if >396.3 pg/mL (adjusted HR 2.03, 95% CI 1.07–3.85, $p=0.03$). The areas under the curve (a measure of discrimination, that is, the ability of αKlotho to correctly classify those with and without the outcome) for 1/serum αKlotho to predicate the composite outcome (doubling S_{Cr}, ESRD, or death) were 0.81, 0.78, and 0.72 at 12, 24, and 36 months (Kim et al., 2013; Table 2). Further studies are needed to corroborate these findings.

Taken together, CKD is a state of αKlotho deficiency and dysfunctional vitamin D–FGF23–αKlotho pathways. Plasma αKlotho level positively correlates with eGFR and negatively with S_{Cr} and FGF23 and is a promising biomarker for the prediction of adverse outcomes (eg, CKD progression, CV morbidity, and death) in CKD patients. Therefore, αKlotho may represent a novel therapy for CKD patients that needs to be further investigated.

7. αKlotho AS A PROMISING TREATMENT STRATEGY FOR CKD

The kidney is confirmed as the major site that contributes to circulating αKlotho (Hu, Shi, Zhang, et al., 2015; Lindberg et al., 2014) and dysfunctional or decreased number of αKlotho-producing cells contribute to αKlotho deficiency leading to accelerated aging (Hu, Shi, Zhang, et al., 2015; Lindberg et al., 2014). Therefore, αKlotho deficiency may not only be a pathogenic intermediate for accelerating CKD progression but also a main promoter of complications such as secondary hyperparathyroidism and CV disease in CKD. Conceivably, any therapy that restores or stimulates endogenous αKlotho or administration of exogenous αKlotho might provide a novel treatment strategy in CKD.

7.1 Epigenetic regulation of αKlotho Expression

αKlotho deficiency in the kidney of hypomorphic αKlotho-deficient mice was thought to result from interruption of the promoter of *αKlotho* gene by the exogenous transgene (Kuro-o et al., 1997). But recent data do not entirely support this notion because there was aberrant αKlotho promoter methylation in *kl/kl* mice. The in vitro study showed that *αKlotho* gene promoter methylation reduced promoter activity by 30–40%, whereas DNA demethylating agents increased *αKlotho* expression 1.5- to 3.0-fold (Azuma et al., 2012). Similarly, uremic toxins—indoxyl sulfate or *p*-cresyl sulfate—induced hypermethylation of the *αKlotho* gene, and decreased αKlotho expression in renal tubules and kidney cell line, which can be reversed by demethylation of the *αKlotho* gene. Therefore, hypermethylation may be one of the mechanisms of *αKlotho* gene expression inhibition in CKD (Chen, Zhang, et al., 2013; Sun et al., 2012; Young & Wu, 2012). In addition, promoter histone acetylation was also proposed as a possible mechanism for αKlotho silencing in many types of cancer cell lines (Rubinek et al., 2012; Xie et al., 2013). Furthermore, TNF and TNF-like weak inducer of apoptosis (TWEAK)-induced downregulation of αKlotho expression in the kidney and kidney cell lines can be blunted by inhibition of histone deacetylase (Moreno et al., 2011). Therefore, demethylating agents and deacetylase inhibitors may be agents to reactivate αKlotho expression in the kidney and consequently increase circulating αKlotho.

7.2 Reactivation of Endogenous αKlotho Expression Independently of Epigenetics

To date, several categories of drugs in the market including peroxisome proliferator-activated receptors-gamma (PPAR-γ) agonists (Chen, Cheng, Ku, & Lin, 2014; Yang et al., 2009; Zhang et al., 2008; Zhang & Zheng, 2008), angiotensin II type I receptor antagonists (Karalliedde, Maltese, Hill, Viberti, & Gnudi, 2013; Yoon et al., 2011; Zhou et al., 2010), 3-hydroxy-3-methylglutaryl CoA (HMG-CoA) reductase inhibitors (statin; Narumiya et al., 2004), and vitamin D active derivatives (de Borst et al., 2011; Forster et al., 2011; Lau et al., 2012; Lim et al., 2012; Ritter et al., 2015) have been shown to upregulate αKlotho expression in vivo and in vitro. The effect of upregulating αKlotho is definitely not associated with their well-identified original pharmacological targets, and pharmacological mechanisms remain to be explored.

7.3 Administration of Soluble αKlotho Protein

αKlotho gene delivery is shown to effectively rescue many phenotypes observed in αKlotho-deficient mice (Shiraki-Iida et al., 2000), attenuating the progression of hypertension and kidney damage in spontaneous hypertensive rats (Wang & Sun, 2009, 2014), improving kidney function in acute kidney injury (Sugiura et al., 2005), ameliorating angiotensin II-induced kidney injury (Mitani et al., 2002), improving endothelial function (Saito et al., 2000), and protecting from uremic cardiomyopathy (Xie et al., 2012, 2015). Although gene therapy is effective in animal studies, its safety is still questionable and human clinical application is not in the proximity. There are only few clinical trials testing gene therapy in specific human diseases including genetic diseases (Williams, 2014) and some types of cancers (Heller & Heller, 2015).

In contrast, administration of exogenous αKlotho protein is more direct, safer, and an easier modality to restore endocrine αKlotho deficiency. Animal studies have already provided prove-of-concept encouraging data that soluble αKlotho protein administration is safe and effective (Hu, Shi, Zhang, Quinones, et al., 2010; Shi et al., 2015). Soluble αKlotho protein attenuates kidney damage and preserves kidney function in an ischemia-reperfusion injury model causing acute kidney injury, which is a state of acute αKlotho deficiency (Hu, Shi, Zhang, Quinones, et al., 2010). Furthermore, αKlotho protein inhibited renal fibrosis in a UUO kidney injury model (Doi et al., 2011), and effectively extended the life span of homozygous αKlotho-deficient mice, ameliorating premature aging-related phenotypes (Chen, Kuro, et al., 2013). Although there are no clinical data showing αKlotho protein administration to CKD patients, the preclinical data clearly support the therapeutic potential of soluble αKlotho protein in CKD patients.

8. CONCLUSION AND FUTURE DIRECTIONS

Although progress has been made in understanding that αKlotho is not only an aging suppressor involved in longevity and aging, but also a renoprotective factor which has a central impact on renal physiology and renal pathophysiology, the utility of αKlotho in clinical practice is still under development (Table 3). First, αKlotho could serve as an early and sensitive biomarker of CKD, although its specificity and its prognostic value require further exploration in humans. Similarly, whether urine or serum αKlotho is a better biomarker remains to be confirmed. Second, exogenous αKlotho supplementation may represent a novel therapy to retard or block progressive CKD, post-AKI transition to CKD, as well as preventing and reversing CV complications associated with renal disease as shown in animal studies. The therapeutic efficacy of αKlotho in kidney disease has been unequivocally demonstrated in animal models. One needs to validate the efficacy of αKlotho therapy in different stages of kidney disease.

Although the effects of αKlotho protein on the kidney will invariably be pleiotropic, how αKlotho exerts its renoprotection is largely inconclusive. High levels of FGF23 have been found to be associated with high mortality and morbidity in CKD (Arnlov et al., 2013; Desjardins et al., 2012; Faul et al., 2011; Fliser et al., 2007; Grabner et al., 2015; Greenhill, 2011; Guo & Yuan, 2015; Hanks, Casazza, Judd, Jenny, & Gutierrez, 2015; Hasegawa et al., 2010; Krupp & Madhivanan, 2014;

Table 3 Potential Applications of αKlotho in Chronic Kidney Disease

Biomarker		**Therapeutic Agent**	
Diagnostic	Early detection of CKD	Prevention	• CKD progression • CV comorbidity and complications
Prognostic	Prediction of progression to ESRD	Treatment	• Anti-Pi toxicity • Attenuation of CV complications

CKD, chronic kidney disease; *CV*, cardiovascular event; *ESRD*, end-stage renal disease.

Mencke et al., 2015; Mirza, Larsson, Melhus, Lind, & Larsson, 2009; Razzaque, 2009a, 2009b; Rotondi et al., 2015; Sawires et al., 2015; Silswal et al., 2014; Silver, Rodriguez, & Slatopolsky, 2012; Sinha et al., 2015; Wolf, 2010; Wright et al., 2014; Zhang, Yan, Zhu, & Ni, 2015; Zhang, Yang, et al., 2015). Given that administration of exogenous αKlotho has a favorable effect on CKD animals in terms of improvement of renal function, better maintenance of phosphate homeostasis, and attenuation of vascular calcification and cardiac hypertrophy, whether synergistic utilization of FGF23 antagonist or inhibitor and αKlotho can enhance αKlotho therapeutic efficacy needs to be tested.

Some therapeutic modalities including ACE inhibition, HMG-CoA reductase inhibition, vitamin D derivatives, and antioxidants could sustain or increase endogenous αKlotho expression. It is conceivable that a multi-targeted regimen including αKlotho protein, FGF23 antagonist, and endogenous αKlotho inducers may enhance the beneficial effect of αKlotho on retardation of CKD progression and prevention or reduction of CV morbidity and mortality in CKD. The immediate challenge is to test whether human CKD resembles the rodent counterpart and if so, how to more efficiently increase αKlotho levels in patients with CKD, either by stimulating endogenous αKlotho or by administering recombinant αKlotho. Finally, αKlotho has shown to be effective in experimental CKD models such as renal ablation (Xie et al., 2015), hypertensive kidney damage (Tang et al., 2011; Wang & Sun, 2009), UUO (Doi et al., 2011; Guan et al., 2014; Satoh et al., 2012; Sugiura et al., 2012), but few experimental data showed its effect on CKD secondary to diabetes or other glomerular diseases which account for large part of the CKD population. More animal studies and clinical observational studies are required to consider the administration of exogenous αKlotho protein a potential novel therapeutic strategy for CKD.

ACKNOWLEDGMENTS

The authors appreciate Dr. Orson Moe for helpful interactions in the editing of this chapter. The research work in MC Hu's research laboratory is in part supported by the NIH (R01-DK091392, R01-DK092461), the George M. O'Brien Kidney Research Center (P30-DK-07938), the Charles and Jane Pak Foundation, and the Pak Center Innovative Research Support. J.A.N. is in part supported by the Ben J. Lipps Research Fellowship Program of American Society of Nephrology Foundation for Kidney Research and the Truelson Fellowship Fund at the Charles and Jane Pak Center of Mineral Metabolism and Clinical Research.

REFERENCES

Adema, A. Y., Vervloet, M. G., Blankenstein, M. A., & Heijboer, A. C. (2015). α-Klotho is unstable in human urine. *Kidney International, 88*(6), 1442–1444.

Adijiang, A., Shimizu, H., Higuchi, Y., Nishijima, F., & Niwa, T. (2011). Indoxyl sulfate reduces klotho expression and promotes senescence in the kidneys of hypertensive rats. *Journal of Renal Nutrition, 21*(1), 105–109.

Akimoto, T., Kimura, T., Watanabe, Y., Ishikawa, N., Iwazu, Y., Saito, O., et al. (2013). The impact of nephrectomy and renal transplantation on serum levels of soluble Klotho protein. *Transplantation Proceedings, 45*(1), 134–136.

Akimoto, T., Yoshizawa, H., Watanabe, Y., Numata, A., Yamazaki, T., Takeshima, E., et al. (2012). Characteristics of urinary and serum soluble Klotho protein in patients with different degrees of chronic kidney disease. *BMC Nephrology, 13*, 155.

Andrukhova, O., Smorodchenko, A., Egerbacher, M., Streicher, C., Zeitz, U., Goetz, R., et al. (2014). FGF23 promotes renal calcium reabsorption through the TRPV5 channel. *The EMBO Journal, 33*(3), 229–246.

Ardura, J. A., Rayego-Mateos, S., Ramila, D., Ruiz-Ortega, M., & Esbrit, P. (2010). Parathyroid hormone-related protein promotes epithelial-mesenchymal transition. *Journal of the American Society of Nephrology, 21*(2), 237–248.

Arnlov, J., Carlsson, A. C., Sundstrom, J., Ingelsson, E., Larsson, A., Lind, L., et al. (2013). Serum FGF23 and risk of cardiovascular events in relation to mineral metabolism and cardiovascular pathology. *Clinical Journal of the American Society of Nephrology, 8*(5), 781–786.

Asai, O., Nakatani, K., Tanaka, T., Sakan, H., Imura, A., Yoshimoto, S., et al. (2012). Decreased renal alpha-Klotho expression in early diabetic nephropathy in humans and mice and its possible role in urinary calcium excretion. *Kidney International, 81*(6), 539–547.

Azuma, M., Koyama, D., Kikuchi, J., Yoshizawa, H., Thasinas, D., Shiizaki, K., et al. (2012). Promoter methylation confers kidney-specific expression of the Klotho gene. *The FASEB Journal, 26*(10), 4264–4274.

Barker, S. L., Pastor, J., Carranza, D., Quinones, H., Griffith, C., Goetz, R., et al. (2015). The demonstration of alphaKlotho deficiency in human chronic kidney disease with a novel synthetic antibody. *Nephrology, Dialysis, Transplantation, 30*(2), 223–233.

Ben-Dov, I. Z., Galitzer, H., Lavi-Moshayoff, V., Goetz, R., Kuro-o, M., Mohammadi, M., et al. (2007). The parathyroid is a target organ for FGF23 in rats. *The Journal of Clinical Investigation, 117*(12), 4003–4008.

Bian, A., Neyra, J. A., Zhan, M., & Hu, M. C. (2015). Klotho, stem cells, and aging. *Clinical Interventions in Aging, 10*, 1233–1243.

Bian, A., Xing, C., & Hu, M. C. (2014). Alpha Klotho and phosphate homeostasis. *Journal of Endocrinological Investigation, 37*(11), 1121–1126.

Bloch, L., Sineshchekova, O., Reichenbach, D., Reiss, K., Saftig, P., Kuro-o, M., et al. (2009). Klotho is a substrate for alpha-, beta- and gamma-secretase. *FEBS Letters, 583*(19), 3221–3224.

Block, G. A., Hulbert-Shearon, T. E., Levin, N. W., & Port, F. K. (1998). Association of serum phosphorus and calcium × phosphate product with mortality risk in chronic hemodialysis patients: A national study. *American Journal of Kidney Diseases, 31*(4), 607–617.

Camilli, T. C., Xu, M., O'Connell, M. P., Chien, B., Frank, B. P., Subaran, S., et al. (2011). Loss of Klotho during melanoma progression leads to increased filamin cleavage, increased Wnt5A expression, and enhanced melanoma cell motility. *Pigment Cell & Melanoma Research, 24*(1), 175–186.

Canalejo, R., Canalejo, A., Martinez-Moreno, J. M., Rodriguez-Ortiz, M. E., Estepa, J. C., Mendoza, F. J., et al. (2010). FGF23 fails to inhibit uremic parathyroid glands. *Journal of the American Society of Nephrology*, *21*(7), 1125–1135.

Cannata-Andia, J. B., & Rodriguez-Garcia, M. (2002). Hyperphosphataemia as a cardiovascular risk factor—How to manage the problem. *Nephrology, Dialysis, Transplantation*, *17*(Suppl. 11), 16–19.

Cano, F. J., Freundlich, M., Ceballos, M. L., Rojo, A. P., Azocar, M. A., Delgado, I. O., et al. (2014). Longitudinal FGF23 and Klotho axis characterization in children treated with chronic peritoneal dialysis. *Clinical Kidney Journal*, 7(5), 457–463.

Carracedo, J., Buendia, P., Merino, A., Madueno, J. A., Peralbo, E., Ortiz, A., et al. (2012). Klotho modulates the stress response in human senescent endothelial cells. *Mechanisms of Ageing and Development*, *133*(11–12), 647–654.

Carracedo, J., Buendia, P., Merino, A., Soriano, S., Esquivias, E., Martin-Malo, A., et al. (2013). Cellular senescence determines endothelial cell damage induced by uremia. *Experimental Gerontology*, *48*(8), 766–773.

Chade, A. R., Zhu, X., Mushin, O. P., Napoli, C., Lerman, A., & Lerman, L. O. (2006). Simvastatin promotes angiogenesis and prevents microvascular remodeling in chronic renal ischemia. *The FASEB Journal*, *20*(10), 1706–1708.

Chen, L. J., Cheng, M. F., Ku, P. M., & Lin, J. W. (2014). Rosiglitazone increases cerebral klotho expression to reverse baroreflex in type 1-like diabetic rats. *BioMed Research International*, *2014*, 309151.

Chen, T. H., Kuro, O. M., Chen, C. H., Sue, Y. M., Chen, Y. C., Wu, H. H., et al. (2013). The secreted Klotho protein restores phosphate retention and suppresses accelerated aging in Klotho mutant mice. *European Journal of Pharmacology*, *698*(1–3), 67–73.

Chen, C. D., Li, H., Liang, J., Hixson, K., Zeldich, E., & Abraham, C. R. (2015). The anti-aging and tumor suppressor protein klotho enhances differentiation of a human oligodendrocytic hybrid cell line. *Journal of Molecular Neuroscience*, *55*(1), 76–90.

Chen, C. D., Podvin, S., Gillespie, E., Leeman, S. E., & Abraham, C. R. (2007). Insulin stimulates the cleavage and release of the extracellular domain of Klotho by ADAM10 and ADAM17. *Proceedings of the National Academy of Sciences of the United States of America*, *104*(50), 19796–19801.

Chen, C. D., Tung, T. Y., Liang, J., Zeldich, E., Tucker Zhou, T. B., Turk, B. E., et al. (2014). Identification of cleavage sites leading to the shed form of the anti-aging protein klotho. *Biochemistry*, *53*(34), 5579–5587.

Chen, J., Zhang, X., Zhang, H., Lin, J., Zhang, C., Wu, Q., et al. (2013). Elevated Klotho promoter methylation is associated with severity of chronic kidney disease. *PloS One*, *8*(11), e79856.

Cheng, M. F., Chen, L. J., Niu, H. S., Yang, T. T., Lin, K. C., & Cheng, J. T. (2015). Signals mediating Klotho-induced neuroprotection in hippocampal neuronal cells. *Acta Neurobiologiae Experimentalis (Wars)*, *75*(1), 60–71.

Clements, M. E., Chaber, C. J., Ledbetter, S. R., & Zuk, A. (2013). Increased cellular senescence and vascular rarefaction exacerbate the progression of kidney fibrosis in aged mice following transient ischemic injury. *PloS One*, *8*(8), e70464.

Davidovich, E., Davidovits, M., Peretz, B., Shapira, J., & Aframian, D. J. (2009). The correlation between dental calculus and disturbed mineral metabolism in paediatric patients with chronic kidney disease. *Nephrology, Dialysis, Transplantation*, *24*(8), 2439–2445.

de Borst, M. H., Vervloet, M. G., ter Wee, P. M., & Navis, G. (2011). Cross talk between the renin-angiotensin-aldosterone system and vitamin D-FGF-23-klotho in chronic kidney disease. *Journal of the American Society of Nephrology*, *22*(9), 1603–1609.

de Oliveira, R. M. (2006). Klotho RNAi induces premature senescence of human cells via a p53/p21 dependent pathway. *FEBS Letters*, *580*(24), 5753–5758.

Degaspari, S., Tzanno-Martins, C. B., Fujihara, C. K., Zatz, R., Branco-Martins, J. P., Viel, T. A., et al. (2015). Altered KLOTHO and NF-kappaB-TNF-alpha signaling are correlated with nephrectomy-induced cognitive impairment in rats. *PloS One, 10*(5), e0125271.

Delmez, J. A., & Slatopolsky, E. (1992). Hyperphosphatemia: Its consequences and treatment in patients with chronic renal disease. *American Journal of Kidney Diseases, 19*, 303–317.

Desjardins, L., Liabeuf, S., Renard, C., Lenglet, A., Lemke, H. D., Choukroun, G., et al. (2012). FGF23 is independently associated with vascular calcification but not bone mineral density in patients at various CKD stages. *Osteoporosis International, 23*(7), 2017–2025.

Devaraj, S., Syed, B., Chien, A., & Jialal, I. (2012). Validation of an immunoassay for soluble Klotho protein: Decreased levels in diabetes and increased levels in chronic kidney disease. *American Journal of Clinical Pathology, 137*(3), 479–485.

D'Hoore, E., Neirynck, N., Schepers, E., Vanholder, R., Verbeke, F., Van Thielen, M., et al. (2015). Chronic kidney disease progression is mainly associated with non-recovery of acute kidney injury. *Journal of Nephrology, 28*(6), 709–716.

Di Marco, G. S., Hausberg, M., Hillebrand, U., Rustemeyer, P., Wittkowski, W., Lang, D., et al. (2008). Increased inorganic phosphate induces human endothelial cell apoptosis in vitro. *American Journal of Physiology. Renal Physiology, 294*(6), F1381–F1387.

Dmitrieva, N. I., & Burg, M. B. (2007). High NaCl promotes cellular senescence. *Cell Cycle, 6*(24), 3108–3113.

Doi, S., Yorioka, N., Doi, S., Zou, Y., Togao, O., Pastor, J. V., et al. (2011). Klotho inhibits transforming growth factor-beta1 (TGF-beta1) signaling and suppresses renal fibrosis and cancer metastasis in mice. *The Journal of Biological Chemistry, 286*(10), 8655–8665.

Donate-Correa, J., Mora-Fernandez, C., Martinez-Sanz, R., Muros-de-Fuentes, M., Perez, H., Meneses-Perez, B., et al. (2013). Expression of FGF23/KLOTHO system in human vascular tissue. *International Journal of Cardiology, 165*(1), 179–183.

Eckardt, K. U., Coresh, J., Devuyst, O., Johnson, R. J., Kottgen, A., Levey, A. S., et al. (2013). Evolving importance of kidney disease: From subspecialty to global health burden. *Lancet, 382*(9887), 158–169.

Emami Aleagha, M. S., Siroos, B., Ahmadi, M., Balood, M., Palangi, A., Haghighi, A. N., et al. (2015). Decreased concentration of Klotho in the cerebrospinal fluid of patients with relapsing-remitting multiple sclerosis. *Journal of Neuroimmunology, 281*, 5–8.

Fahrleitner-Pammer, A., Herberth, J., Browning, S. R., Obermayer-Pietsch, B., Wirnsberger, G., Holzer, H., et al. (2008). Bone markers predict cardiovascular events in chronic kidney disease. *Journal of Bone and Mineral Research, 23*(11), 1850–1858.

Fang, Y., Ginsberg, C., Sugatani, T., Monier-Faugere, M. C., Malluche, H., & Hruska, K. A. (2014). Early chronic kidney disease-mineral bone disorder stimulates vascular calcification. *Kidney International, 85*(1), 142–150.

Faul, C., Amaral, A. P., Oskouei, B., Hu, M. C., Sloan, A., Isakova, T., et al. (2011). FGF23 induces left ventricular hypertrophy. *The Journal of Clinical Investigation, 121*(11), 4393–4408.

Ferenbach, D. A., & Bonventre, J. V. (2015). Mechanisms of maladaptive repair after AKI leading to accelerated kidney ageing and CKD. *Nature Reviews. Nephrology, 11*(5), 264–276.

Fernandez-Martin, J. L., Martinez-Camblor, P., Dionisi, M. P., Floege, J., Ketteler, M., London, G., et al. (2015). Improvement of mineral and bone metabolism markers is associated with better survival in haemodialysis patients: The COSMOS study. *Nephrology, Dialysis, Transplantation, 30*(9), 1542–1551.

Fliser, D., Kollerits, B., Neyer, U., Ankerst, D. P., Lhotta, K., Lingenhel, A., et al. (2007). Fibroblast growth factor 23 (FGF23) predicts progression of chronic kidney disease: The Mild to Moderate Kidney Disease (MMKD) Study. *Journal of the American Society of Nephrology, 18*(9), 2600–2608.

Fliser, D., Seiler, S., Heine, G. H., & Ketteler, M. (2012). Measurement of serum soluble Klotho levels in CKD 5D patients: Useful tool or dispensable biomarker? *Nephrology, Dialysis, Transplantation*, *27*(5), 1702–1703.

Floege, J., Kim, J., Ireland, E., Chazot, C., Drueke, T., de Francisco, A., et al. (2011). Serum iPTH, calcium and phosphate, and the risk of mortality in a European haemodialysis population. *Nephrology, Dialysis, Transplantation*, *26*(6), 1948–1955.

Foley, R. N., Parfrey, P. S., Harnett, J. D., Kent, G. M., Martin, C. J., Murray, D. C., et al. (1995). Clinical and echocardiographic disease in patients starting end-stage renal disease therapy. *Kidney International*, *47*(1), 186–192.

Foley, R. N., Parfrey, P. S., & Sarnak, M. J. (1998). Clinical epidemiology of cardiovascular disease in chronic renal disease. *American Journal of Kidney Diseases*, *32*(5 Suppl. 3), S112–S119.

Forster, R. E., Jurutka, P. W., Hsieh, J. C., Haussler, C. A., Lowmiller, C. L., Kaneko, I., et al. (2011). Vitamin D receptor controls expression of the anti-aging klotho gene in mouse and human renal cells. *Biochemical and Biophysical Research Communications*, *414*(3), 557–562.

Foster, M. C., Inker, L. A., Levey, A. S., Selvin, E., Eckfeldt, J., Juraschek, S. P., et al. (2013). Novel filtration markers as predictors of all-cause and cardiovascular mortality in US adults. *American Journal of Kidney Diseases*, *62*(1), 42–51.

Fufaa, G. D., Weil, E. J., Nelson, R. G., Hanson, R. L., Bonventre, J. V., Sabbisetti, V., et al. (2015). Association of urinary KIM-1, L-FABP, NAG and NGAL with incident end-stage renal disease and mortality in American Indians with type 2 diabetes mellitus. *Diabetologia*, *58*(1), 188–198.

Fukino, K., Suzuki, T., Saito, Y., Shindo, T., Amaki, T., Kurabayashi, M., et al. (2002). Regulation of angiogenesis by the aging suppressor gene klotho. *Biochemical and Biophysical Research Communications*, *293*(1), 332–337.

Galitzer, H., Ben-Dov, I. Z., Silver, J., & Naveh-Many, T. (2010). Parathyroid cell resistance to fibroblast growth factor 23 in secondary hyperparathyroidism of chronic kidney disease. *Kidney International*, 77(3), 211–218.

Gattineni, J., & Baum, M. (2010). Regulation of phosphate transport by fibroblast growth factor 23 (FGF23): Implications for disorders of phosphate metabolism. *Pediatric Nephrology*, *25*(4), 591–601.

Glassock, R. J., Pecoits-Filho, R., & Barberato, S. H. (2009). Left ventricular mass in chronic kidney disease and ESRD. *Clinical Journal of the American Society of Nephrology*, *4*(Suppl. 1), S79–S91.

Go, A. S., Chertow, G. M., Fan, D., McCulloch, C. E., & Hsu, C. Y. (2004). Chronic kidney disease and the risks of death, cardiovascular events, and hospitalization. *The New England Journal of Medicine*, *351*(13), 1296–1305.

Goetz, R., Nakada, Y., Hu, M. C., Kurosu, H., Wang, L., Nakatani, T., et al. (2010). Isolated C-terminal tail of FGF23 alleviates hypophosphatemia by inhibiting FGF23-FGFR-Klotho complex formation. *Proceedings of the National Academy of Sciences of the United States of America*, *107*(1), 407–412.

Grabner, A., Amaral, A. P., Schramm, K., Singh, S., Sloan, A., Yanucil, C., et al. (2015). Activation of cardiac fibroblast growth factor receptor 4 causes left ventricular hypertrophy. *Cell Metabolism*, *22*(6), 1020–1032.

Greenhill, C. (2011). Risk factors: Levels of FGF23 predict outcomes in advanced CKD. *Nature Reviews. Nephrology*, 7(11), 606.

Gross, M. L., & Ritz, E. (2008). Hypertrophy and fibrosis in the cardiomyopathy of uremia—Beyond coronary heart disease. *Seminars in Dialysis*, *21*(4), 308–318.

Guan, X., Nie, L., He, T., Yang, K., Xiao, T., Wang, S., et al. (2014). Klotho suppresses renal tubulo-interstitial fibrosis by controlling basic fibroblast growth factor-2 signalling. *The Journal of Pathology*, *234*(4), 560–572.

Guo, Y. C., & Yuan, Q. (2015). Fibroblast growth factor 23 and bone mineralisation. *International Journal of Oral Science*, 7(1), 8–13.

Gutiérrez, O. M. (2010). Fibroblast growth factor 23 and disordered vitamin D metabolism in chronic kidney disease: Updating the "trade-off" hypothesis. *Clinical Journal of the American Society of Nephrology*, *5*(9), 1710–1716.

Gutiérrez, O. M., Januzzi, J. L., Isakova, T., Laliberte, K., Smith, K., Collerone, G., et al. (2009). Fibroblast growth factor 23 and left ventricular hypertrophy in chronic kidney disease. *Circulation*, *119*(19), 2545–2552.

Gutiérrez, O. M., Mannstadt, M., Isakova, T., Rauh-Hain, J. A., Tamez, H., Shah, A., et al. (2008). Fibroblast growth factor 23 and mortality among patients undergoing hemodialysis. *The New England Journal of Medicine*, *359*(6), 584–592.

Hanks, L. J., Casazza, K., Judd, S. E., Jenny, N. S., & Gutierrez, O. M. (2015). Associations of fibroblast growth factor-23 with markers of inflammation, insulin resistance and obesity in adults. *PloS One*, *10*(3), e0122885.

Haruna, Y., Kashihara, N., Satoh, M., Tomita, N., Namikoshi, T., Sasaki, T., et al. (2007). Amelioration of progressive renal injury by genetic manipulation of Klotho gene. *Proceedings of the National Academy of Sciences of the United States of America*, *104*(7), 2331–2336.

Hasegawa, H., Nagano, N., Urakawa, I., Yamazaki, Y., Iijima, K., Fujita, T., et al. (2010). Direct evidence for a causative role of FGF23 in the abnormal renal phosphate handling and vitamin D metabolism in rats with early-stage chronic kidney disease. *Kidney International*, *78*(10), 975–980.

Heller, R., & Heller, L. C. (2015). Gene electrotransfer clinical trials. *Advances in Genetics*, *89*, 235–262.

Hemmelgarn, B. R., Manns, B. J., Lloyd, A., James, M. T., Klarenbach, S., Quinn, R. R., et al. (2010). Relation between kidney function, proteinuria, and adverse outcomes. *JAMA*, *303*(5), 423–429.

Heydarkhan-Hagvall, S., Helenius, G., Johansson, B. R., Li, J. Y., Mattsson, E., & Risberg, B. (2003). Co-culture of endothelial cells and smooth muscle cells affects gene expression of angiogenic factors. *Journal of Cellular Biochemistry*, *89*(6), 1250–1259.

Hruska, K. A., Saab, G., Mathew, S., & Lund, R. (2007). Renal osteodystrophy, phosphate homeostasis, and vascular calcification. *Seminars in Dialysis*, *20*(4), 309–315.

Hu, M. C. (2016). Klotho connects intermedin1-53 to suppression of vascular calcification in chronic kidney disease. *Kidney International*, *89*(3), 534–537.

Hu, M. C., Kuro-o, M., & Moe, O. W. (2012). Secreted klotho and chronic kidney disease. *Advances in Experimental Medicine and Biology*, *728*, 126–157.

Hu, M. C., Kuro-o, M., & Moe, O. W. (2013a). Klotho and chronic kidney disease. *Contributions to Nephrology*, *180*, 47–63.

Hu, M. C., Kuro-o, M., & Moe, O. W. (2013b). Renal and extrarenal actions of Klotho. *Seminars in Nephrology*, *33*(2), 118–129.

Hu, M. C., Kuro-o, M., & Moe, O. W. (2014). alphaKlotho and vascular calcification: An evolving paradigm. *Current Opinion in Nephrology and Hypertension*, *23*(4), 331–339.

Hu, M. C., Shi, M., Cho, H. J., Adams-Huet, B., Paek, J., Hill, K., et al. (2015a). Klotho and phosphate are modulators of pathologic uremic cardiac remodeling. *Journal of the American Society of Nephrology*, *26*(6), 1290–1302.

Hu, M. C., Shi, M., Cho, H. J., Zhang, J., Pavlenco, A., Liu, S., et al. (2013). The erythropoietin receptor is a downstream effector of Klotho-induced cytoprotection. *Kidney International*, *84*(3), 468–481.

Hu, M. C., Shi, M., Zhang, J., Addo, T., Cho, H. J., Barker, S. L., et al. (2015b). Renal production, uptake, and handling of circulating αKlotho. *Journal of the American Society of Nephrology*, *27*(1), 79–90.

Hu, M. C., Shi, M., Zhang, J., Pastor, J., Nakatani, T., Lanske, B., et al. (2010a). Klotho: A novel phosphaturic substance acting as an autocrine enzyme in the renal proximal tubule. *The FASEB Journal*, *24*(9), 3438–3450.

Hu, M. C., Shi, M., Zhang, J., Quinones, H., Griffith, C., Kuro-o, M., et al. (2011). Klotho deficiency causes vascular calcification in chronic kidney disease. *Journal of the American Society of Nephrology*, *22*(1), 124–136.

Hu, M. C., Shi, M., Zhang, J., Quinones, H., Kuro, O. M., & Moe, O. W. (2010b). Klotho deficiency is an early biomarker of renal ischemia-reperfusion injury and its replacement is protective. *Kidney International*, *78*(12), 1240–1251.

Hu, M. C., Shiizaki, K., Kuro-o, M., & Moe, O. W. (2013). Fibroblast growth factor 23 and Klotho: Physiology and pathophysiology of an endocrine network of mineral metabolism. *Annual Review of Physiology*, *75*, 503–533.

Ichikawa, S., Imel, E. A., Kreiter, M. L., Yu, X., Mackenzie, D. S., Sorenson, A. H., et al. (2007). A homozygous missense mutation in human KLOTHO causes severe tumoral calcinosis. *The Journal of Clinical Investigation*, *117*(9), 2684–2691.

Ikushima, M., Rakugi, H., Ishikawa, K., Maekawa, Y., Yamamoto, K., Ohta, J., et al. (2006). Anti-apoptotic and anti-senescence effects of Klotho on vascular endothelial cells. *Biochemical and Biophysical Research Communications*, *339*(3), 827–832.

Imura, A., Iwano, A., Tohyama, O., Tsuji, Y., Nozaki, K., Hashimoto, N., et al. (2004). Secreted Klotho protein in sera and CSF: Implication for post-translational cleavage in release of Klotho protein from cell membrane. *FEBS Letters*, *565*(1–3), 143–147.

Isakova, T., Gutierrez, O. M., Chang, Y., Shah, A., Tamez, H., Smith, K., et al. (2009). Phosphorus binders and survival on hemodialysis. *Journal of the American Society of Nephrology*, *20*(2), 388–396.

Ito, S., Fujimori, T., Hayashizaki, Y., & Nabeshima, Y. (2002). Identification of a novel mouse membrane-bound family 1 glycosidase-like protein, which carries an atypical active site structure. *Biochimica et Biophysica Acta*, *1576*(3), 341–345.

Ito, S., Kinoshita, S., Shiraishi, N., Nakagawa, S., Sekine, S., Fujimori, T., et al. (2000). Molecular cloning and expression analyses of mouse betaklotho, which encodes a novel Klotho family protein. *Mechanisms of Development*, *98*(1–2), 115–119.

Iwano, M., Plieth, D., Danoff, T. M., Xue, C., Okada, H., & Neilson, E. G. (2002). Evidence that fibroblasts derive from epithelium during tissue fibrosis. *The Journal of Clinical Investigation*, *110*(3), 341–350.

Ix, J. H., Katz, R., Kestenbaum, B. R., de Boer, I. H., Chonchol, M., Mukamal, K. J., et al. (2012). Fibroblast growth factor-23 and death, heart failure, and cardiovascular events in community-living individuals: CHS (Cardiovascular Health Study). *Journal of the American College of Cardiology*, *60*(3), 200–207.

Jennings, P., Koppelstaetter, C., Aydin, S., Abberger, T., Wolf, A. M., Mayer, G., et al. (2007). Cyclosporine A induces senescence in renal tubular epithelial cells. *American Journal of Physiology Renal Physiology*, *293*(3), F831–F838.

Jie, K. E., Zaikova, M. A., Bergevoet, M. W., Westerweel, P. E., Rastmanesh, M., Blankestijn, P. J., et al. (2010). Progenitor cells and vascular function are impaired in patients with chronic kidney disease. *Nephrology, Dialysis, Transplantation*, *25*(6), 1875–1882.

Jimbo, R., Kawakami-Mori, F., Mu, S., Hirohama, D., Majtan, B., Shimizu, Y., et al. (2014). Fibroblast growth factor 23 accelerates phosphate-induced vascular calcification in the absence of Klotho deficiency. *Kidney International*, *85*(5), 1103–1111.

Jono, S., McKee, M. D., Murry, C. E., Shioi, A., Nishizawa, Y., Mori, K., et al. (2000). Phosphate regulation of vascular smooth muscle cell calcification. *Circulation Research*, *87*(7), E10–E17.

Kailong, L., Du, X., Yani, H., Lin, Z., Jvrong, Y., Ruihua, S., et al. (2007). P53-Rb signaling pathway is involved in tubular cell senescence in renal ischemia/reperfusion injury. *Biocell*, *31*(2), 213–223.

Kaisar, M., Isbel, N., & Johnson, D. W. (2007). Cardiovascular disease in patients with chronic kidney disease. A clinical review. *Minerva Urologica e Nefrologica, 59*(3), 281–297.
Kalantar-Zadeh, K., Shah, A., Duong, U., Hechter, R. C., Dukkipati, R., & Kovesdy, C. P. (2010). Kidney bone disease and mortality in CKD: Revisiting the role of vitamin D, calcimimetics, alkaline phosphatase, and minerals. *Kidney International. Supplement, 117*, S10–S21.
Kalluri, R., & Neilson, E. G. (2003). Epithelial-mesenchymal transition and its implications for fibrosis. *The Journal of Clinical Investigation, 112*(12), 1776–1784.
Karalliedde, J., Maltese, G., Hill, B., Viberti, G., & Gnudi, L. (2013). Effect of renin-angiotensin system blockade on soluble Klotho in patients with type 2 diabetes, systolic hypertension, and albuminuria. *Clinical Journal of the American Society of Nephrology, 8*(11), 1899–1905.
Kato, Y., Arakawa, E., Kinoshita, S., Shirai, A., Furuya, A., Yamano, K., et al. (2000). Establishment of the anti-Klotho monoclonal antibodies and detection of Klotho protein in kidneys. *Biochemical and Biophysical Research Communications, 267*(2), 597–602.
Kestenbaum, B., & Belozeroff, V. (2007). Mineral metabolism disturbances in patients with chronic kidney disease. *European Journal of Clinical Investigation, 37*(8), 607–622.
Kestenbaum, B., Sampson, J. N., Rudser, K. D., Patterson, D. J., Seliger, S. L., Young, B., et al. (2005). Serum phosphate levels and mortality risk among people with chronic kidney disease. *Journal of the American Society of Nephrology, 16*(2), 520–528.
Khan, S. (2007). Vitamin D deficiency and secondary hyperparathyroidism among patients with chronic kidney disease. *The American Journal of the Medical Sciences, 333*(4), 201–207.
Kim, H. R., Nam, B. Y., Kim, D. W., Kang, M. W., Han, J. H., Lee, M. J., et al. (2013). Circulating alpha-klotho levels in CKD and relationship to progression. *American Journal of Kidney Diseases, 61*(6), 899–909.
King, G. D., Rosene, D. L., & Abraham, C. R. (2012). Promoter methylation and age-related downregulation of Klotho in rhesus monkey. *Age (Dordrecht, Netherlands), 34*(6), 1405–1419.
Kitagawa, M., Sugiyama, H., Morinaga, H., Inoue, T., Takiue, K., Ogawa, A., et al. (2013). A decreased level of serum soluble Klotho is an independent biomarker associated with arterial stiffness in patients with chronic kidney disease. *PloS One, 8*(2), e56695.
Koh, N., Fujimori, T., Nishiguchi, S., Tamori, A., Shiomi, S., Nakatani, T., et al. (2001). Severely reduced production of klotho in human chronic renal failure kidney. *Biochemical and Biophysical Research Communications, 280*(4), 1015–1020.
Kooman, J. P., Broers, N. J., Usvyat, L., Thijssen, S., van der Sande, F. M., Cornelis, T., et al. (2013). Out of control: Accelerated aging in uremia. *Nephrology, Dialysis, Transplantation, 28*(1), 48–54.
Krajisnik, T., Olauson, H., Mirza, M. A., Hellman, P., Akerstrom, G., Westin, G., et al. (2010). Parathyroid Klotho and FGF-receptor 1 expression decline with renal function in hyperparathyroid patients with chronic kidney disease and kidney transplant recipients. *Kidney International, 78*(10), 1024–1032.
Kramann, R., Tanaka, M., & Humphreys, B. D. (2014). Fluorescence microangiography for quantitative assessment of peritubular capillary changes after AKI in mice. *Journal of the American Society of Nephrology, 25*(9), 1924–1931.
Krupp, K., & Madhivanan, P. (2014). FGF23 and risk of all-cause mortality and cardiovascular events: A meta-analysis of prospective cohort studies. *International Journal of Cardiology, 176*(3), 1341–1342.
Kuro-o, M., Matsumura, Y., Aizawa, H., Kawaguchi, H., Suga, T., Utsugi, T., et al. (1997). Mutation of the mouse klotho gene leads to a syndrome resembling ageing. *Nature, 390*(6655), 45–51.
Kurosu, H., Yamamoto, M., Clark, J. D., Pastor, J. V., Nandi, A., Gurnani, P., et al. (2005). Suppression of aging in mice by the hormone Klotho. *Science, 309*(5742), 1829–1833.

Kusaba, T., Okigaki, M., Matui, A., Murakami, M., Ishikawa, K., Kimura, T., et al. (2010). Klotho is associated with VEGF receptor-2 and the transient receptor potential canonical-1 Ca2+ channel to maintain endothelial integrity. *Proceedings of the National Academy of Sciences of the United States of America*, *107*(45), 19308–19313.

Lanske, B., & Razzaque, M. S. (2007). Vitamin D and aging: Old concepts and new insights. *The Journal of Nutritional Biochemistry*, *18*(12), 771–777.

Lau, W. L., Leaf, E. M., Hu, M. C., Takeno, M. M., Kuro-o, M., Moe, O. W., et al. (2012). Vitamin D receptor agonists increase klotho and osteopontin while decreasing aortic calcification in mice with chronic kidney disease fed a high phosphate diet. *Kidney International*, *82*(12), 1261–1270.

Lee, J., Jeong, D. J., Kim, J., Lee, S., Park, J. H., Chang, B., et al. (2010). The anti-aging gene KLOTHO is a novel target for epigenetic silencing in human cervical carcinoma. *Molecular Cancer*, *9*, 109.

Lee, S. R., Lee, S. H., Moon, J. Y., Park, J. Y., Lee, D., Lim, S. J., et al. (2010). Repeated administration of bone marrow-derived mesenchymal stem cells improved the protective effects on a remnant kidney model. *Renal Failure*, *32*(7), 840–848.

Levey, A. S., de Jong, P. E., Coresh, J., El Nahas, M., Astor, B. C., Matsushita, K., et al. (2011). The definition, classification, and prognosis of chronic kidney disease: A KDIGO Controversies Conference report. *Kidney International*, *80*(1), 17–28.

Li, L., & Clevers, H. (2010). Coexistence of quiescent and active adult stem cells in mammals. *Science*, *327*(5965), 542–545.

Lim, K., Groen, A., Molostvov, G., Lu, T., Lilley, K. S., Snead, D., et al. (2015). α-Klotho expression in human tissues. *The Journal of Clinical Endocrinology and Metabolism*, *100*(10), E1308–E1318.

Lim, K., Lu, T. S., Molostvov, G., Lee, C., Lam, F. T., Zehnder, D., et al. (2012). Vascular Klotho deficiency potentiates the development of human artery calcification and mediates resistance to fibroblast growth factor 23. *Circulation*, *125*(18), 2243–2255.

Lindberg, K., Amin, R., Moe, O. W., Hu, M. C., Erben, R. G., Ostman Wernerson, A., et al. (2014). The kidney is the principal organ mediating klotho effects. *Journal of the American Society of Nephrology*, *25*(10), 2169–2175.

Lindberg, K., Olauson, H., Amin, R., Ponnusamy, A., Goetz, R., Taylor, R. F., et al. (2013). Arterial klotho expression and FGF23 effects on vascular calcification and function. *PloS One*, *8*(4), e60658.

Lips, P. (2001). Vitamin D deficiency and secondary hyperparathyroidism in the elderly: Consequences for bone loss and fractures and therapeutic implications. *Endocrine Reviews*, *22*(4), 477–501.

Liu, Y. (2010). New insights into epithelial-mesenchymal transition in kidney fibrosis. *Journal of the American Society of Nephrology*, *21*(2), 212–222.

Liu, H., Fergusson, M. M., Castilho, R. M., Liu, J., Cao, L., Chen, J., et al. (2007). Augmented Wnt signaling in a mammalian model of accelerated aging. *Science*, *317*(5839), 803–806.

Liu, S., Tang, W., Zhou, J., Stubbs, J. R., Luo, Q., Pi, M., et al. (2006). Fibroblast growth factor 23 is a counter-regulatory phosphaturic hormone for vitamin D. *Journal of the American Society of Nephrology*, *17*(5), 1305–1315.

Liu, K. D., Yang, W., Anderson, A. H., Feldman, H. I., Demirjian, S., Hamano, T., et al. (2013). Urine neutrophil gelatinase-associated lipocalin levels do not improve risk prediction of progressive chronic kidney disease. *Kidney International*, *83*(5), 909–914.

Lomashvili, K. A., Cobbs, S., Hennigar, R. A., Hardcastle, K. I., & O'Neill, W. C. (2004). Phosphate-induced vascular calcification: Role of pyrophosphate and osteopontin. *Journal of the American Society of Nephrology*, *15*(6), 1392–1401.

London, G. M., Marchais, S. J., Guerin, A. P., & Metivier, F. (2005). Arteriosclerosis, vascular calcifications and cardiovascular disease in uremia. *Current Opinion in Nephrology and Hypertension*, *14*(6), 525–531.

Maekawa, Y., Ishikawa, K., Yasuda, O., Oguro, R., Hanasaki, H., Kida, I., et al. (2009). Klotho suppresses TNF-alpha-induced expression of adhesion molecules in the endothelium and attenuates NF-kappaB activation. *Endocrine*, *35*(3), 341–346.

Masuda, H., Chikuda, H., Suga, T., Kawaguchi, H., & Kuro-o, M. (2005). Regulation of multiple ageing-like phenotypes by inducible klotho gene expression in klotho mutant mice. *Mechanisms of Ageing and Development*, *126*(12), 1274–1283.

Mathew, S., Tustison, K. S., Sugatani, T., Chaudhary, L. R., Rifas, L., & Hruska, K. A. (2008). The mechanism of phosphorus as a cardiovascular risk factor in CKD. *Journal of the American Society of Nephrology*, *19*(6), 1092–1105.

Matsumura, Y., Aizawa, H., Shiraki-Iida, T., Nagai, R., Kuro-o, M., & Nabeshima, Y. (1998). Identification of the human klotho gene and its two transcripts encoding membrane and secreted klotho protein. *Biochemical and Biophysical Research Communications*, *242*(3), 626–630.

Mencke, R., Harms, G., Mirkovic, K., Struik, J., van Ark, J., van Loon, E., et al. (2015). Membrane-bound Klotho is not expressed endogenously in healthy or uremic human vascular tissue. *Cardiovascular Research*, *108*(2), 220–231.

Merchant, M. L., Niewczas, M. A., Ficociello, L. H., Lukenbill, J. A., Wilkey, D. W., Li, M., et al. (2013). Plasma kininogen and kininogen fragments are biomarkers of progressive renal decline in type 1 diabetes. *Kidney International*, *83*(6), 1177–1184.

Mirza, M. A., Larsson, A., Melhus, H., Lind, L., & Larsson, T. E. (2009). Serum intact FGF23 associate with left ventricular mass, hypertrophy and geometry in an elderly population. *Atherosclerosis*, *207*(2), 546–551.

Mitani, H., Ishizaka, N., Aizawa, T., Ohno, M., Usui, S., Suzuki, T., et al. (2002). In vivo klotho gene transfer ameliorates angiotensin II-induced renal damage. *Hypertension*, *39*(4), 838–843.

Mohandas, R., & Segal, M. S. (2010). Endothelial progenitor cells and endothelial vesicles—What is the significance for patients with chronic kidney disease? *Blood Purification*, *29*(2), 158–162.

Moreno, J. A., Izquierdo, M. C., Sanchez-Nino, M. D., Suarez-Alvarez, B., Lopez-Larrea, C., Jakubowski, A., et al. (2011). The inflammatory cytokines TWEAK and TNFalpha reduce renal klotho expression through NFkappaB. *Journal of the American Society of Nephrology*, *22*(7), 1315–1325.

Morishita, K., Shirai, A., Kubota, M., Katakura, Y., Nabeshima, Y., Takeshige, K., et al. (2001). The progression of aging in klotho mutant mice can be modified by dietary phosphorus and zinc. *The Journal of Nutrition*, *131*(12), 3182–3188.

Mu, W., Long, D. A., Ouyang, X., Agarwal, A., Cruz, P. E., Roncal, C. A., et al. (2009). Angiostatin overexpression is associated with an improvement in chronic kidney injury by an anti-inflammatory mechanism. *American Journal of Physiology. Renal Physiology*, *296*(1), F145–F152.

Nagai, R., Saito, Y., Ohyama, Y., Aizawa, H., Suga, T., Nakamura, T., et al. (2000). Endothelial dysfunction in the klotho mouse and downregulation of klotho gene expression in various animal models of vascular and metabolic diseases. *Cellular and Molecular Life Sciences*, *57*(5), 738–746.

Nagano, N., Miyata, S., Abe, M., Kobayashi, N., Wakita, S., Yamashita, T., et al. (2006). Effect of manipulating serum phosphorus with phosphate binder on circulating PTH and FGF23 in renal failure rats. *Kidney International*, *69*(3), 531–537.

Nakamura, T., Saito, Y., Ohyama, Y., Masuda, H., Sumino, H., Kuro-o, M., et al. (2002). Production of nitric oxide, but not prostacyclin, is reduced in klotho mice. *Japanese Journal of Pharmacology*, *89*(2), 149–156.

Nakano-Kurimoto, R., Ikeda, K., Uraoka, M., Nakagawa, Y., Yutaka, K., Koide, M., et al. (2009). Replicative senescence of vascular smooth muscle cells enhances the calcification through initiating the osteoblastic transition. *American Journal of Physiology. Heart and Circulatory Physiology*, *297*(5), H1673–H1684.

Narumiya, H., Sasaki, S., Kuwahara, N., Irie, H., Kusaba, T., Kameyama, H., et al. (2004). HMG-CoA reductase inhibitors up-regulate anti-aging klotho mRNA via RhoA inactivation in IMCD3 cells. *Cardiovascular Research*, *64*(2), 331–336.

Navarro-Gonzalez, J. F., Donate-Correa, J., Muros de Fuentes, M., Perez-Hernandez, H., Martinez-Sanz, R., & Mora-Fernandez, C. (2014). Reduced Klotho is associated with the presence and severity of coronary artery disease. *Heart*, *100*(1), 34–40.

Nishida, Y., Taketani, Y., Yamanaka-Okumura, H., Imamura, F., Taniguchi, A., Sato, T., et al. (2006). Acute effect of oral phosphate loading on serum fibroblast growth factor 23 levels in healthy men. *Kidney International*, *70*(12), 2141–2147.

Niwa, T., & Shimizu, H. (2012). Indoxyl sulfate induces nephrovascular senescence. *Journal of Renal Nutrition*, *22*(1), 102–106.

Obi, Y., Hamano, T., & Isaka, Y. (2015). Prevalence and prognostic implications of vitamin D deficiency in chronic kidney disease. *Disease Markers*, *2015*, 868961.

Ohnishi, M., Nakatani, T., Lanske, B., & Razzaque, M. S. (2009). Reversal of mineral ion homeostasis and soft-tissue calcification of klotho knockout mice by deletion of vitamin D 1alpha-hydroxylase. *Kidney International*, *75*(11), 1166–1172.

Olson, E. R., Naugle, J. E., Zhang, X., Bomser, J. A., & Meszaros, J. G. (2005). Inhibition of cardiac fibroblast proliferation and myofibroblast differentiation by resveratrol. *American Journal of Physiology. Heart and Circulatory Physiology*, *288*(3), H1131–H1138.

Ozeki, M., Fujita, S., Kizawa, S., Morita, H., Sohmiya, K., Hoshiga, M., et al. (2014). Association of serum levels of FGF23 and alpha-Klotho with glomerular filtration rate and proteinuria among cardiac patients. *BMC Nephrology*, *15*, 147.

Panesso, M. C., Shi, M., Cho, H. J., Paek, J., Ye, J., Moe, O. W., et al. (2014). Klotho has dual protective effects on cisplatin-induced acute kidney injury. *Kidney International*, *85*(4), 855–870.

Panichi, V., Bigazzi, R., Paoletti, S., Mantuano, E., Beati, S., Marchetti, V., et al. (2010). Impact of calcium, phosphate, PTH abnormalities and management on mortality in hemodialysis: Results from the RISCAVID study. *Journal of Nephrology*, *23*(5), 556–562.

Park, M. Y., Herrmann, S. M., Saad, A., Eirin, A., Tang, H., Lerman, A., et al. (2015). Biomarkers of kidney injury and klotho in patients with atherosclerotic renovascular disease. *Clinical Journal of the American Society of Nephrology*, *10*(3), 443–451.

Patel, S., Barron, J. L., Mirzazedeh, M., Gallagher, H., Hyer, S., Cantor, T., et al. (2011). Changes in bone mineral parameters, vitamin D metabolites, and PTH measurements with varying chronic kidney disease stages. *Journal of Bone and Mineral Metabolism*, *29*(1), 71–79.

Pavik, I., Jaeger, P., Ebner, L., Wagner, C. A., Petzold, K., Spichtig, D., et al. (2013). Secreted Klotho and FGF23 in chronic kidney disease Stage 1 to 5: A sequence suggested from a cross-sectional study. *Nephrology, Dialysis, Transplantation*, *28*(2), 352–359.

Polichnowski, A. J., Lan, R., Geng, H., Griffin, K. A., Venkatachalam, M. A., & Bidani, A. K. (2014). Severe renal mass reduction impairs recovery and promotes fibrosis after AKI. *Journal of the American Society of Nephrology*, *25*(7), 1496–1507.

Ponte, B., Trombetti, A., Hadaya, K., Ernandez, T., Fumeaux, D., Iselin, C., et al. (2014). Acute and long term mineral metabolism adaptation in living kidney donors: A prospective study. *Bone*, *62*, 36–42.

Pontoriero, G., Cozzolino, M., Locatelli, F., & Brancaccio, D. (2010). CKD patients: The dilemma of serum PTH levels. *Nephron. Clinical Practice*, *116*(4), c263–c268.

Ravani, P., Tripepi, G., Malberti, F., Testa, S., Mallamaci, F., & Zoccali, C. (2005). Asymmetrical dimethylarginine predicts progression to dialysis and death in patients with chronic kidney disease: A competing risks modeling approach. *Journal of the American Society of Nephrology*, *16*(8), 2449–2455.

Ravikumar, P., Ye, J., Zhang, J., Pinch, S. N., Hu, M. C., Kuro-o, M., et al. (2014). alpha-Klotho protects against oxidative damage in pulmonary epithelia. *American Journal of Physiology. Lung Cellular and Molecular Physiology*, *307*(7), L566–L575.

Razzaque, M. S. (2009a). FGF23-mediated regulation of systemic phosphate homeostasis: Is Klotho an essential player? *American Journal of Physiology. Renal Physiology*, *296*(3), F470–F476.

Razzaque, M. S. (2009b). Does FGF23 toxicity influence the outcome of chronic kidney disease? *Nephrology, Dialysis, Transplantation*, *24*(1), 4–7.

Razzaque, M. S. (2012). FGF23, klotho and vitamin D interactions: What have we learned from in vivo mouse genetics studies? *Advances in Experimental Medicine and Biology*, *728*, 84–91.

Rebholz, C. M., Grams, M. E., Coresh, J., Selvin, E., Inker, L. A., Levey, A. S., et al. (2015a). Serum fibroblast growth factor-23 is associated with incident kidney disease. *Journal of the American Society of Nephrology*, *26*(1), 192–200.

Rebholz, C. M., Grams, M. E., Matsushita, K., Selvin, E., & Coresh, J. (2015b). Change in novel filtration markers and risk of ESRD. *American Journal of Kidney Diseases*, *66*(1), 47–54.

Recio-Mayoral, A., Banerjee, D., Streather, C., & Kaski, J. C. (2011). Endothelial dysfunction, inflammation and atherosclerosis in chronic kidney disease—A cross-sectional study of predialysis, dialysis and kidney-transplantation patients. *Atherosclerosis*, *216*(2), 446–451.

Reinders, M. E., Rabelink, T. J., & Briscoe, D. M. (2006). Angiogenesis and endothelial cell repair in renal disease and allograft rejection. *Journal of the American Society of Nephrology*, *17*(4), 932–942.

Rimes-Stigare, C., Frumento, P., Bottai, M., Martensson, J., Martling, C. R., Walther, S. M., et al. (2015). Evolution of chronic renal impairment and long-term mortality after de novo acute kidney injury in the critically ill; a Swedish multi-centre cohort study. *Critical Care*, *19*, 221.

Ritter, C. S., Zhang, S., Delmez, J., Finch, J. L., & Slatopolsky, E. (2015). Differential expression and regulation of Klotho by paricalcitol in the kidney, parathyroid, and aorta of uremic rats. *Kidney International*, *87*(6), 1141–1152.

Rotondi, S., Pasquali, M., Tartaglione, L., Muci, M. L., Mandanici, G., Leonangeli, C., et al. (2015). Soluble alpha-Klotho serum levels in chronic kidney disease. *International Journal of Endocrinology*, *2015*, 872193.

Rubinek, T., Shulman, M., Israeli, S., Bose, S., Avraham, A., Zundelevich, A., et al. (2012). Epigenetic silencing of the tumor suppressor klotho in human breast cancer. *Breast Cancer Research and Treatment*, *133*(2), 649–657.

Sabbisetti, V. S., Waikar, S. S., Antoine, D. J., Smiles, A., Wang, C., Ravisankar, A., et al. (2014). Blood kidney injury molecule-1 is a biomarker of acute and chronic kidney injury and predicts progression to ESRD in type I diabetes. *Journal of the American Society of Nephrology*, *25*(10), 2177–2186.

Saito, Y., Nakamura, T., Ohyama, Y., Suzuki, T., Iida, A., Shiraki-Iida, T., et al. (2000). In vivo klotho gene delivery protects against endothelial dysfunction in multiple risk factor syndrome. *Biochemical and Biophysical Research Communications*, *276*(2), 767–772.

Saito, Y., Yamagishi, T., Nakamura, T., Ohyama, Y., Aizawa, H., Suga, T., et al. (1998). Klotho protein protects against endothelial dysfunction. *Biochemical and Biophysical Research Communications*, *248*(2), 324–329.

Sakan, H., Nakatani, K., Asai, O., Imura, A., Tanaka, T., Yoshimoto, S., et al. (2014). Reduced renal alpha-Klotho expression in CKD patients and its effect on renal phosphate handling and vitamin D metabolism. *PloS One*, *9*(1), e86301.

Sastre, C., Rubio-Navarro, A., Buendía, I., Gómez-Guerrero, C., Blanco, J., Mas, S., et al. (2013). Hyperlipidemia-associated renal damage decreases Klotho expression in kidneys from ApoE knockout mice. *PloS One*, *8*(12), e83713.

Satoh, M., Nagasu, H., Morita, Y., Yamaguchi, T. P., Kanwar, Y. S., & Kashihara, N. (2012). Klotho protects against mouse renal fibrosis by inhibiting Wnt signaling. *American Journal of Physiology. Renal Physiology*, *303*(12), F1641–F1651.

Sawires, H. K., Essam, R. M., Morgan, M. F., & Mahmoud, R. A. (2015). Serum klotho: Relation to fibroblast growth factor-23 and other regulators of phosphate metabolism in children with chronic kidney disease. *Nephron*, *129*(4), 293–299.

Schmitt, R., Susnik, N., & Melk, A. (2015). Molecular aspects of renal senescence. *Current Opinion in Organ Transplantation*, *20*(4), 412–416.

Scholze, A., Liu, Y., Pedersen, L., Xia, S., Roth, H. J., Hocher, B., et al. (2014). Soluble alpha-klotho and its relation to kidney function and fibroblast growth factor-23. *The Journal of Clinical Endocrinology and Metabolism*, *99*(5), E855–E861.

Schwedhelm, E., & Boger, R. H. (2011). The role of asymmetric and symmetric dimethylarginines in renal disease. *Nature Reviews. Nephrology*, 7(5), 275–285.

Scialla, J. J., Lau, W. L., Reilly, M. P., Isakova, T., Yang, H. Y., Crouthamel, M. H., et al. (2013). Fibroblast growth factor 23 is not associated with and does not induce arterial calcification. *Kidney International*, *83*(6), 1159–1168.

Seiler, S., Rogacev, K. S., Roth, H. J., Shafein, P., Emrich, I., Neuhaus, S., et al. (2014). Associations of FGF-23 and sKlotho with cardiovascular outcomes among patients with CKD stages 2-4. *Clinical Journal of the American Society of Nephrology*, *9*(6), 1049–1058.

Semba, R. D., Moghekar, A. R., Hu, J., Sun, K., Turner, R., Ferrucci, L., et al. (2014). Klotho in the cerebrospinal fluid of adults with and without Alzheimer's disease. *Neuroscience Letters*, *558*, 37–40.

Shi, M., Flores, B., Gillings, N., Bian, A., Cho, H. J., Yan, S., et al. (2015). αKlotho mitigates progression of acute kidney injury to chronic kidney disease through activation of autophagy. *Journal of the American Society of Nephrology*, pii: ASN.2015060613 [Epub ahead of print].

Shimada, T., Takeshita, Y., Murohara, T., Sasaki, K., Egami, K., Shintani, S., et al. (2004). Angiogenesis and vasculogenesis are impaired in the precocious-aging klotho mouse. *Circulation*, *110*(9), 1148–1155.

Shimada, T., Urakawa, I., Isakova, T., Yamazaki, Y., Epstein, M., Wesseling-Perry, K., et al. (2010). Circulating fibroblast growth factor 23 in patients with end-stage renal disease treated by peritoneal dialysis is intact and biologically active. *The Journal of Clinical Endocrinology and Metabolism*, *95*(2), 578–585.

Shimamura, Y., Hamada, K., Inoue, K., Ogata, K., Ishihara, M., Kagawa, T., et al. (2012). Serum levels of soluble secreted alpha-Klotho are decreased in the early stages of chronic kidney disease, making it a probable novel biomarker for early diagnosis. *Clinical and Experimental Nephrology*, *16*(5), 722–729.

Shimizu, H., Bolati, D., Adijiang, A., Adelibieke, Y., Muteliefu, G., Enomoto, A., et al. (2011). Indoxyl sulfate downregulates renal expression of Klotho through production of ROS and activation of nuclear factor-κB. *American Journal of Nephrology*, *33*(4), 319–324.

Shiraki-Iida, T., Aizawa, H., Matsumura, Y., Sekine, S., Iida, A., Anazawa, H., et al. (1998). Structure of the mouse klotho gene and its two transcripts encoding membrane and secreted protein. *FEBS Letters*, *424*(1–2), 6–10.

Shiraki-Iida, T., Iida, A., Nabeshima, Y., Anazawa, H., Nishikawa, S., Noda, M., et al. (2000). Improvement of multiple pathophysiological phenotypes of klotho (kl/kl) mice by adenovirus-mediated expression of the klotho gene. *The Journal of Gene Medicine*, *2*(4), 233–242.

Shroff, R., Aitkenhead, H., Costa, N., Trivelli, A., Litwin, M., Picca, S., et al. (2016). Normal 25-hydroxyvitamin D levels are associated with less proteinuria and attenuate renal failure progression in children with CKD. *Journal of the American Society of Nephrology*, *27*(1), 314–322.

Shuto, E., Taketani, Y., Tanaka, R., Harada, N., Isshiki, M., Sato, M., et al. (2009). Dietary phosphorus acutely impairs endothelial function. *Journal of the American Society of Nephrology*, *20*(7), 1504–1512.

Silswal, N., Touchberry, C. D., Daniel, D. R., McCarthy, D. L., Zhang, S., Andresen, J., et al. (2014). FGF23 directly impairs endothelium-dependent vasorelaxation by increasing superoxide levels and reducing nitric oxide bioavailability. *American Journal of Physiology. Endocrinology and Metabolism*, *307*(5), E426–E436.

Silver, J., & Naveh-Many, T. (2010). FGF23 and the parathyroid glands. *Pediatric Nephrology*, *25*(11), 2241–2245.

Silver, J., Rodriguez, M., & Slatopolsky, E. (2012). FGF23 and PTH—Double agents at the heart of CKD. *Nephrology, Dialysis, Transplantation*, *27*(5), 1715–1720.

Sinha, M. D., Turner, C., Booth, C. J., Waller, S., Rasmussen, P., Goldsmith, D. J., et al. (2015). Relationship of FGF23 to indexed left ventricular mass in children with non-dialysis stages of chronic kidney disease. *Pediatric Nephrology*, *30*(10), 1843–1852.

Siomou, E., & Stefanidis, C. J. (2012). FGF-23 in children with CKD: A new player in the development of CKD-mineral and bone disorder. *Nephrology, Dialysis, Transplantation*, *27*(12), 4259–4262.

Sitia, S., Tomasoni, L., Atzeni, F., Ambrosio, G., Cordiano, C., Catapano, A., et al. (2010). From endothelial dysfunction to atherosclerosis. *Autoimmunity Reviews*, *9*(12), 830–834.

Six, I., Okazaki, H., Gross, P., Cagnard, J., Boudot, C., Maizel, J., et al. (2014). Direct, acute effects of Klotho and FGF23 on vascular smooth muscle and endothelium. *PloS One*, *9*(4), e93423.

Slatopolsky, E., & Bricker, N. S. (1973). The role of phosphorus restrict ion in the prevention of secondary hyperparathyroidism in chronic renal disease. *Kidney International*, *4*, 141–146.

Small, D. M., Bennett, N. C., Roy, S., Gabrielli, B. G., Johnson, D. W., & Gobe, G. C. (2012). Oxidative stress and cell senescence combine to cause maximal renal tubular epithelial cell dysfunction and loss in an in vitro model of kidney disease. *Nephron. Experimental Nephrology*, *122*(3–4), 123–130.

Smith, R. C., O'Bryan, L. M., Farrow, E. G., Summers, L. J., Clinkenbeard, E. L., Roberts, J. L., et al. (2012). Circulating αKlotho influences phosphate handling by controlling FGF23 production. *The Journal of Clinical Investigation*, *122*(12), 4710–4715.

Snyder, J. J., Foley, R. N., & Collins, A. J. (2009). Prevalence of CKD in the United States: A sensitivity analysis using the National Health and Nutrition Examination Survey (NHANES) 1999–2004. *American Journal of Kidney Diseases*, *53*(2), 218–228.

Stenvinkel, P., & Larsson, T. E. (2013). Chronic kidney disease: A clinical model of premature aging. *American Journal of Kidney Diseases*, *62*(2), 339–351.

Sugiura, H., Yoshida, T., Shiohira, S., Kohei, J., Mitobe, M., Kurosu, H., et al. (2012). Reduced Klotho expression level in kidney aggravates renal interstitial fibrosis. *American Journal of Physiology. Renal Physiology*, *302*(10), F1252–F1264.

Sugiura, H., Yoshida, T., Tsuchiya, K., Mitobe, M., Nishimura, S., Shirota, S., et al. (2005). Klotho reduces apoptosis in experimental ischaemic acute renal failure. *Nephrology, Dialysis, Transplantation*, *20*(12), 2636–2645.

Sun, C. Y., Chang, S. C., & Wu, M. S. (2012). Suppression of Klotho expression by protein-bound uremic toxins is associated with increased DNA methyltransferase expression and DNA hypermethylation. *Kidney International*, *81*(7), 640–650.

Sun, S., Cheng, B., Sun, P. G., Wu, X. H., Wu, Q. Q., & He, P. (2015). RTEF-1 protects against oxidative damage induced by H2O2 in human umbilical vein endothelial cells through Klotho activation. *Experimental Biology and Medicine (Maywood, NJ)*, *240*(12), 1606–1613.

Taddei, S., Nami, R., Bruno, R. M., Quatrini, I., & Nuti, R. (2011). Hypertension, left ventricular hypertrophy and chronic kidney disease. *Heart Failure Reviews*, *16*(6), 615–620.

Takeshita, K., Fujimori, T., Kurotaki, Y., Honjo, H., Tsujikawa, H., Yasui, K., et al. (2004). Sinoatrial node dysfunction and early unexpected death of mice with a defect of klotho gene expression. *Circulation*, *109*(14), 1776–1782.

Tang, R., Zhou, Q. L., Ao, X., Peng, W. S., Veeraragoo, P., & Tang, T. F. (2011). Fosinopril and losartan regulate klotho gene and nicotinamide adenine dinucleotide phosphate oxidase expression in kidneys of spontaneously hypertensive rats. *Kidney & Blood Pressure Research*, *34*(5), 350–357.

Taniyama, Y., & Morishita, R. (2010). Does therapeutic angiogenesis overcome CKD? *Hypertension Research*, *33*(2), 114–115.

Tohyama, O., Imura, A., Iwano, A., Freund, J. N., Henrissat, B., Fujimori, T., et al. (2004). Klotho is a novel beta-glucuronidase capable of hydrolyzing steroid beta-glucuronides. *The Journal of Biological Chemistry*, *279*(11), 9777–9784.

Tsirpanlis, G. (2008). Cellular senescence, cardiovascular risk, and CKD: A review of established and hypothetical interconnections. *American Journal of Kidney Diseases*, *51*(1), 131–144.

Tsujikawa, H., Kurotaki, Y., Fujimori, T., Fukuda, K., & Nabeshima, Y. (2003). Klotho, a gene related to a syndrome resembling human premature aging, functions in a negative regulatory circuit of vitamin D endocrine system. *Molecular Endocrinology*, *17*(12), 2393–2403.

Tumur, Z., & Niwa, T. (2009). Indoxyl sulfate inhibits nitric oxide production and cell viability by inducing oxidative stress in vascular endothelial cells. *American Journal of Nephrology*, *29*(6), 551–557.

Tumur, Z., Shimizu, H., Enomoto, A., Miyazaki, H., & Niwa, T. (2010). Indoxyl sulfate upregulates expression of ICAM-1 and MCP-1 by oxidative stress-induced NF-kappaB activation. *American Journal of Nephrology*, *31*(5), 435–441.

Tyralla, K., & Amann, K. (2003). Morphology of the heart and arteries in renal failure. *Kidney International. Supplement*, *84*, S80–S83.

Utsugi, T., Ohno, T., Ohyama, Y., Uchiyama, T., Saito, Y., Matsumura, Y., et al. (2000). Decreased insulin production and increased insulin sensitivity in the klotho mutant mouse, a novel animal model for human aging. *Metabolism*, *49*(9), 1118–1123.

van Ballegooijen, A. J., Rhee, E. P., Elmariah, S., de Boer, I. H., & Kestenbaum, B. (2016). Renal clearance of mineral metabolism biomarkers. *Journal of the American Society of Nephrology*, *27*(2), 392–397.

Van, T. V., Watari, E., Taketani, Y., Kitamura, T., Shiota, A., Tanaka, T., et al. (2012). Dietary phosphate restriction ameliorates endothelial dysfunction in adenine-induced kidney disease rats. *Journal of Clinical Biochemistry and Nutrition*, *51*(1), 27–32.

Venkatachalam, M. A., Weinberg, J. M., Kriz, W., & Bidani, A. K. (2015). Failed tubule recovery, AKI-CKD transition, and kidney disease progression. *Journal of the American Society of Nephrology*, *26*(8), 1765–1776.

Verbeke, F., Van Biesen, W., & Vanholder, R. (2011). The role of collagen metabolism in CKD-associated arterial senescence: Underestimated and underappreciated. *Nephrology, Dialysis, Transplantation*, *26*(9), 2726–2728.

Vervloet, M. G., Adema, A. Y., Larsson, T. E., & Massy, Z. A. (2014). The role of klotho on vascular calcification and endothelial function in chronic kidney disease. *Seminars in Nephrology*, *34*(6), 578–585.

Wan, M., Smith, C., Shah, V., Gullet, A., Wells, D., Rees, L., et al. (2013). Fibroblast growth factor 23 and soluble klotho in children with chronic kidney disease. *Nephrology, Dialysis, Transplantation*, *28*(1), 153–161.

Wang, Y., Chen, L., Huang, G., He, D., He, J., Xu, W., et al. (2013). Klotho sensitizes human lung cancer cell line to cisplatin via PI3k/Akt pathway. *PloS One*, *8*(2), e57391.

Wang, Y., & Sun, Z. (2009). Klotho gene delivery prevents the progression of spontaneous hypertension and renal damage. *Hypertension*, *54*(4), 810–817.

Wang, Y., & Sun, Z. (2014). Antiaging gene Klotho regulates endothelin-1 levels and endothelin receptor subtype B expression in kidneys of spontaneously hypertensive rats. *Journal of Hypertension, 32*(8), 1629–1636.

Weber, T. J., Liu, S., Indridason, O. S., & Quarles, L. D. (2003). Serum FGF23 levels in normal and disordered phosphorus homeostasis. *Journal of Bone and Mineral Research, 18*(7), 1227–1234.

Weissman, I. L. (2000). Stem cells: Units of development, units of regeneration, and units in evolution. *Cell, 100*(1), 157–168.

Wesseling-Perry, K. (2015). Defective skeletal mineralization in pediatric CKD. *Current Osteoporosis Reports, 13*(2), 98–105.

Westerweel, P. E., Hoefer, I. E., Blankestijn, P. J., de Bree, P., Groeneveld, D., van Oostrom, O., et al. (2007). End-stage renal disease causes an imbalance between endothelial and smooth muscle progenitor cells. *American Journal of Physiology. Renal Physiology, 292*(4), F1132–F1140.

Williams, D. A. (2014). Curing genetic disease with gene therapy. *Transactions of the American Clinical and Climatological Association, 125*, 122–128. discussion 8–9.

Wolf, M. (2010). Forging forward with 10 burning questions on FGF23 in kidney disease. *Journal of the American Society of Nephrology, 21*(9), 1427–1435.

Wolf, M. (2012). Update on fibroblast growth factor 23 in chronic kidney disease. *Kidney International, 82*(7), 737–747.

Wright, C. B., Dong, C., Stark, M., Silverberg, S., Rundek, T., Elkind, M. S., et al. (2014). Plasma FGF23 and the risk of stroke: The Northern Manhattan Study (NOMAS). *Neurology, 82*(19), 1700–1706.

Xie, J., Cha, S. K., An, S. W., Kuro, O. M., Birnbaumer, L., & Huang, C. L. (2012). Cardioprotection by Klotho through downregulation of TRPC6 channels in the mouse heart. *Nature Communications, 3*, 1238.

Xie, J., Yoon, J., An, S. W., Kuro-o, M., & Huang, C. L. (2015). Soluble Klotho protects against uremic cardiomyopathy independently of fibroblast growth factor 23 and phosphate. *Journal of the American Society of Nephrology, 26*(5), 1150–1160.

Xie, B., Zhou, J., Yuan, L., Ren, F., Liu, D. C., Li, Q., et al. (2013). Epigenetic silencing of Klotho expression correlates with poor prognosis of human hepatocellular carcinoma. *Human Pathology, 44*(5), 795–801.

Yamada, S., Tatsumoto, N., Tokumoto, M., Noguchi, H., Ooboshi, H., Kitazono, T., et al. (2015). Phosphate binders prevent phosphate-induced cellular senescence of vascular smooth muscle cells and vascular calcification in a modified, adenine-based uremic rat model. *Calcified Tissue International, 96*(4), 347–358.

Yamagishi, T., Saito, Y., Nakamura, T., Takeda, S., Kanai, H., Sumino, H., et al. (2001). Troglitazone improves endothelial function and augments renal klotho mRNA expression in Otsuka Long-Evans Tokushima Fatty (OLETF) rats with multiple atherogenic risk factors. *Hypertension Research, 24*(6), 705–709.

Yamazaki, Y., Imura, A., Urakawa, I., Shimada, T., Murakami, J., Aono, Y., et al. (2010). Establishment of sandwich ELISA for soluble alpha-Klotho measurement: Age-dependent change of soluble alpha-Klotho levels in healthy subjects. *Biochemical and Biophysical Research Communications, 398*(3), 513–518.

Yang, H. C., Deleuze, S., Zuo, Y., Potthoff, S. A., Ma, L. J., & Fogo, A. B. (2009). The PPARgamma agonist pioglitazone ameliorates aging-related progressive renal injury. *Journal of the American Society of Nephrology, 20*(11), 2380–2388.

Yang, H., & Fogo, A. B. (2010). Cell senescence in the aging kidney. *Journal of the American Society of Nephrology, 21*(9), 1436–1439.

Yang, K., Nie, L., Huang, Y., Zhang, J., Xiao, T., Guan, X., et al. (2012). Amelioration of uremic toxin indoxyl sulfate-induced endothelial cell dysfunction by Klotho protein. *Toxicology Letters, 215*(2), 77–83.

Yang, K., Wang, C., Nie, L., Zhao, X., Gu, J., Guan, X., et al. (2015). Klotho protects against indoxyl sulphate-induced myocardial hypertrophy. *Journal of the American Society of Nephrology, 26*(10), 2434–2446.

Yao, Y., Jumabay, M., Ly, A., Radparvar, M., Cubberly, M. R., & Bostrom, K. I. (2013). A role for the endothelium in vascular calcification. *Circulation Research, 113*(5), 495–504.

Yilmaz, M. I., Saglam, M., Caglar, K., Cakir, E., Sonmez, A., Ozgurtas, T., et al. (2006). The determinants of endothelial dysfunction in CKD: Oxidative stress and asymmetric dimethylarginine. *American Journal of Kidney Diseases, 47*(1), 42–50.

Yoon, H. E., Ghee, J. Y., Piao, S., Song, J. H., Han, D. H., Kim, S., et al. (2011). Angiotensin II blockade upregulates the expression of Klotho, the anti-ageing gene, in an experimental model of chronic cyclosporine nephropathy. *Nephrology, Dialysis, Transplantation, 26*(3), 800–813.

Yoon, H. E., Lim, S. W., Piao, S. G., Song, J. H., Kim, J., & Yang, C. W. (2012). Statin upregulates the expression of klotho, an anti-aging gene, in experimental cyclosporine nephropathy. *Nephron Experimental Nephrology, 120*(4), e123–e133.

Young, G. H., & Wu, V. C. (2012). KLOTHO methylation is linked to uremic toxins and chronic kidney disease. *Kidney International, 81*(7), 611–612.

Zeisberg, M., & Duffield, J. S. (2010). Resolved: EMT produces fibroblasts in the kidney. *Journal of the American Society of Nephrology, 21*(8), 1247–1253.

Zeisberg, M., Hanai, J., Sugimoto, H., Mammoto, T., Charytan, D., Strutz, F., et al. (2003). BMP-7 counteracts TGF-beta1-induced epithelial-to-mesenchymal transition and reverses chronic renal injury. *Nature Medicine, 9*(7), 964–968.

Zhang, H., Li, Y., Fan, Y., Wu, J., Zhao, B., Guan, Y., et al. (2008). Klotho is a target gene of PPAR-gamma. *Kidney International, 74*(6), 732–739.

Zhang, M., Yan, J., Zhu, M., & Ni, Z. (2015). Fibroblast growth factor 23 predicts coronary calcification and poor prognosis in patients with chronic kidney disease stages 3-5D. *Annals of Clinical and Laboratory Science, 45*(1), 17–22.

Zhang, L. N., Yang, G., Cheng, C., Shen, C., Cui, Y. Y., Zhang, J., et al. (2015). Plasma FGF23 levels and heart rate variability in patients with stage 5 CKD. *Osteoporosis International, 26*(1), 395–405.

Zhang, L., Yang, Y., Tang, Y., Zhao, Y., Cao, Y., Su, B., et al. (2013). Recovery from AKI following multiple wasp stings: A case series. *Clinical Journal of the American Society of Nephrology, 8*(11), 1850–1856.

Zhang, R., & Zheng, F. (2008). PPAR-gamma and aging: One link through klotho? *Kidney International, 74*(6), 702–704.

Zhou, L., Li, Y., Zhou, D., Tan, R. J., & Liu, Y. (2013). Loss of Klotho contributes to kidney injury by derepression of Wnt/beta-catenin signaling. *Journal of the American Society of Nephrology, 24*(5), 771–785.

Zhou, Q., Lin, S., Tang, R., Veeraragoo, P., Peng, W., & Wu, R. (2010). Role of fosinopril and valsartan on Klotho gene expression induced by angiotensin II in rat renal tubular epithelial cells. *Kidney & Blood Pressure Research, 33*(3), 186–192.

CHAPTER TWELVE

Deficiency of Soluble α-Klotho as an Independent Cause of Uremic Cardiomyopathy

J. Xie, Y.-L. Wu, C.-L. Huang[1]
University of Texas Southwestern Medical Center, Dallas, TX, United States
[1]Corresponding author: e-mail address: chou-long.huang@utsouthwestern.edu

Contents

Abstract

Cardiovascular disease (CVD) is the major cause of mortality for patients with chronic kidney disease (CKD). Cardiac hypertrophy, occurring in up to 95% patients with CKD (also known as uremic cardiomyopathy), increases their risk for cardiovascular death. Many CKD-specific risk factors of uremic cardiomyopathy have been recognized, such as secondary hyperparathyroidism, indoxyl sulfate (IS)/*p*-cresyl, and vitamin D deficiency. However, several randomized controlled trials have recently shown that these risk factors have little impact on the mortality of CVD. Klotho is a type 1 membrane protein predominantly produced in the kidney, and CKD is known to be a Klotho-deficient state. Because of its important role in FGF23 and phosphate metabolism, Klotho is believed to affect cardiac growth and function indirectly through FGF23 and phosphate. Recent

Vitamins and Hormones, Volume 101
ISSN 0083-6729
http://dx.doi.org/10.1016/bs.vh.2016.02.010

studies showed that soluble Klotho protects the heart against stress-induced cardiac hypertrophy by inhibiting TRPC6 channel-mediated abnormal Ca^{2+} signaling in the heart, and the decreased level of circulating soluble Klotho in CKD is an important cause of uremic cardiomyopathy independent of FGF23 and phosphate. These new evidence suggested that Klotho is an independent contributing factor for uremic cardiomyopathy and a possible new target for treatment of this disease.

1. INTRODUCTION

Chronic kidney disease (CKD) affects approximately 10% of the general population (Coresh et al., 2007; Go, Chertow, Fan, McCulloch, & Hsu, 2004). CKD patients are more likely to die from cardiovascular disease (CVD) than to reach end stage kidney disease and receive dialysis. In healthy individuals, the frequency of cardiovascular mortality is predominantly a function of age. In contrast, young adults with advanced CKD on dialysis have ~300-fold higher rate of CVD. Age-associated increases in CV mortality in dialysis patient population are relatively less pronounced (Foley, Parfrey, & Sarnak, 1998). Left ventricular hypertrophy (LVH) is very common among CKD patients and contributes to CV mortality by causing arrhythmia, sudden death, or congestive heart failure. LVH has traditionally been defined by echocardiogram, but recent studies using magnetic resonance imaging (MRI) show that LVH is more prevalent in CKD patients than previously thought (Mark et al., 2006). Many studies have shown that among the various cardiovascular complications in CKD patients, LVH plays a major role (Glassock, Pecoits-Filho, & Barberato, 2009; Go et al., 2004; Gross & Ritz, 2008; Taddei, Nami, Bruno, Quatrini, & Nuti, 2011). Furthermore, LVH is one of the most prevalent manifestations of CVD (Park et al., 2012), and is predictive of an increased mortality and cardiovascular death in advanced CKD patients (Foley, Parfrey, Harnett, Kent, Murray, et al., 1995; Parfrey et al., 1996; Stack & Saran, 2002). LVH in CKD, also known as uremic cardiomyopathy, can be due to conventional risk factors, such as hypertension and anemia or CKD-specific factors (Harnett, Kent, Barre, Taylor, & Parfrey, 1994; Levin et al., 1999). The role of many other uremia-specific cardiovascular risk factors remains unclear. Nevertheless, whether in human or animal data, LVH occurs even in the absence of hypertension (Wolf et al., 2000). Randomized trials have shown that correction of anemia alone fails to result in reduction in left ventricular mass (Cianciaruso, Ravani, Barrett, Levin, & investigators, I.-E., 2008;

Foley et al., 2000; Roger et al., 2004). Discovery of new CKD-specific risk factors for cardiac hypertrophy is urgent in targeting novel therapies for reducing the burden of CVD in CKD.

2. CLINICAL MANIFESTATIONS OF UREMIC CARDIOMYOPATHY

2.1 Definition of Uremic Cardiomyopathy

In the 1980–1990s, echocardiographic studies identified common abnormalities of cardiac structure and function in ESRD patients (Foley, Parfrey, Harnett, Kent, Martin, et al., 1995; Huting, Kramer, Schutterle, & Wizemann, 1988; London et al., 1987). These abnormalities include LVH, left ventricular (LV) dilation, and LV systolic dysfunction. Nevertheless, further study by contrast-enhanced cardiac MRI in ESRD patients revealed that LVH may be relatively unique in CKD patients caused by uremia-specific factors, while LV dilation and systolic dysfunction may be due to underlying (possibly silent) ischemic heart disease (Mark et al., 2006). Thus, LVH is frequently considered as a hallmark of "uremic cardiomyopathy."

2.2 Epidemiology of Uremic Cardiomyopathy

The prevalence of cardiac hypertrophy is markedly increased in patients with CKD. More than 75% of patients with CKD already had established LVH at the time of dialysis initiation (Foley, Parfrey, Harnett, Kent, Martin, et al., 1995; Stewart et al., 2005). A cross-sectional analysis from the Chronic Renal Insufficiency Cohort (CRIC) reported a progressive increase of prevalent LVH diagnosed by echocardiography from 32% among patients with an estimated glomerular filtration rate (eGFR) $\geq$60 mL/min per 1.73 m^2 to 75% in those with an eGFR $<$30 mL/min per 1.73 m^2 (Park et al., 2012). Recent longitudinal echocardiographic data showed CKD stage 4 and 5 patients had greater increase in LV mass index over 1 year compared with patients in stage 3a and 3b (Cai, Lu, Lu, & Wang, 2014). These data suggested that worsening kidney function may be an important contributing factor for LVH in CKD.

LVH is clinically important because it is associated with increases in the incidence of heart failure (HF), ventricular arrhythmias, death following acute myocardial infarction, sudden cardiac death, aortic root dilation, and stoke in general population. In ESRD patients, LVH was associated with $>$2-fold increase in the risk of stroke and a 60% increase in the risk of sudden death (Krane et al., 2009). In a cross-sectional study, LVH was also linked to increased QT interval and dispersal, and with minor rhythm abnormalities, providing a link with the high risk of sudden death in ESRD patients (Stewart et al., 2005).

2.3 Mechanism of Cardiac Hypertrophy in CKD Patients

2.3.1 New CKD-Specific Risk for Cardiac Hypertrophy Should Be Proposed

The pathophysiology of uremic cardiomyopathy has three components: (1) LV pressure overload, (2) LV volume overload, and (3) CKD (or uremia)-specific factors. LV pressure overload frequently results from hypertension and arteriosclerosis, and occasionally from aortic stenosis. Pressure overload causes increased wall stress during systole, rising cardiomyocyte proliferation and wall thickening with preservation, or reduction of cavity volume (concentric hypertrophy). LV volume overload occurs as a result of the presence of high output states like anemia of chronic disease, an arteriovenous fistula, and hypervolemia. Volume overload stretches the ventricular wall, engendering cardiomyocyte elongation, which thus enlarges the internal dimensions of the left ventricular cavity (eccentric hypertrophy). These conventional risk factors, such as hypertension, volume overload, and anemia, are important in the pathogenesis of cardiac hypertrophy in CKD (Glassock et al., 2009; Gross & Ritz, 2008). Yet, many CKD patients even without underlying diabetes and ischemic heart disease develop cardiac hypertrophy despite intensive control of blood pressure, anemia, and fluid retention. Thus, risk factors specific to CKD also play important roles. Several CKD-specific risk factors have been proposed (Glassock et al., 2009; Gross & Ritz, 2008). Among these include secondary hyperparathyroidism, vitamin D deficiency, retention of protein-bound uremic toxins (IS, etc.), phosphate retention, increases in circulating FGF23 levels, and decreases in circulating soluble Klotho (Fig. 1).

2.3.2 Secondary Hyperparathyroidism and Uremic Cardiomyopathy

The implication that elevated parathyroid hormone (PTH) levels may be causative for cardiac hypertrophy in CKD was first proposed in early 1980s by Massry and colleagues (Bogin, Massry, & Harary, 1981). They showed that PTH increases Ca^{2+} entry in rat ventricular myocytes, involving L-type Ca^{2+} channels. Additionally, receptors for PTH and PTHrP are present in cardiomyocytes. Parathyroidectomy alleviates uremic cardiomyopathy in animal model of CKD and infusion of PTH aggravates the disease

Mechanisms of cardiac hypertrophy in CKD

Conventional:	Hypertension Volume overload Anemia
CKD-specific:	Hyperparathyroidism Vitamin D deficiency Indoxyl sulfate Phosphate retention Elevated FGF23 Klotho deficiency, etc.

Fig. 1 Mechanisms of cardiac hypertrophy in CKD.

(Massry & Fadda, 1993). Interestingly, patients with primary hyperparathyroidism also have increased CVD and mortality (Andersson, Rydberg, & Willenheimer, 2004). Whether this is related to hypercalcemia-induced hypertension, cardiac, vascular and/or valvular calcification, etc., has never been clearly settled. In general, PTH is not considered a major prohypertrophic hormone for cardiomyocytes.

In the largest clinical trial involving HD patients ever, the EVOLVE study examined the effect of suppression of PTH on CVD (EVOLVE Trail Investigators et al., 2012). Three thousand eight hundred eighty-three patients with moderate-to-severe secondary hyperparathyroidism undergoing hemodialysis (median level of intact PTH, 693 pg/mL [10th–90th percentile, 363–1694]) were randomly assigned to receive either cinacalcet or placebo. All patients were eligible to receive conventional therapy, including phosphate binders, vitamin D sterols, or both. The patients were followed for up to 64 months. The primary composite end point was the time until death, myocardial infarction, hospitalization for unstable angina, HF, or a peripheral vascular event. The primary analysis was performed on the basis of the intention-to-treat principle. The median duration of study-drug exposure was 21.2 months in the cinacalcet group, vs 17.5 months in the placebo group. The primary composite end point was reached in 938 of 1948 patients (48.2%) in the cinacalcet group and 952 of 1935 patients (49.2%) in the placebo group (relative hazard in the cinacalcet group vs the placebo group, 0.93; 95% confidence interval, 0.85–1.02; $p=0.11$). Hypocalcemia and gastrointestinal adverse events were significantly more frequent in patients receiving cinacalcet. In an unadjusted intention-to-treat analysis, cinacalcet did not significantly reduce the risk of death or major cardiovascular events in patients with moderate-to-severe secondary hyperparathyroidism who were undergoing dialysis. After adjustment for baseline characteristics, there was a nominally significant 12% reduction in risk. The conclusion of the trial is considered as nondefinitive.

2.3.3 Vitamin D Deficiency and Uremic Cardiomyopathy

Vitamin D deficiency is known to be associated with decreased cardiovascular-related morbidity and mortality, possibly by modifying cardiac structure and function, yet firm evidence for either remains lacking. In experimental animal model, active vitamin D attenuates cardiac hypertrophy (Bodyak et al., 2007; Freundlich et al., 2014; Panizo et al., 2013). However, a double-blind, randomized placebo-controlled (PRIMO) trial among 227 patients with stage 3 and 4 CKD, mild-to-moderate LVH, and preserved left ventricular ejection fraction, was conducted to determine the effects of an active vitamin D compound, paricalcitol, on left ventricular mass over 48 weeks (Thadhani et al., 2012). The eGFR of study patients ranged between 15 and 60 mL/min/1.73 m^2. Change in left ventricular mass index over 48 weeks was assessed by cardiovascular MRI. Secondary end points included echocardiographic changes in left ventricular diastolic function. It was found that 48 weeks therapy with paricalcitol did not alter left ventricular mass index or improve certain measures of diastolic dysfunction in patients with CKD. Some may criticize that the negative findings in the PRIMO study was due to the low LV mass and absence of LV hypertrophy. However, another similar prospective, double-blind, randomized placebo-controlled (OPERA) trial has shown similar result in Hong Kong (Wang et al., 2014). Sixty subjects enrolled by echocardiographic criteria of LV hypertrophy were randomly assigned to receive either oral paricalcitol (1 μg) one time daily ($n=30$) or matching placebo ($n=30$) for 52 weeks. Again, the primary end point by cardiac MRI was change in LV mass indexed by body surface area, which did not differ between groups (median [interquartile range], −2.59 [−6.13 to 0.32] g/m^2 with paricalcitol vs −4.85 [−9.89 to 1.10] g/m^2 with placebo). Changes in LV volume and ejection fraction were also not different between the two groups. Notably, paricalcitol treatment significantly reduced intact PTH ($p<0.001$), whereas induced more hypercalcemia ($p<0.001$). In this trial, 52 weeks of treatment with oral paricalcitol significantly improved secondary hyperparathyroidism but had no impact on LV structure in patients with advanced CKD.

2.3.4 Indoxyl Sulfate/p-Cresyl Sulfate and Uremic Cardiomyopathy

Indoxyl sulfate (IS) and *p*-cresyl sulfate (*p*-CS) both originate from bacterial protein fermentation in the large intestine. Colonic microbiota degrades tryptophan to indole. Further hydroxylation results in 3-hydroxy-indole, the majority of which is sulfonated to IS. In parallel, fermentation of tyrosine results in *p*-cresol and ultimately *p*-CS (Meijers & Evenepoel, 2011). Most *p*-CS and IS in circulation noncovalently bind to albumin and compete for the same albumin-binding sites. Being bound to albumin, IS and *p*-CS are not easily cleared by glomerular filtration. Normally, they are cleared from the systemic circulation predominantly by tubular secretion via organic anion transporters. In individuals with normal GFR, the circulating levels of free IS are below detection and of total (free and bound) ~0.5 mg/L. The level of the total IS may be up to ~20 mg/L in uremia. IS/*p*-CS have been implicated in causing many uremia-associated complications including oxidative stress, inflammation, fibrosis, endothelial dysfunction, cardiac dysfunction, etc. (Lekawanvijit, Kompa, Wang, Kelly, & Krum, 2012). In a small cross-section survey of 86 patients with mild to severe CKD, LVH was positively related to serum IS level ($r=0.69$, $p<0.001$) (Yang et al., 2015).

AST-120 is a carbonaceous oral adsorbent first approved for use in Japan in 1991 and later on in other Asian and European countries. It decreases serum IS/PS levels by binding and excretion through gastrointestinal tract. It has been used in several regions including Japan as an agent to retard the progression of CKD. CAP-KD study is an RCT study of 460 CKD patients examining the effect of AST-120 on progression of CKD (Akizawa et al., 2009). Composite primary end point includes doubling of serum creatinine (sCr) level, increase in sCr level to 6.0 mg/dL or more, need for dialysis or transplantation, or death. During 56 weeks, numbers of primary end-point events (43 for control vs 42 for AST-120) and event-free survival ($p=0.9$) did not differ between groups. Gastrointestinal adverse events were less common in the control group than the AST-120 group (2 vs 32 events). Estimated creatinine clearance decreased more in the control group than in the AST-120 group (−15% per year vs −12% per year, relative to the baseline value; [corrected] $p=0.001$). However, the difference was not significant. Thus, AST-120 did not substantially slow the progression of kidney disease in patients with moderate-to-severe CKD during 1 year.

This study was recently followed by a larger (EPPIC I and II Trial) multinational (North/Central America & Europe) RCT study of 2035 moderate-to-severe CKD patients to examine the effect of AST-120 on the progression of CKD, defined as initiation of dialysis, kidney transplantation, and sCr doubling (Schulman et al., 2015). Again, until accrual of 291 primary end points reached for each trial, the time to primary end point was similar between the AST-120 and the placebo groups in both trials (EPPIC-1: HR, 1.03; 95% CI, 0.84–1.27; $p=0.78$) (EPPIC-2: HR, 0.91; 95% CI, 0.74–1.12; $p=0.37$). All cause-mortality was also no difference between the treatment and placebo group in each trial (EPPIC I: 9.8% vs 8.4%, $p=0.45$; EPPIC II: 9.8% vs 12.3%, $p=0.19$). Therefore, AST-120 had no impact on CKD progression and all-cause mortality by these trials.

2.3.5 Phosphate Retention and Uremic Cardiomyopathy

Studies in model organism and small animals have provided evidence that dietary phosphate intake inversely correlates with life span independently of caloric intake. Direct examination of phosphate intake on human life span is impossible. However, several observational studies have examined relationship between hyperphosphatemia and CVD and cardiovascular modality in populations that have shortened life span, for example, individuals with prior events of acute coronary syndrome or with CKD. A post hoc analysis of data from the Cholesterol And Recurrent Events (CARE) study, baseline serum phosphate levels from 4127 fasting participants who were categorized into four groups: <2.5, 2.5–3.4, 3.5–3.9, and ≥4 mg/dL (Tonelli et al., 2005). Patients were followed up for a median of 59.7 months. Cox proportional-hazards models were used to examine the association between serum phosphate and adverse clinical outcomes after adjustment for potential confounders. During nearly 60 months of follow-up, 375 participants died. The analysis revealed a significant association between baseline serum phosphate level and the age-, race-, and sex-adjusted risk of all-cause death.

Block et al. examined the relationship between mineral metabolism and mortality in dialysis patients (Block, Hulbert-Shearon, Levin, & Port, 1998). Multivariable-adjusted relative risks of death were calculated for categories of serum phosphorus, calcium, calcium × phosphorus product, and intact PTH using proportional hazards regression. They found that hyperphosphatemia and hyperparathyroidism were significantly associated with all-cause, cardiovascular, and fracture-related hospitalization. Hyperphosphatemia is an important risk factor for vascular calcification, which may contribute to cardiac hypertrophy, by causing vascular calcification, decreasing vascular compliance, and thus systolic hypertension (Glassock et al., 2009; Gross & Ritz, 2008). Whether phosphate is a direct toxin for the heart and causes cardiac hypertrophy independently of vascular and hemodynamic factors are not known.

3. MINERAL METABOLISM DISORDER OF CKD

3.1 Klotho and Its Role in Phosphate Metabolism

Klotho is a type-1 membrane protein predominantly produced in the kidney that exerts some antiaging function (Kuro-o et al., 1997). Mice homozygous for hypomorphic *klotho* (*kl*) allele, due to disruption of the promoter region by a randomly inserted transgene, show multiple phenotypes resembling premature human aging, including shortened lifespan, growth retardation, skin atrophy, hypogonadism, hyperphosphatemia, and tissue calcification (Kuro-o et al., 1997). Conversely, overexpression of *klotho* transgene results in extended lifespan and increased resistance to oxidative stress (Kurosu et al., 2005). The extracellular domain of Klotho consisting of the homologous KL1 and KL2 repeats can be cleaved and released into the extracellular fluid such as blood, urine, and cerebrospinal fluid (Imura et al., 2004; Kuro-o et al., 1997; Kurosu et al., 2005). Thus, Klotho exists in two different forms in vivo: a full-length membranous Klotho and a shed soluble Klotho ectodomain, each with own separate function (Imura et al., 2004; Kuro-o et al., 1997; Kurosu et al., 2005) (Fig. 2).

Soluble Klotho has been shown to regulate ion transport and growth factor signaling acting as a paracrine or endocrine factor (Cha et al., 2008;

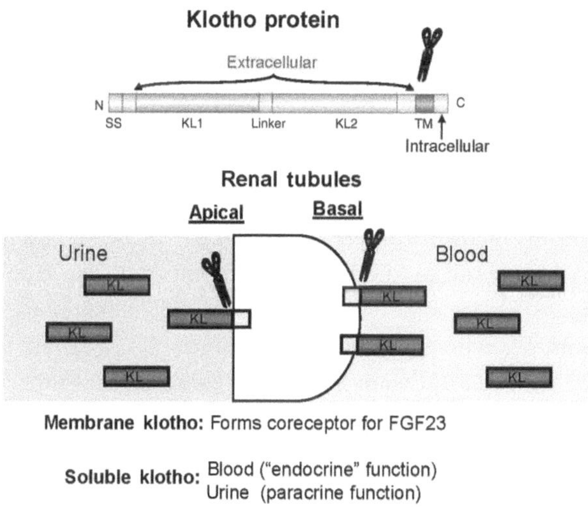

Fig. 2 Klotho protein and shedding model.

Kurosu et al., 2005). Membranous Klotho interacts with fibroblast growth factor receptor (FGFR) and functions as coreceptor for the ligand FGF23 (Kurosu et al., 2006; Urakawa et al., 2006). Activation of membranous Klotho–FGFR coreceptor complex by FGF23 suppresses renal synthesis of 1,25-$(OH)_2$-vitamin D (therefore decreasing gastrointestinal phosphate absorption) and inhibits renal phosphate reabsorption by sodium-phosphate cotransporter (Consortium, A, 2000; Kurosu et al., 2006; Razzaque, 2009; Urakawa et al., 2006). Recent studies have indicated that regulation of phosphate metabolism by FGF23 via Klotho and FGFR coreceptor complex plays a critical role in the aging suppression by Klotho. Mice homozygous for *fgf23* deletion have profound hyperphosphatemia and the same premature aging phenotypes as in homozygous *klotho*-hypomorphic mice (Razzaque, 2009). Dietary phosphate restriction rescues premature aging and death in both *fgf23* knockout and homozygous *klotho*-hypomorphic mice (Morishita et al., 2001; Nakatani et al., 2009; Razzaque, 2009).

3.2 Role of FGF23 and Klotho in CKD-MBD

FGF23 is a bone-derived endocrine hormone that potently lowers serum phosphate levels by suppressing 1,25-$(OH)_2$-vitamin D synthesis (thus decreasing gut absorption of phosphate) and inhibiting renal phosphate reabsorption mediated by Na-phosphate cotransporter (Consortium, A, 2000). Serum FGF23 levels rise in early stages of CKD and increase progressively in advanced stages to the levels 100s and 1000s fold of normal serum levels (Fliser et al., 2007; Razzaque, 2009). The mechanism for the increase in FGF23 levels in CKD remains incompletely understood, but phosphate retention plays an important role. In early stages of CKD, renal resistance to FGF23 due to decrease in the abundance of membranous Klotho in the diseased kidney (thus with impairment of its FGF23 coreceptor function and elevation of 1,25-dihydroxy vitamin D levels) may play a key role. As CKD progresses, phosphate retention as the result of decreased renal excretion further stimulates FGF23 production from the bone (Kuro-o, 2013; Wolf, 2012). Increases in circulating FGF23 levels suppress 1,25-$(OH)_2$-vitamin D, which causes an increase in PTH levels by releasing inhibition by 1,25-$(OH)_2$-vitamin D in the parathyroid glands. In the early stages of CKD, rises in FGF23 and PTH levels coupled with the decreases in 1,25-$(OH)_2$-vitamin D keep serum phosphate levels from rising beyond the normal range (Fig. 3). Thus, rise of FGF23 in CKD is considered an adaptive mechanism to combat phosphate retention (Kuro-o, 2013;

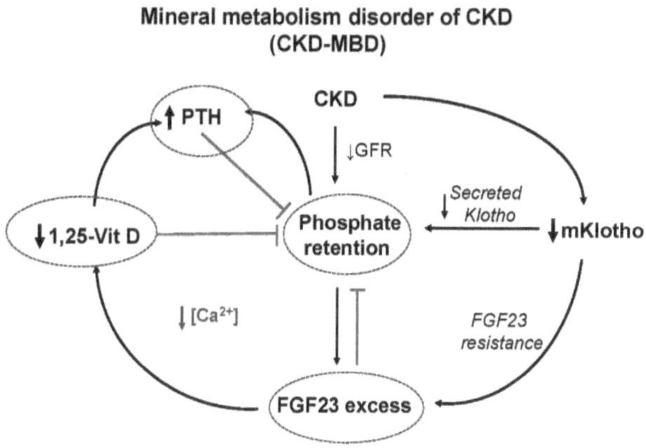

Fig. 3 Role of FGF23 and Klotho in CKD.

Razzaque, 2009; Wolf, 2012). Consistent with this notion, many studies have found that circulating FGF23 levels strongly correlate with the stages of CKD and its cardiovascular complications (Faul et al., 2011; Gutierrez et al., 2009; Kuro-o, 2013; Wolf, 2012). Recently, a study reported that rise in FGF23 levels in CKD may be maladaptive; that is, it stimulates cardiomyocyte growth to induce cardiac hypertrophy (Faul et al., 2011). However, other studies have not found evidence supporting the notion that FGF23 directly induces cardiac hypertrophy (Agarwal et al., 2014).

4. DECREASE IN CIRCULATING SOLUBLE KLOTHO CONTRIBUTES TO UREMIC CARDIOMYOPATHY: EVIDENCE AND MECHANISM

4.1 Klotho-Deficiency Aggravates Stress-Induced and Uremic Cardiomyopathy, Independently of Hypertension, Hyperphosphatemia, and Excess FGF23

Since soluble Klotho level is decreased in aging and in late stage CKD, while cardiomyopathy is a major consequence of aging and CKD, it is natural to ask whether Klotho deficiency is a contributing factor of age-related and/or uremic cardiomyopathy. Xie et al. started to address this question by showing Klotho deficiency aggravates pathological cardiac hypertrophy in Klotho-deficient mice (Xie et al., 2012). Overstimulation by isoproterenol (ISO), a well-accepted experimental model of stress-induced cardiac hypertrophy, induced pathological hypertrophy in wild type (*WT*) mice as reflected by increase in heart weight indices (heart weight normalized to body weight or tibia length), and the ISO-induced cardiac hypertrophy was aggravated in homozygous *klotho*-hypomorphic (*kl/kl*) mice. This aggravated cardiomyopathy was further measured and supported by the increased heart size (histology and MRI imaging), elevated hypertrophic marker gene expression (qRT-PCR), and decreased cardiac ejection fraction (MRI). It is interesting to note that *klotho*-deficient (*kl/kl*) and heterozygous (*kl*/+) mice were previously reported to develop cardiomyopathy, supposedly via elevated FGF23 levels (Faul et al., 2011). In contrast, Xie et al. did not observe significant cardiac hypertrophy in *kl*/+ mice and *kl/kl* mice rescued with phosphate-restricted diet. Klotho deficiency only aggravates stress-induced cardiomyopathy in their study. Another report (Agarwal et al., 2014) showing Klotho-deficient mice do not develop spontaneous cardiac hypertrophy seems to support the notion that stress-induced condition is needed for Klotho to exhibit its cardiac protection effect.

To further analyze Klotho's role in uremic cardiomyopathy caused by CKD, Xie et al. generated CKD by performing 5/6 nephrectomy in *WT* and *kl*/+ mice. 4–5 weeks after the surgery, *WT* and *kl*/+ mice developed similar degree of CKD with classic risk factors, such as blood pressure, creatinine clearance, serum Pi, and hematocrit, all at similar levels. Serum soluble Klotho level was decreased to ~50% and ~13% of normal level, in *WT* CKD and *kl*/+ CKD mice, respectively. The CKD mice developed cardiac hypertrophy and cardiac fibrosis. The uremic cardiomyopathy was dramatically more severe in the *kl*/+ CKD mice compared to the *WT* CKD mice, as analyzed by the heart weight indices, cardiac MRI, and histology. Blood pressure and serum Pi were normalized to normal levels by treatment with antihypertensive drugs and phosphate-restriction diet. Normalization of BP only partially rescued cardiac hypertrophy in CKD, and did not ameliorate the difference in cardiomyopathy between *WT* CKD and *kl*/+ CKD mice. Normalization of serum Pi also corrected serum FGF23 levels in both *WT* CKD and *kl*/+ CKD mice, but did not significantly alter cardiac hypertrophic response in either group. Therefore, these results suggested that Klotho-deficiency aggravates uremic cardiomyopathy independently of hypertension, hyperphosphatemia, and excess FGF23 (Xie, Yoon, An, Kuro-o, & Huang, 2015).

4.2 Transgenic *klotho*-Expression Attenuates Stress-Induced and Uremic Cardiomyopathy

More evidence supporting that Klotho is an independent cause of uremic cardiomyopathy also came from studies on *klotho*-overexpressing transgenic animals with stress-induced cardiomyopathy (ISO model) and uremic cardiomyopathy (CKD model). A transgenic mouse line overexpressing *klotho* globally (*KL-Tg*) has about twofold increase on the circulating level of soluble Klotho. The Klotho overexpression did not cause detectable changes in heart mass and systemic blood pressure at baseline, nor did it alter serum phosphate or FGF23 levels, but it blunted the ISO-induced cardiac hypertrophic responses (Xie et al., 2012). In the CKD mice developing uremic cardiomyopathy, transient Klotho-expressing was achieved by intravenous delivery of *klotho* transgene. The transient expression of transgenic *klotho* in CKD mice attenuated cardiac hypertrophy, when compared with CKD mice with vector DNA injection. The blood pressure, creatinine clearance, BUN, serum Pi, FGF23, and hematocrit were not altered by this *klotho* transgenic expression (Xie et al., 2015).

4.3 Klotho Protects the Heart by Downregulation of TRPC6

4.3.1 TPRC6 Channel and Cardiomyopathy

Calcineurin-NFAT pathway is one of the most critical regulatory pathways in the development of cardiomyopathy. The calmodulin-dependent serine–threonine protein phosphatase calcineurin is activated by increases in intracellular calcium. Activated calcineurin then dephosphorylates a transcription factor, nuclear factor of activated T cells (NFAT), and causes its nuclear translocation and binding to the regulator regions of cardiac genes. In conjunction with other transcription factors, NFAT induces cardiac gene expression and promotes hypertrophic growth and remodeling (Vega, Bassel-Duby, & Olson, 2003).

Triggered by extracellular stimuli, Ca^{2+} can either be released from intracellular organelles or entered across the plasma membrane to increase intracellular Ca^{2+} level. The TRPC family channels are Ca^{2+}-permeable cation channels expressed in the plasma membrane of many tissues including the heart (Nilius, Owsianik, Voets, & Peters, 2007). Evidence indicates that Ca^{2+} influx through cardiac TRPC channels including TRPC1, 3, 4, 5, and 6, is important in calcineurin signaling and hypertrophic growth of hearts. Their expression is increased in hypertrophic hearts stimulated by various stresses and their downregulation protects against cardiac hypertrophy (Rowell, Koitabashi, & Kass, 2010). TRPC6 contains NFAT-responsive elements in its promoter, which have a pivotal role in amplifying and sustaining gene expression through a feed-forward circuit (Eder & Molkentin, 2011; Kuwahara et al., 2006). Thus, TRPC6 is an important modulator of cardiac hypertrophy and a potential target for treatment.

4.3.2 *Klotho Downregulates TRPC6 in Heart: Genetics Evidence*

A possible regulatory relation between Klotho and TRPC6 was investigated by Xie et al. (Xie et al., 2012, 2015). Klotho-deficiency in mice aggravated ISO-induced cardiac hypertrophy, and also the increases of TRPC6 expression in heart. Conversely, Klotho overexpression in mice attenuated the ISO-induced increases in cardiac *Trpc6* mRNA expression. Deletion of *Trpc6* in mice partially protected against ISO-induced cardiac remodeling. In a *Klotho/Trpc6* double-knockout mouse model, absence of TRPC6 completely prevented the exaggerated ISO-induced cardiac hypertrophy caused by Klotho deficiency. Mice with cardiac-specific overexpression of TRPC6 (*C6-Tg*) develop spontaneous cardiac hypertrophy and have shortened life span (Kuwahara et al., 2006). When they were crossed with transgenic mice-overexpressing *klotho* (*Kl-Tg*), the resulting double transgenic mice (*C6-Tg*, *Kl-Tg*) had ameliorated cardiac hypertrophy and improved survival curve, compared to *C6-Tg* mice. Also, in uremic cardiomyopathy induced by CKD, *Trpc6* gene expression and channel activity were upregulated. This *Trpc6* upregulation was further aggravated in *klotho*-deficient CKD heart, compared with *WT* CKD heart. These genetic studies suggested that Klotho protects the heart by inhibiting cardiac TRPC6.

4.3.3 *Soluble Klotho Inhibits TRPC6 Activity by Downregulating Its Exocytosis*

In vitro assays have also been performed to verify that Klotho inhibits TRPC6 activity in cardiomyocytes and to investigate the underlying mechanism. TRPC6 channel activity was measured in freshly isolated ventricular myocytes by whole-cell patch-clamp recording. It was identified as current increase above baseline in *WT* cardiomyocytes after ISO treatment, and in *Trpc6*-overexpressing cardiomyocytes. The current increase by ISO treatment was eliminated in *Trpc6*-knockout hearts. The ISO-induced increase in currents was prevented in cardiomyocytes of *klotho*-overexpressing transgenic mice, similar to the pattern observed in *Trpc6*-knockout cardiomyocytes. Acute addition of soluble Klotho to culture media decreased TRPC6-mediated currents in *WT* cardiomyocytes after ISO treatment and that in *Trpc6*-overexpressing cardiomyocytes. These results confirmed that soluble Klotho directly inhibits TRPC6 activity in cardiac myocytes. Soluble Klotho also inhibits TRPC6 currents in HEK cells expressing recombinant TRPC6. Biotinylation assays of the TRPC6-expressing HEK cells revealed that the Klotho treatment decreased cell surface abundance of TRPC6. Blocking exocytosis by v-SNARE inhibitor tetanus toxin decreased TRPC6 currents in these cells, and prevented further inhibition by soluble Klotho, suggesting that the major action of Klotho on TRPC6 is by blocking exocytosis. On the other hand, blocking endocytosis using a dominant-negative dynamin did not affect the ability of Klotho to inhibit TRPC6 (Xie et al., 2012).

It has been reported that cell surface abundance of TRPC channels is regulated by exocytosis, which is under the control of phosphoinositide-3-kinase (PI3K) signaling pathways (Bezzerides, Ramsey, Kotecha, Greka, & Clapham, 2004). Soluble Klotho is known to inhibit PI3K signaling by insulin and insulin-like growth factors (IGF), which contributes to the antiaging and tumor suppression effects of Klotho (Château, Araiz, Descamps, & Galas, 2010; Kurosu et al., 2005; Wolf et al., 2008). When analyzed in TRPC6-expressing HEK cells, IGF1 promoted TRPC6 currents, while addition of soluble Klotho inhibited this IGF1-stimulated TRPC6 current. Moreover, PI3K inhibitor wortmannin decreased TRPC6 currents stimulated by IGF1, and abrogated the TRPC6-inhibiting effect of Klotho. In ISO-treated cardiac myocytes, the TRPC6 current was decreased by treatment of wortmannin to the similar levels of that was treated by soluble Klotho and exocytosis blocker tetanus toxin. Any combination of the three did not further decrease TRPC6 current. These results indicate that soluble Klotho inhibits IGF1- and PI3K-dependent exocytosis of TRPC6 (Xie et al., 2012).

5. CONCLUSION

Recent new evidences have shown that soluble Klotho inhibits a cardiac Ca^{2+} channel TRPC6 by downregulating IGF1- and PI3K-dependent TRPC6 exocytosis, and protects the heart against stress-induced pathological hypertrophy and remodeling (Fig. 4). Decreased level of circulating soluble Klotho in CKD is an important cause of uremic cardiomyopathy independent of FGF23 and phosphate, opening new avenues for treatment of this disease.

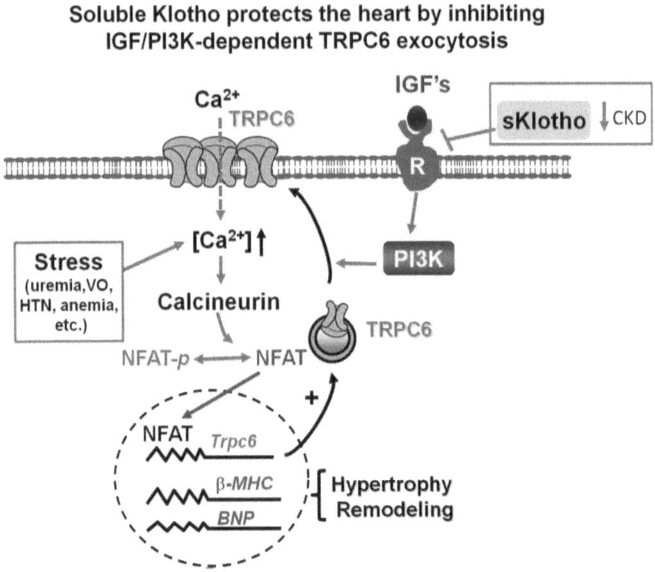

Fig. 4 Soluble Klotho protects the heart by inhibiting IGF/PI3K-dependent TRPC6 exocytosis. Stresses (ISO overstimulation, uremia, etc.) cause an abnormal intracellular Ca^{2+} signaling, which activates calcineurin and NFAT, thereby inducing cardiac hypertrophy and remodeling, as well as *Trpc6* gene expression. Upregulation of TRPC6 provides a feed-forward loop that amplifies and sustains the pathological cardiac responses. IGF1 activates PI3K to promote exocytosis of TRPC6. Soluble Klotho (sKlotho) inhibits IGF1 activation of PI3K, partly by direct interactions with the receptors. Decreased production of soluble Klotho in CKD leads to upregulation of TRPC6 activity under stress and contributes to cardiac hypertrophy.

REFERENCES

Agarwal, I., Ide, N., Ix, J. H., Kestenbaum, B., Lanske, B., Schiller, N. B., … Mukamal, K. J. (2014). Fibroblast growth factor-23 and cardiac structure and function. *Journal of the American Heart Association. 3*(1). e000584. http://dx.doi.org/10.1161/JAHA.113.000584.

Akizawa, T., Asano, Y., Morita, S., Wakita, T., Onishi, Y., Fukuhara, S., … Group, C.-K. S. (2009). Effect of a carbonaceous oral adsorbent on the progression of CKD: A multicenter, randomized, controlled trial. *American Journal of Kidney Diseases, 54*(3), 459–467. http://dx.doi.org/10.1053/j.ajkd.2009.05.011.

Andersson, P., Rydberg, E., & Willenheimer, R. (2004). Primary hyperparathyroidism and heart disease—A review. *European Heart Journal, 25*(20), 1776–1787. http://dx.doi.org/10.1016/j.ehj.2004.07.010.

Bezzerides, V. J., Ramsey, I. S., Kotecha, S., Greka, A., & Clapham, D. E. (2004). Rapid vesicular translocation and insertion of TRP channels. *Nature Cell Biology, 6*, 709–720.

Block, G. A., Hulbert-Shearon, T. E., Levin, N. W., & Port, F. K. (1998). Association of serum phosphorus and calcium x phosphate product with mortality risk in chronic hemodialysis patients: A national study. *American Journal of Kidney Diseases, 31*(4), 607–617.

Bodyak, N., Ayus, J. C., Achinger, S., Shivalingappa, V., Ke, Q., Chen, Y. S., ... Kang, P. M. (2007). Activated vitamin D attenuates left ventricular abnormalities induced by dietary sodium in Dahl salt-sensitive animals. *Proceedings of the National Academy of Sciences of the United States of America, 104*(43), 16810–16815. http://dx.doi.org/10.1073/pnas.0611202104.

Bogin, E., Massry, S. G., & Harary, I. (1981). Effect of parathyroid hormone on rat heart cells. *The Journal of Clinical Investigation, 67*(4), 1215–1227.

Cai, Q. Z., Lu, X. Z., Lu, Y., & Wang, A. Y. (2014). Longitudinal changes of cardiac structure and function in CKD (CASCADE study). *Journal of the American Society of Nephrology, 25*(7), 1599–1608. http://dx.doi.org/10.1681/ASN.2013080899.

Cha, S. K., Ortega, B., Kurosu, H., Rosenblatt, K. P., Kuro, O. M., & Huang, C. L. (2008). Removal of sialic acid involving Klotho causes cell-surface retention of TRPV5 channel via binding to galectin-1. *Proceedings of the National Academy of Sciences of the United States of America, 105*(28), 9805–9810. http://dx.doi.org/10.1073/pnas.0803223105.

Château, M. T., Araiz, C., Descamps, S., & Galas, S. (2010). Klotho interferes with a novel FGF-signalling pathway and insulin/Igf-like signalling to improve longevity and stress resistance in *Caenorhabditis elegans*. *Aging, 2*, 567–581.

Cianciaruso, B., Ravani, P., Barrett, B. J., Levin, A., & Investigators, I.-E. (2008). Italian randomized trial of hemoglobin maintenance to prevent or delay left ventricular hypertrophy in chronic kidney disease. *Journal of Nephrology, 21*(6), 861–870.

Consortium, A (2000). Autosomal dominant hypophosphataemic rickets is associated with mutations in FGF23. *Nature Genetics, 26*(3), 345–348. http://dx.doi.org/10.1038/81664.

Coresh, J., Selvin, E., Stevens, L. A., Manzi, J., Kusek, J. W., Eggers, P., ... Levey, A. S. (2007). Prevalence of chronic kidney disease in the United States. *JAMA, 298*(17), 2038–2047. http://dx.doi.org/10.1001/jama.298.17.2038.

Eder, P., & Molkentin, J. D. (2011). TRPC6 channels as effectors of cardiac hypertrophy. *Circulation, 108*, 265–272.

EVOLVE Trail Investigators, E.T., Chertow, G. M., Block, G. A., Correa-Rotter, R., Drueke, T. B., Floege, J., ... Parfrey, P. S. (2012). Effect of cinacalcet on cardiovascular disease in patients undergoing dialysis. *The New England Journal of Medicine, 367*(26), 2482–2494. http://dx.doi.org/10.1056/NEJMoa1205624.

Faul, C., Amaral, A. P., Oskouei, B., Hu, M. C., Sloan, A., Isakova, T., ... Wolf, M. (2011). FGF23 induces left ventricular hypertrophy. *The Journal of Clinical Investigation, 121*(11), 4393–4408. http://dx.doi.org/10.1172/JCI46122.

Fliser, D., Kollerits, B., Neyer, U., Ankerst, D. P., Lhotta, K., Lingenhel, A., ... Riegler, P. (2007). Fibroblast growth factor 23 (FGF23) predicts progression of chronic kidney disease: The Mild to Moderate Kidney Disease (MMKD) Study. *Journal of the American Society of Nephrology, 18*(9), 2600–2608. http://dx.doi.org/10.1681/ASN.2006080936.

Foley, R. N., Parfrey, P. S., Harnett, J. D., Kent, G. M., Martin, C. J., Murray, D. C., & Barre, P. E. (1995a). Clinical and echocardiographic disease in patients starting end-stage renal disease therapy. *Kidney International, 47*(1), 186–192.

Foley, R. N., Parfrey, P. S., Harnett, J. D., Kent, G. M., Murray, D. C., & Barre, P. E. (1995b). The prognostic importance of left ventricular geometry in uremic cardiomyopathy. *Journal of the American Society of Nephrology, 5*(12), 2024–2031.

Foley, R. N., Parfrey, P. S., Morgan, J., Barre, P. E., Campbell, P., Cartier, P., ... Wells, G. A. (2000). Effect of hemoglobin levels in hemodialysis patients with asymptomatic cardiomyopathy. *Kidney International, 58*(3), 1325–1335. http://dx.doi.org/10.1046/j.1523-1755.2000.00289.x.

Foley, R. N., Parfrey, P. S., & Sarnak, M. J. (1998). Clinical epidemiology of cardiovascular disease in chronic renal disease. *American Journal of Kidney Diseases, 32*(5 Suppl. 3), S112–S119.

Freundlich, M., Li, Y. C., Quiroz, Y., Bravo, Y., Seeherunvong, W., Faul, C., … Rodriguez-Iturbe, B. (2014). Paricalcitol downregulates myocardial renin-angiotensin and fibroblast growth factor expression and attenuates cardiac hypertrophy in uremic rats. *American Journal of Hypertension*, *27*(5), 720–726. http://dx.doi.org/10.1093/ajh/hpt177.

Glassock, R. J., Pecoits-Filho, R., & Barberato, S. H. (2009). Left ventricular mass in chronic kidney disease and ESRD. *Clinical Journal of the American Society of Nephrology*, *4*(Suppl. 1), S79–S91. http://dx.doi.org/10.2215/CJN.04860709.

Go, A. S., Chertow, G. M., Fan, D., McCulloch, C. E., & Hsu, C. Y. (2004). Chronic kidney disease and the risks of death, cardiovascular events, and hospitalization. *The New England Journal of Medicine*, *351*(13), 1296–1305. http://dx.doi.org/10.1056/NEJMoa041031.

Gross, M. L., & Ritz, E. (2008). Hypertrophy and fibrosis in the cardiomyopathy of uremia—Beyond coronary heart disease. *Seminars in Dialysis*, *21*(4), 308–318. http://dx.doi.org/10.1111/j.1525-139X.2008.00454.x.

Gutierrez, O. M., Januzzi, J. L., Isakova, T., Laliberte, K., Smith, K., Collerone, G., … Wolf, M. (2009). Fibroblast growth factor 23 and left ventricular hypertrophy in chronic kidney disease. *Circulation*, *119*(19), 2545–2552. http://dx.doi.org/10.1161/CIRCULATIONAHA.108.844506.

Harnett, J. D., Kent, G. M., Barre, P. E., Taylor, R., & Parfrey, P. S. (1994). Risk factors for the development of left ventricular hypertrophy in a prospectively followed cohort of dialysis patients. *Journal of the American Society of Nephrology*, *4*(7), 1486–1490.

Huting, J., Kramer, W., Schutterle, G., & Wizemann, V. (1988). Analysis of left-ventricular changes associated with chronic hemodialysis. A noninvasive follow-up study. *Nephron*, *49*(4), 284–290.

Imura, A., Iwano, A., Tohyama, O., Tsuji, Y., Nozaki, K., Hashimoto, N., … Nabeshima, Y. (2004). Secreted Klotho protein in sera and CSF: Implication for post-translational cleavage in release of Klotho protein from cell membrane. *FEBS Letters*, *565*(1–3), 143–147. http://dx.doi.org/10.1016/j.febslet.2004.03.090.

Krane, V., Heinrich, F., Meesmann, M., Olschewski, M., Lilienthal, J., Angermann, C., … Dialysis Study, I. (2009). Electrocardiography and outcome in patients with diabetes mellitus on maintenance hemodialysis. *Clinical Journal of the American Society of Nephrology*, *4*(2), 394–400. http://dx.doi.org/10.2215/CJN.02020408.

Kuro-o, M. (2013). Klotho, phosphate and FGF-23 in ageing and disturbed mineral metabolism. *Nature Reviews. Nephrology*, *9*(11), 650–660. http://dx.doi.org/10.1038/nrneph.2013.111.

Kuro-o, M., Matsumura, Y., Aizawa, H., Kawaguchi, H., Suga, T., Utsugi, T., … Nabeshima, Y. I. (1997). Mutation of the mouse klotho gene leads to a syndrome resembling ageing. *Nature*, *390*(6655), 45–51. http://dx.doi.org/10.1038/36285.

Kurosu, H., Ogawa, Y., Miyoshi, M., Yamamoto, M., Nandi, A., Rosenblatt, K. P., … Kuro-o, M. (2006). Regulation of fibroblast growth factor-23 signaling by klotho. *The Journal of Biological Chemistry*, *281*(10), 6120–6123. http://dx.doi.org/10.1074/jbc.C500457200.

Kurosu, H., Yamamoto, M., Clark, J. D., Pastor, J. V., Nandi, A., Gurnani, P., … Kuro-o, M. (2005). Suppression of aging in mice by the hormone Klotho. *Science*, *309*(5742), 1829–1833. http://dx.doi.org/10.1126/science.1112766.

Kuwahara, K., et al. (2006). TRPC6 fulfills a calcineurin signaling circuit during pathologic cardiac remodeling. *The Journal of Clinical Investigation*, *116*, 3114–3126.

Lekawanvijit, S., Kompa, A. R., Wang, B. H., Kelly, D. J., & Krum, H. (2012). Cardiorenal syndrome: The emerging role of protein-bound uremic toxins. *Circulation Research*, *111*(11), 1470–1483. http://dx.doi.org/10.1161/CIRCRESAHA.112.278457.

Levin, A., Thompson, C. R., Ethier, J., Carlisle, E. J., Tobe, S., Mendelssohn, D., … Singer, J. (1999). Left ventricular mass index increase in early renal disease: Impact of decline in hemoglobin. *American Journal of Kidney Diseases*, *34*(1), 125–134.

London, G. M., Fabiani, F., Marchais, S. J., de Vernejoul, M. C., Guerin, A. P., Safar, M. E., … Llach, F. (1987). Uremic cardiomyopathy: An inadequate left ventricular hypertrophy. *Kidney International*, *31*(4), 973–980.

Mark, P. B., Johnston, N., Groenning, B. A., Foster, J. E., Blyth, K. G., Martin, T. N., … Jardine, A. G. (2006). Redefinition of uremic cardiomyopathy by contrast-enhanced cardiac magnetic resonance imaging. *Kidney International*, *69*(10), 1839–1845. http://dx.doi.org/10.1038/sj.ki.5000249.

Massry, S. G., & Fadda, G. Z. (1993). Chronic renal failure is a state of cellular calcium toxicity. *American Journal of Kidney Diseases*, *21*(1), 81–86.

Meijers, B. K., & Evenepoel, P. (2011). The gut-kidney axis: Indoxyl sulfate, p-cresyl sulfate and CKD progression. *Nephrology, Dialysis, Transplantation*, *26*(3), 759–761. http://dx.doi.org/10.1093/ndt/gfq818.

Morishita, K., Shirai, A., Kubota, M., Katakura, Y., Nabeshima, Y., Takeshige, K., & Kamiya, T. (2001). The progression of aging in klotho mutant mice can be modified by dietary phosphorus and zinc. *The Journal of Nutrition*, *131*(12), 3182–3188.

Nakatani, T., Sarraj, B., Ohnishi, M., Densmore, M. J., Taguchi, T., Goetz, R., … Razzaque, M. S. (2009). In vivo genetic evidence for klotho-dependent, fibroblast growth factor 23 (Fgf23) -mediated regulation of systemic phosphate homeostasis. *The FASEB Journal*, *23*(2), 433–441. http://dx.doi.org/10.1096/fj.08-114397.

Nilius, B., Owsianik, G., Voets, T., & Peters, J. A. (2007). Transient receptor potential cation channels in disease. *Physiological Reviews*, *87*, 165–217.

Panizo, S., Barrio-Vazquez, S., Naves-Diaz, M., Carrillo-Lopez, N., Rodriguez, I., Fernandez-Vazquez, A., … Cannata-Andia, J. B. (2013). Vitamin D receptor activation, left ventricular hypertrophy and myocardial fibrosis. *Nephrology, Dialysis, Transplantation*, *28*(11), 2735–2744. http://dx.doi.org/10.1093/ndt/gft268.

Parfrey, P. S., Foley, R. N., Harnett, J. D., Kent, G. M., Murray, D. C., & Barre, P. E. (1996). Outcome and risk factors for left ventricular disorders in chronic uraemia. *Nephrology, Dialysis, Transplantation*, *11*(7), 1277–1285.

Park, M., Hsu, C. Y., Li, Y., Mishra, R. K., Keane, M., Rosas, S. E., … Chronic Renal Insufficiency Cohort Study, G (2012). Associations between kidney function and subclinical cardiac abnormalities in CKD. *Journal of the American Society of Nephrology*, *23*(10), 1725–1734. http://dx.doi.org/10.1681/ASN.2012020145.

Razzaque, M. S. (2009). The FGF23-Klotho axis: Endocrine regulation of phosphate homeostasis. *Nature Reviews. Endocrinology*, *5*(11), 611–619. http://dx.doi.org/10.1038/nrendo.2009.196.

Roger, S. D., McMahon, L. P., Clarkson, A., Disney, A., Harris, D., Hawley, C., … Levin, A. (2004). Effects of early and late intervention with epoetin alpha on left ventricular mass among patients with chronic kidney disease (stage 3 or 4): Results of a randomized clinical trial. *Journal of the American Society of Nephrology*, *15*(1), 148–156.

Rowell, J., Koitabashi, N., & Kass, D. A. (2010). TRP-ing up heart and vessels: Canonical transient receptor potential channels and cardiovascular disease. *Journal of Cardiovascular Translational Research*, *3*, 516–524.

Schulman, G., Berl, T., Beck, G. J., Remuzzi, G., Ritz, E., Arita, K., … Shimizu, M. (2015). Randomized placebo-controlled EPPIC trials of AST-120 in CKD. *Journal of the American Society of Nephrology*, *26*(7), 1732–1746. http://dx.doi.org/10.1681/ASN.2014010042.

Stack, A. G., & Saran, R. (2002). Clinical correlates and mortality impact of left ventricular hypertrophy among new ESRD patients in the United States. *American Journal of Kidney Diseases*, *40*(6), 1202–1210. http://dx.doi.org/10.1053/ajkd.2002.36881.

Stewart, G. A., Gansevoort, R. T., Mark, P. B., Rooney, E., McDonagh, T. A., Dargie, H. J., … Jardine, A. G. (2005). Electrocardiographic abnormalities and uremic cardiomyopathy. *Kidney International*, *67*(1), 217–226. http://dx.doi.org/10.1111/j.1523-1755.2005.00072.x.

Taddei, S., Nami, R., Bruno, R. M., Quatrini, I., & Nuti, R. (2011). Hypertension, left ventricular hypertrophy and chronic kidney disease. *Heart Failure Reviews*, *16*(6), 615–620. http://dx.doi.org/10.1007/s10741-010-9197-z.

Thadhani, R., Appelbaum, E., Pritchett, Y., Chang, Y., Wenger, J., Tamez, H., … Solomon, S. D. (2012). Vitamin D therapy and cardiac structure and function in patients with chronic kidney disease: The PRIMO randomized controlled trial. *JAMA*, *307*(7), 674–684. http://dx.doi.org/10.1001/jama.2012.120.

Tonelli, M., Sacks, F., Pfeffer, M., Gao, Z., Curhan, G., & Cholesterol, & Recurrent Events Trial, I (2005). Relation between serum phosphate level and cardiovascular event rate in people with coronary disease. *Circulation*, *112*(17), 2627–2633. http://dx.doi.org/10.1161/CIRCULATIONAHA.105.553198.

Urakawa, I., Yamazaki, Y., Shimada, T., Iijima, K., Hasegawa, H., Okawa, K., … Yamashita, T. (2006). Klotho converts canonical FGF receptor into a specific receptor for FGF23. *Nature*, *444*(7120), 770–774. http://dx.doi.org/10.1038/nature05315.

Vega, R. B., Bassel-Duby, R., & Olson, E. N. (2003). Control of cardiac growth and function by calcineurin signaling. *The Journal of Biological Chemistry*, *278*, 36981–36984.

Wang, A. Y., Fang, F., Chan, J., Wen, Y. Y., Qing, S., Chan, I. H., … Yu, C. M. (2014). Effect of paricalcitol on left ventricular mass and function in CKD—The OPERA trial. *Journal of the American Society of Nephrology*, *25*(1), 175–186. http://dx.doi.org/10.1681/ASN.2013010103.

Wolf, M. (2012). Update on fibroblast growth factor 23 in chronic kidney disease. *Kidney International*, *82*(7), 737–747. http://dx.doi.org/10.1038/ki.2012.176.

Wolf, W. C., Yoshida, H., Agata, J., Chao, L., & Chao, J. (2000). Human tissue kallikrein gene delivery attenuates hypertension, renal injury, and cardiac remodeling in chronic renal failure. *Kidney International*, *58*(2), 730–739. http://dx.doi.org/10.1046/j.1523-1755.2000.00219.

Wolf, I., et al. (2008). Klotho: A tumor suppressor and a modulator of the IGF-1 and FGF pathways in human breast cancer. *Oncogene*, *27*, 7094–7105.

Xie, J., Cha, S. K., An, S. W., Kuro-O, M., Birnbaumer, L., & Huang, C. L. (2012). Cardioprotection by Klotho through downregulation of TRPC6 channels in the mouse heart. *Nature Communications*, *3*, 1238. http://dx.doi.org/10.1038/ncomms2240.

Xie, J., Yoon, J., An, S. W., Kuro-o, M., & Huang, C. L. (2015). Soluble klotho protects against uremic cardiomyopathy independently of fibroblast growth factor 23 and phosphate. *Journal of the American Society of Nephrology*, *26*(5), 1150–1160. http://dx.doi.org/10.1681/ASN.2014040325.

Yang, K., Wang, C., Nie, L., Zhao, X., Gu, J., Guan, X., … Zhao, J. (2015). Klotho protects against indoxyl sulphate-induced myocardial hypertrophy. *Journal of the American Society of Nephrology*, *26*(10), 2434–2446. http://dx.doi.org/10.1681/ASN.2014060543.

INDEX

Note: Page numbers followed by "*f*" indicate figures, and "*t*" indicate tables.

D

E

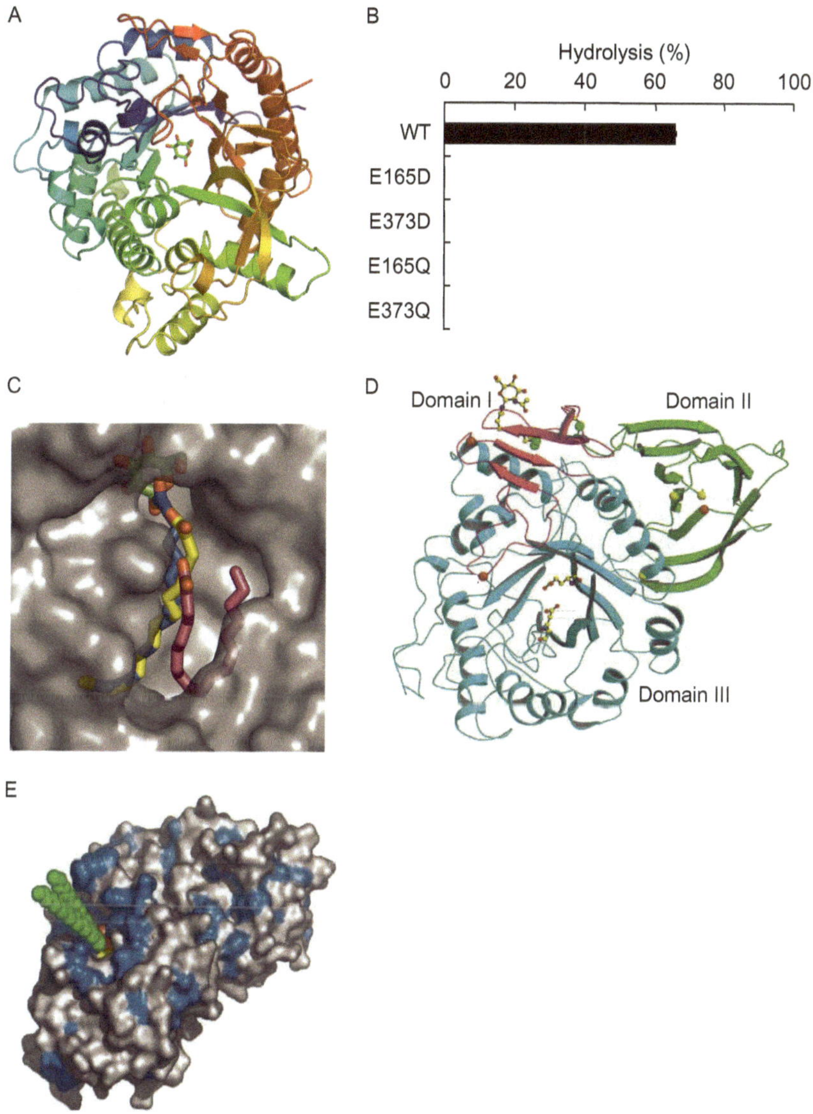

Y. Hayashi and M. Ito, Fig. 3 KLrP and GBA1 X-ray crystal structures. (A) Structure of KLrP in complex with glucose (KLrP/Glc) illustrated using a ribbon diagram. The glucose is shown with carbon atoms in green and oxygen atoms in red. (B) Point mutations of two catalytic residues. The purified wild-type KLrP (WT) and mutants (E165D, E373D, E165Q, and E373Q) were subjected to the neutral GCase assay. For the GCase assay, 100 pmol of C6-NBD-GlcCer was incubated with 50 ng of enzyme in 50 mM MES buffer, pH 6.0, containing 0.25% sodium cholate at 37°C for 30 min. (C) A close-up view of KLrP's substrate-binding cleft. Based on the electron densities, palmitic acid is located at two

(Continued)

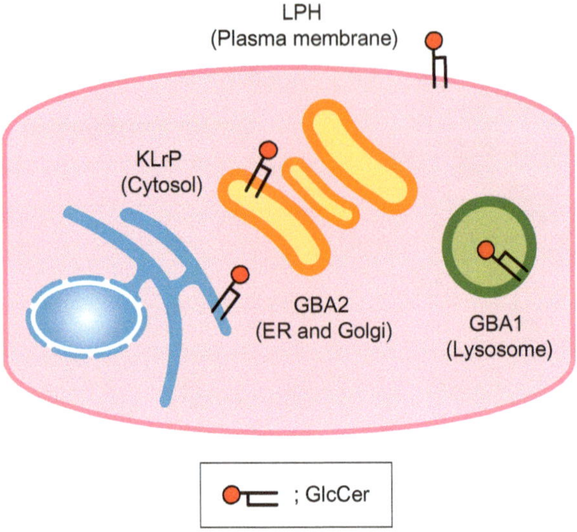

Y. Hayashi and M. Ito, Fig. 4 Cellular localizations for four mammalian GCases. GBA1 is localized to the lysosomes, where it hydrolyzes GlcCer in an acidic environment. GBA2 and KLrP (GBA3) are a nonintegral cytosolic proteins with a neutral pH optimum; however, GBA2, but not KLrP, is strongly associated with the ER/Golgi membranes (Hayashi et al., 2007; Körschen et al., 2013). LPH is exclusively localized to the intestinal epithelial cell microvilli, where it functions as a digestive enzyme.

Y. Hayashi and M. Ito, Fig. 3—Cont'd different positions, form I and form II. *Green*, glucose; *blue*, palmitic acid (form I); *yellow*, palmitic acid (from II); and *purple*, oleic acid. (D and E) The GBA1 structure is illustrated using a ribbon diagram and the docked model of GBA1 with glucose. *Panel (A) Adapted from Hayashi, Y., Okino, N., Kakuta, Y., Shikanai, T., Tani, M., Narimatsu, H., et al. (2007). Klotho-related protein is a novel cytosolic neutral beta-glycosylceramidase.* The Journal of Biological Chemistry, 282, *30889–30900. Panels (D and E) Adapted from Dvir, H., Harel, M., McCarthy, A. A., Toker, L., Silman, I., Futerman, A. H., et al. (2003). X-ray structure of human acid-beta-glucosidase, the defective enzyme in Gaucher disease.* EMBO Reports, 4, *704–709.*

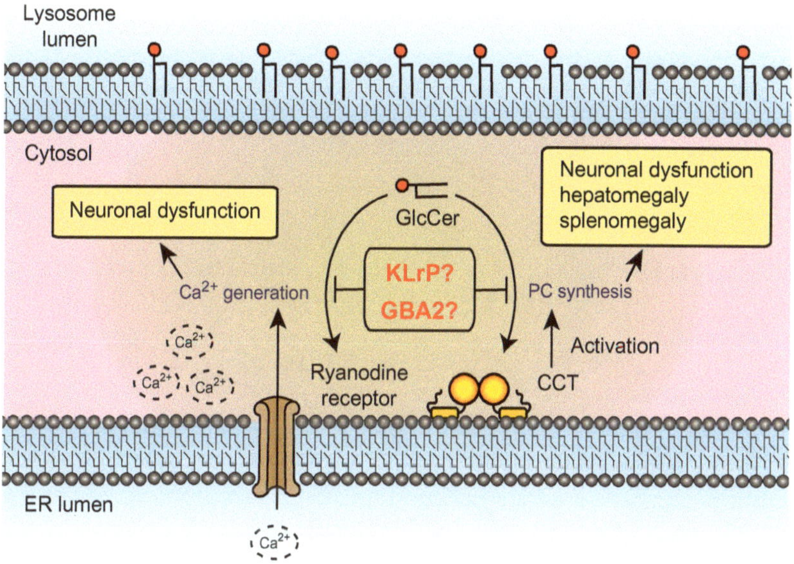

Y. Hayashi and M. Ito, Fig. 5 Working hypothesis for how Gaucher disease (GD) is triggered by GlcCer accumulation (modified from Futerman's model). The GlcCer accumulated in the GD lysosomes escapes to the ER, where the GlcCer enhances an agonist-induced calcium release via activating the ryanodine receptor. GlcCer simultaneously increases PC synthesis through CTP:phosphatidylcholine cytidylyltransferase activation. The disturbances in cellular homeostasis induced by the GlcCer result in neuron dysfunction, hepatomegaly, and splenomegaly in GD patients. Given the cellular localizations of GBA2 and KLrP, both GCases may contribute to hydrolysis of the escaped GlcCer.

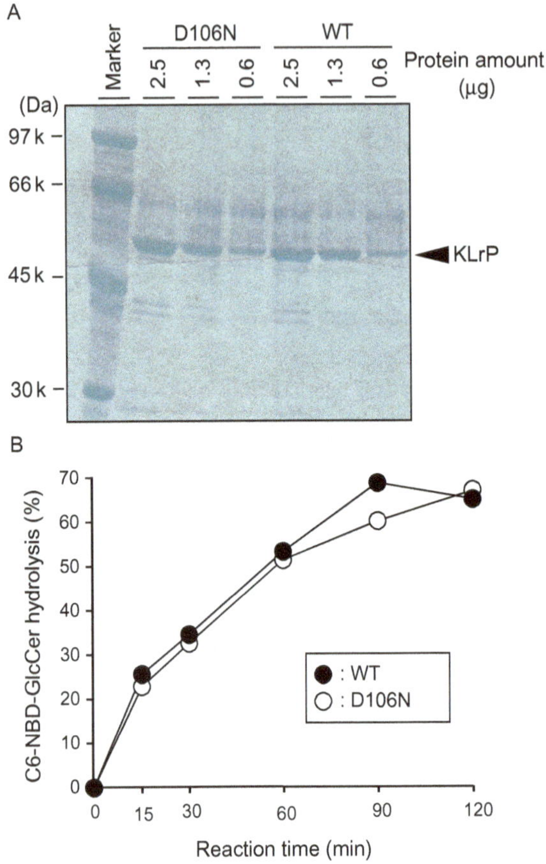

Y. Hayashi and M. Ito, Fig. 6 The effects of a D106N mutation on the GCase activity of KLrP. (A) Purification of the WT and D106N mutation of KLrP. His-tagged KLrP and the D106N mutant were purified using HiTrap Chelating and Superdex200. (B) The GCase activities of the purified WT and D106N mutant. For the GCase assay, 100 pmol of C6-NBD-GlcCer was incubated with 25 ng of enzyme in 50 mM MES buffer, pH 6.0, containing 0.25% sodium cholate at 37°C for several minutes (0, 15, 30, 60, 90, and 120 min). The results represent the average value from two independent experiments.

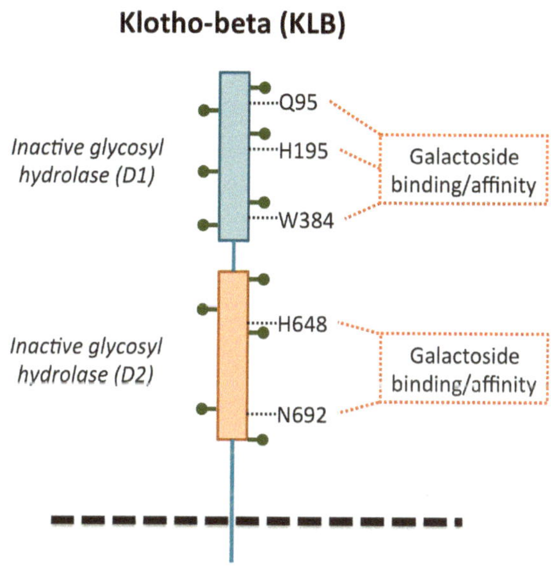

D.M. Kilkenny and J.V. Rocheleau, Fig. 1 KLB structure. KLB transmembrane receptor has a much larger extracellular domain (996 aa) compared to the cytoplasmic (27 aa) region. The extracellular domain contains two inactive glycosyl-hydrolase domains (D1, D2) with residues maintained for binding galactoside sugars. The extracellular region contains 11 putative N-glycosylation sites (indicated by *stick and ball markings*) spread across the D1 and D2 domains.

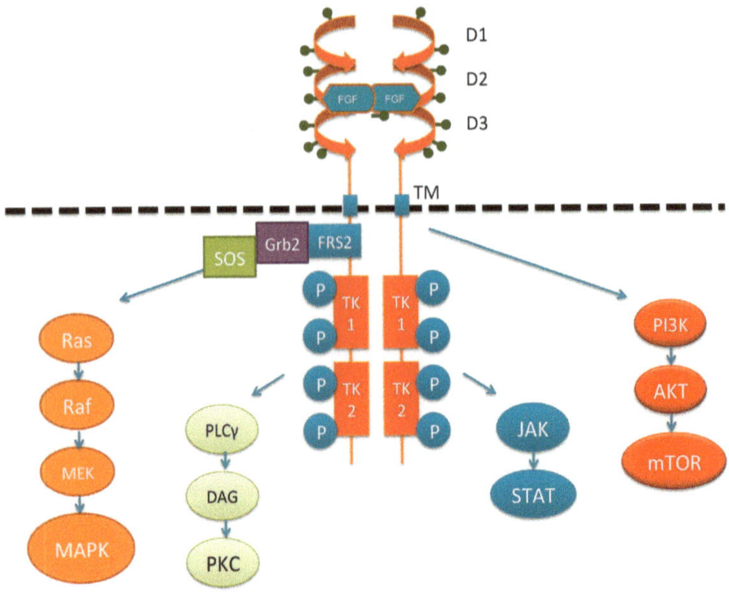

D.M. Kilkenny and J.V. Rocheleau, Fig. 2 FGFR signaling cascades. FGF ligand binding to the extracellular immunoglobulin-like D2–D3 domains of canonical FGFR1, which exhibits seven N-glycosylation sites (indicated by *ball and stick markings*), results in *trans*-activation of the split intracellular tyrosine kinase (TK) domains of a receptor dimer. Receptor activation can initiative a multitude of intracellular signaling cascades, including direct interaction with the FRS2 docking protein, as well as downstream activation of the Ras/MAPK pathway, the PKC pathway, the Jak/STAT pathway, and the PI3K/mTOR pathway.

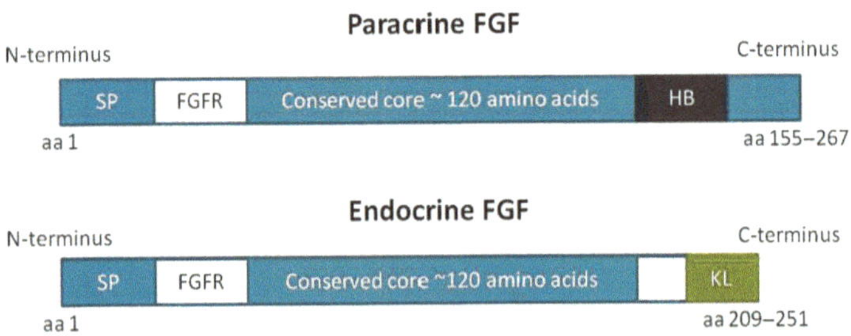

D.M. Kilkenny and J.V. Rocheleau, Fig. 3 Critical domains of paracrine and endocrine FGFs. Despite variability in biological function, paracrine and endocrine FGF peptides share a conserved core sequence of ~120 aa and present as a class of proteins of ~18–24 kDa molecular mass (Table 3). An N-terminal signal peptide (SP) and FGFR binding site (FGFR; directs interaction with high-affinity FGFRs) are also common sequence features. The C-terminal sequence displays some variability and is required for association with low-affinity HSPGs (HB; paracrine ligands) or klotho co-receptors (KL; endocrine ligands).

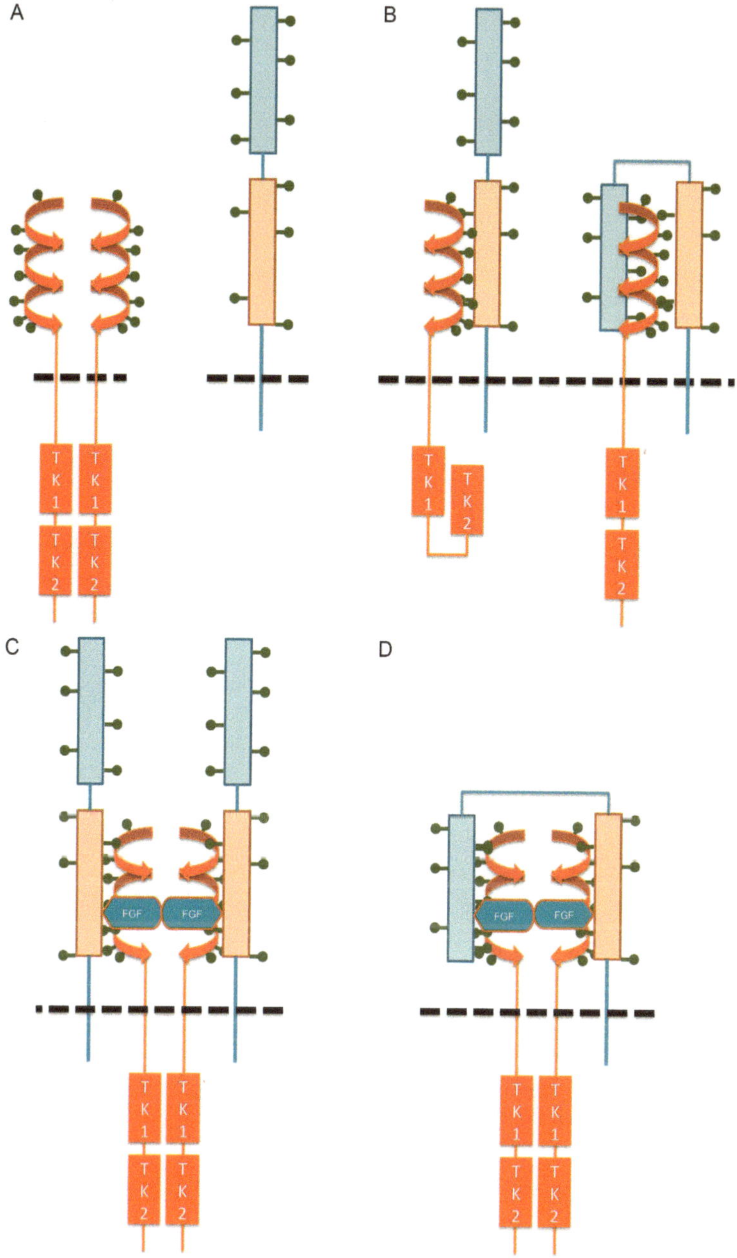

D.M. Kilkenny and J.V. Rocheleau, Fig. 4 Putative receptor stoichiometry. The cartoons represent putative receptor stoichiometry of FGFR1 and KLB in the absence (A, B) and presence (C, D) of FGF21. (A) Our data indicate that FGFR1c diffuses as preformed dimers, while KLB diffuses as a monomer. (B) Co-expression of FGFR1c and KLB results in formation of a 1:1 heterocomplex. This cartoon models two potential interactions highlighting both allosteric conformational changes (*left*) and multiple binding interactions (*right*). (C) A cartoon representation of a 2:2:2 receptor complex formed by FGF21: FGFR1:KLB. This putative model was anticipated after finding a preformed 1:1 heterodimer of FGFR1 and KLB, but has not been observed. (D) A putative model of the 2:2:1 receptor complex of FGF21:FGFR1:KLB. This model indicates putative interaction of one KLB with two FGFR1c receptors in an effort to explain the release of a second KLB when the complex is formed.

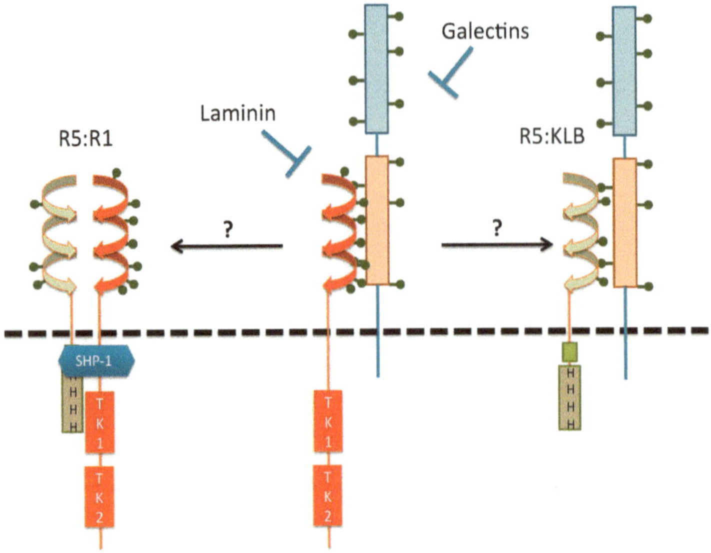

D.M. Kilkenny and J.V. Rocheleau, Fig. 5 Potential regulators of FGFR1:KLB signaling. The formation of FGFR1:KLB heterodimers is potentially diminished by laminin-induced downregulation of FGFR1 expression, as well as association of KLB with the galectin lattice. Given extracellular sequence similarity, the potential heterodimerization of FGFR1c with the novel FGFR5 and KLB is also proposed. We previously identified FGFR5 interaction with SHP-1 which may act as a scaffold protein to differentially activate and/or regulate FGFR1c signaling.

M. Sopjani and M. Dërmaku-Sopjani, Fig. 1 Schematic representation of domain structure of the Klotho (Huang & Moe, 2011; Imura et al., 2007; Kuro-o, 2010). Human Klotho protein is composed of 1014 (1012 in mice) amino acids in length. At its N-terminus it possesses a putative signal sequence (SS), followed by two internal repeats (Kl1 and Kl2) of the extracellular domain that shares sequence homology to the β-glucosidase. A linker (L) is positioned in between two internal repeats. The next segment is a putative transmembrane domain (TM) with a short cytoplasmic domain (CD) at the C-terminus (not drawn to scale).

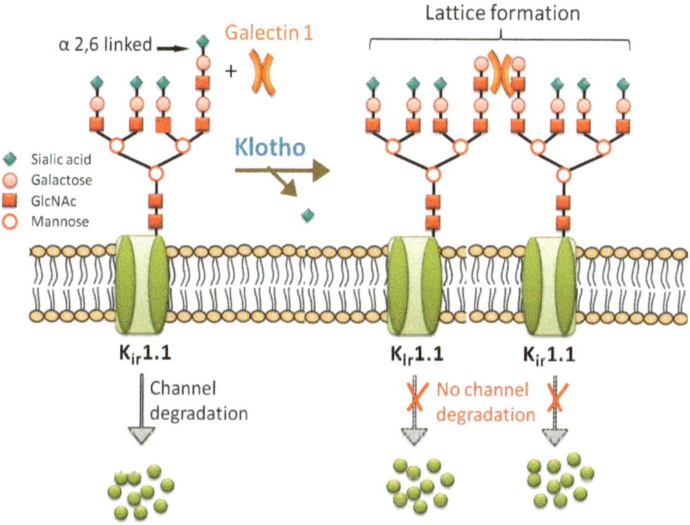

M. Sopjani and M. Dërmaku-Sopjani, Fig. 2 Tentative model showing Klotho function in upregulating the plasma membrane abundance of ion channels K_{ir}1.1 (Cha et al., 2009; Sopjani et al., 2014). Soluble Klotho protein through its enzymatic activity removes the α2,6-specific sialic acids from K_{ir}1.1 (ROMK1) potassium channel N-glycans. By removing sialic acids, which are terminal acidic monosaccharides, it exposes underlying GlcNAc, a ligand-binding galectin-1 at the extracellular surface. This allows binding to galectin-1 and subsequent lattice formation at the extracellular surface between two adjacent ion channels that prevents endocytosis and diverting the channel to degradation. Entrapping of the functional channel on cell membrane leads to increasing K_{ir}1.1-mediated potassium transport on the plasma membrane. Similar model is reported for TRPV5 channel. See text for further details.

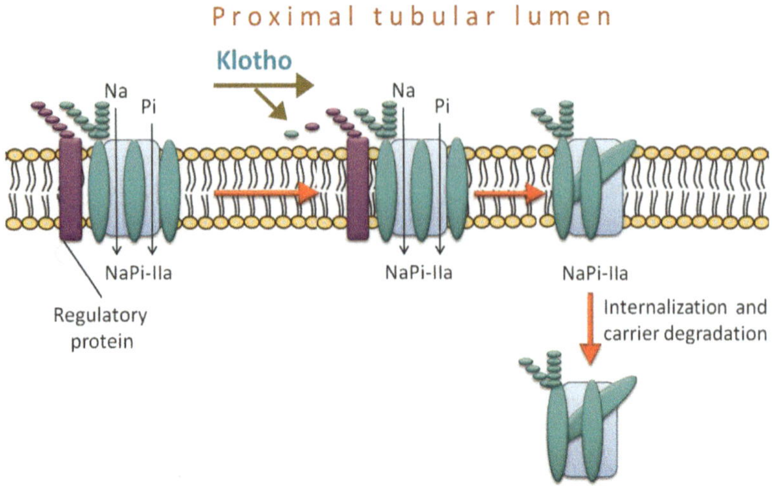

M. Sopjani and M. Dërmaku-Sopjani, Fig. 3 A simple model showing the enzymatic activity of Klotho-dependent regulation of NaPi-IIa carrier protein abundance (Hu et al., 2010; Sopjani et al., 2014) at the plasma membrane. The circulating Klotho inhibits Na^{+}-coupled phosphate transporter NaPi-IIa activity in the apical membrane of the renal proximal tubules. Klotho-sensitive NaPi-IIa inhibition is mediated by the enzymatic activity of Klotho and is realized by deglycosylating extracellular sugar residues on the cell membrane proteins of NaPi-IIa and/or a regulatory protein. This makes NaPi-IIa protein more sensitive to proteases in brush border membrane (BBM) that results in proteolitic degradation in lysosomes. This leads to phosphaturia as a consequence of reduced phosphate transport mediated by NaPi-IIa due to carrier protein internalization from the apical membrane into intracellular pool.

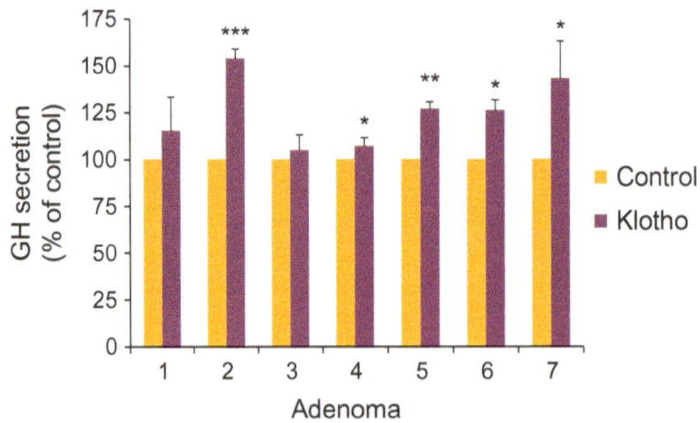

T. Rubinek and D. Modan-Moses, Fig. 1 Klotho increase GH secretion from cultured pituitary adenomas. Seven human GH adenomas were harvested and cultured. Cultures were treated with klotho (0.1 ng/mL) or control vehicle and 24 h later aliquots were removed for GH measurements. Each bar represents GH secretion relative to control treated cells. $^{*}p<0.05$, $^{**}p<0.005$, treatment vs vehicle-treated control. *Shahmoon, S., Rubinfeld, H., Wolf, I., Cohen, Z. R., Hadani, M., Shimon, I., & Rubinek, T. (2014). The aging suppressor klotho: A potential regulator of growth hormone secretion.* American Journal of Physiology. Endocrinology and Metabolism, 307(3), *17, copyright (2014).*

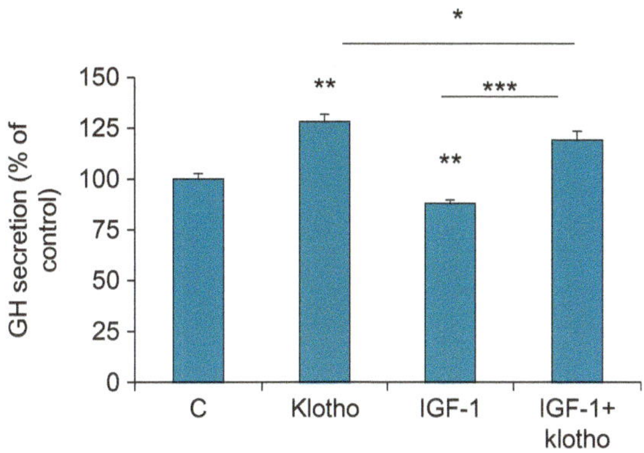

T. Rubinek and D. Modan-Moses, Fig. 2 Klotho partially overrides the inhibitory effect of IGF-1 on GH secretion. Human GH adenoma was harvested, cells were cultured and 48 h later cultures were treated with IGF-1 (50 ng/mL), klotho (0.1 ng/mL), their combination, or treated with control vehicle. Twenty-four hours later aliquots were removed for GH measurements. $*p < 0.05$, $**p < 0.005$, $***p < 0.0005$. *Shahmoon, S., Rubinfeld, H., Wolf, I., Cohen, Z. R., Hadani, M., Shimon, I., & Rubinek, T. (2014). The aging suppressor klotho: A potential regulator of growth hormone secretion.* American Journal of Physiology. Endocrinology and Metabolism, 307(3), *17, copyright (2014).*

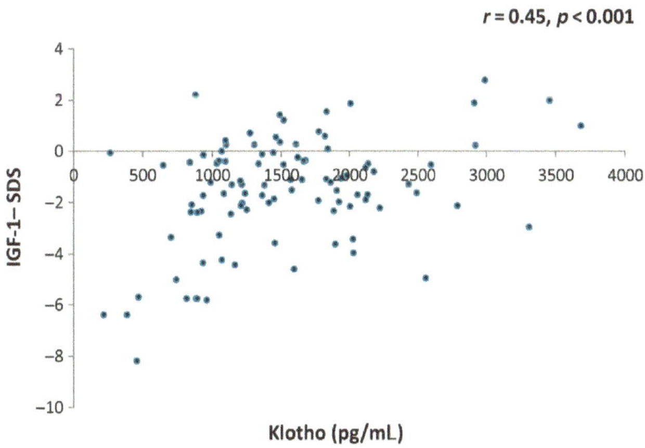

T. Rubinek and D. Modan-Moses, Fig. 3 The correlation between serum klotho levels and IGF-1 in children with short stature. A significant positive correlation was noted ($r = 0.45, p < 0.001$). *Wolf, I., Shahmoon, S., Ben Ami, M., Levy-Shraga, Y., Mazor-Aronovitch, K., Pinhas-Hamiel, O.,…, Modan-Moses, D. (2014). Association between decreased klotho blood levels and organic growth hormone deficiency in children with growth impairment.* PLoS One 9(9), *e107174. http://dx.doi.org/10.1371/journal.pone.0107174, copyright (2014).*

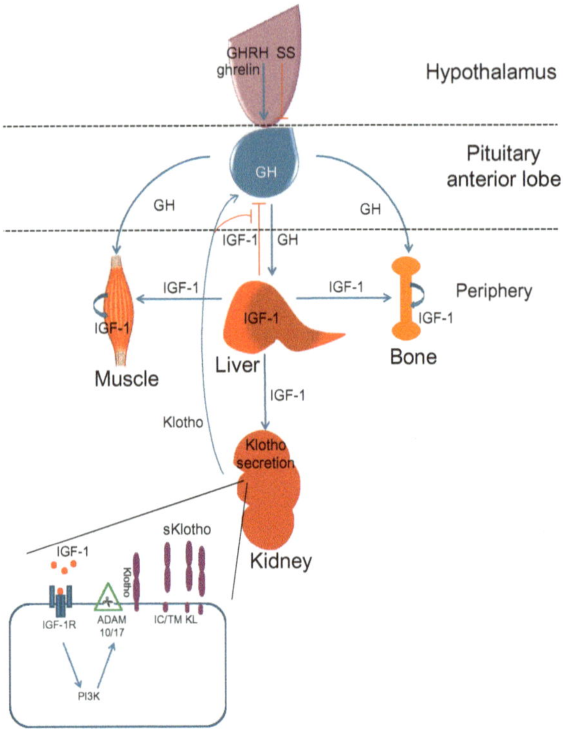

T. Rubinek and D. Modan-Moses, Fig. 4 Regulation of GH/IGF and klotho in the circulation. GH secretion is regulated by hypothalamic factors (GHRH, somatostatin, and ghrelin). GH activates GH receptors (GHR) in the liver, which stimulate synthesis and secretion of IGF-1. GH may also activate GHR in muscles and bones. IGF-1 binds to IGF-1Rs in target tissues like muscles and bones and induces growth responses, and also serves as a negative regulator for GH secretion. Klotho is highly expressed in the kidney and insulin, and IGF-1 trigger its shedding, through the action of the metalloproteases ADAM 10 and ADAM 17 (see inset). Klotho enhances GH secretion from the pituitary directly and by abrogating the inhibitory effect of IGF-1.

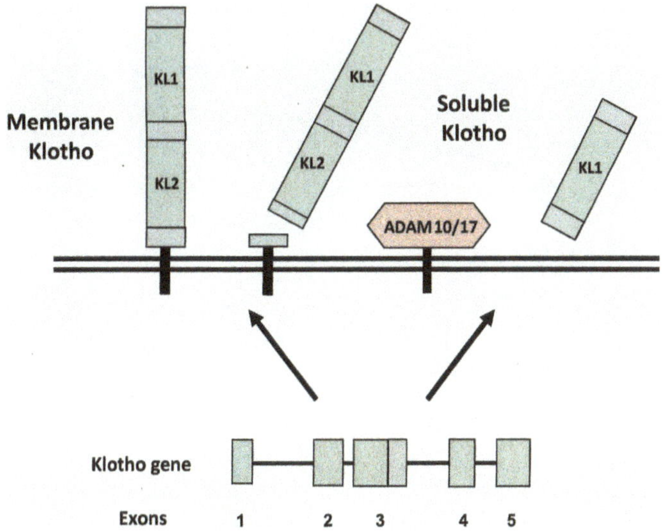

P. Buendía *et al.*, Fig. 1 Klotho structure. Domain structure of human Klotho. KL1 and KL2 indicate first and second Klotho repeat, respectively. The entire extracellular domain consisting of KL1 and KL2 repeats is cleaved by ADAM10 or ADAM17 and released into blood, urine, or cerebrospinal fluid.

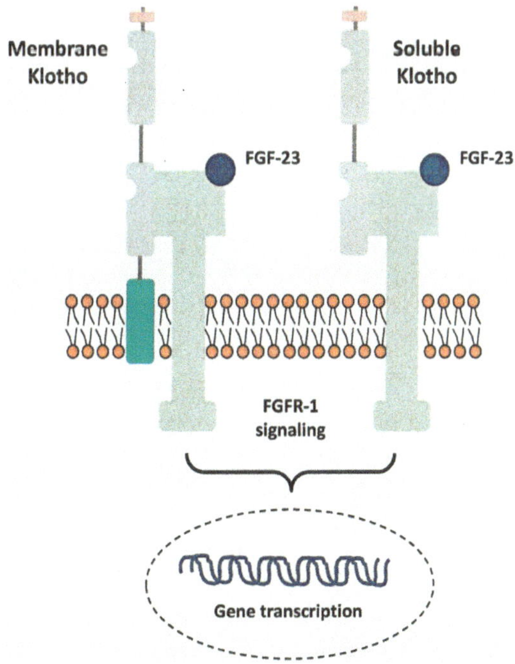

P. Buendía *et al.*, Fig. 3 Klotho-FGF-23 axis. FGF-23 can bind to the Klotho–FGFR1c complex on the surface of the parathyroid cell and thereby triggers downstream signaling.

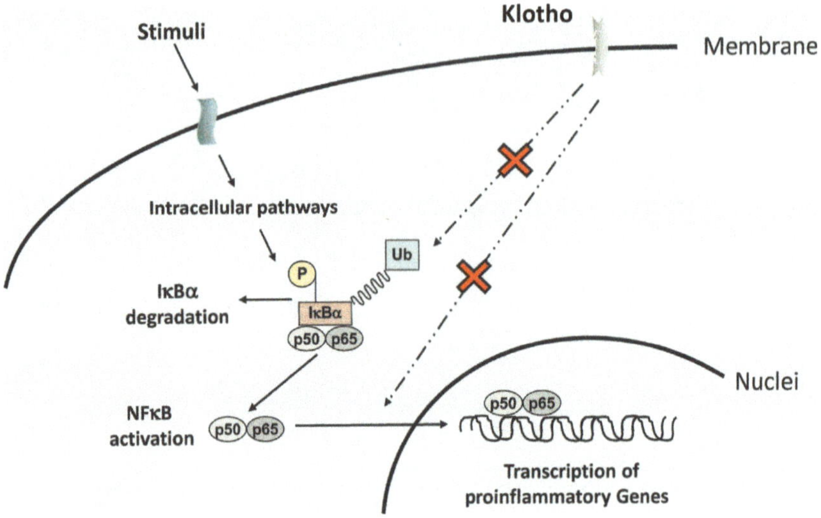

P. Buendía *et al.*, Fig. 4 Klotho and NFκB. Klotho suppresses nuclear NFκB activation and the subsequent transcription of proinflammatory genes.

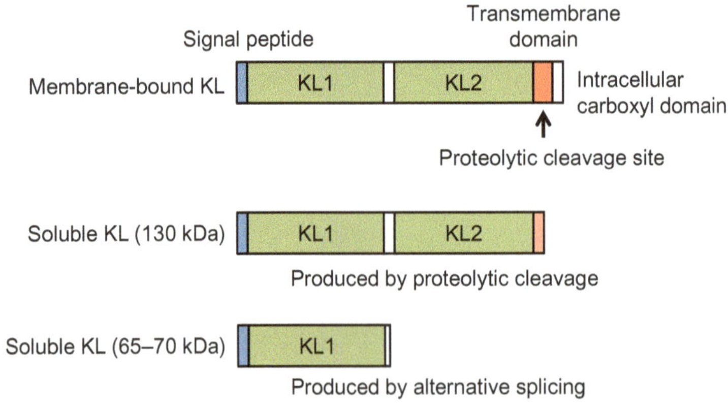

S. Kinoshita and M. Kawai, Fig. 1 Structure of Klotho (KL). KL is composed of two internal repeats (KL1 and KL2), a transmembrane domain and short intracellular carboxyl domain. KL is proteolytically cleaved at the transmembrane domain (*arrow*) to produce its soluble form. The shorter form of soluble KL is also produced by alternative splicing.

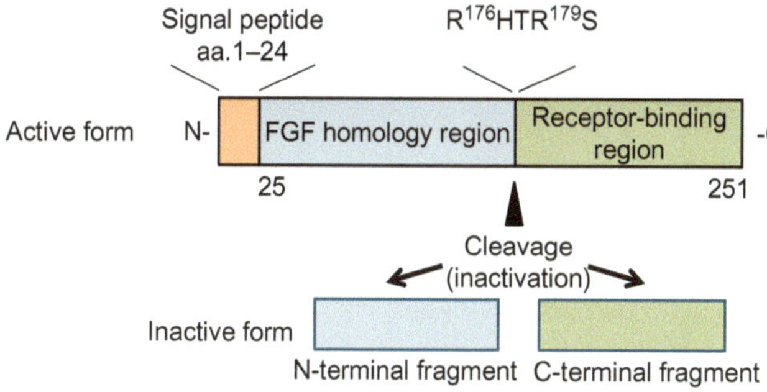

S. Kinoshita and M. Kawai, Fig. 2 Structure of FGF23. Intact FGF23 is proteolytically cleaved to generate inactive forms of FGF23. The site of cleavage is shown in the figure by an *arrowhead.*

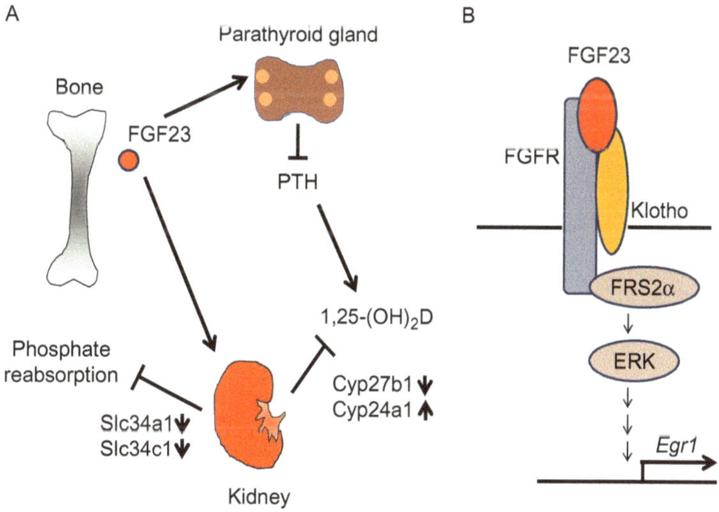

S. Kinoshita and M. Kawai, Fig. 4 Signaling cascades of the FGF23/Klotho pathway. (A) FGF23 produced from osteocytes mainly transduces its signals in the kidney and regulates phosphate reabsorption and the production of the active form of vitamin D. (B) FGF23 requires Klotho as a coreceptor to exert its effects and activate its downstream signaling molecules including FRS2α and ERK.

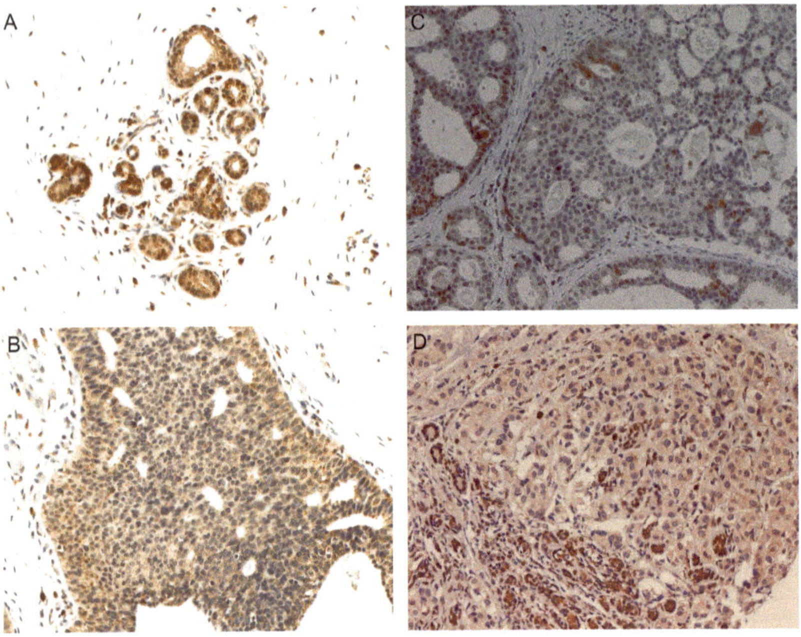

T. Rubinek and I. Wolf, Fig. 1 Klotho expression is downregulated in early stages of the malignant process. Immunohistochemical staining of klotho in mammary tissues. Depicted high klotho expression in normal mammary tissue (A) and reduced expression is observed already in hyperplastic tissue (B), as well as in ductal carcinoma in situ (C) and invasive carcinoma (D).

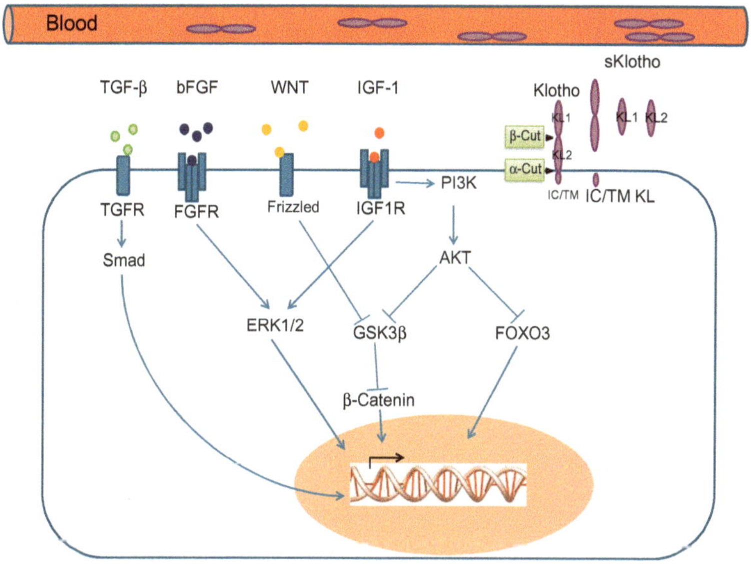

T. Rubinek and I. Wolf, Fig. 2 Signaling pathways affected by klotho in cancer cells. Cancer cells are regulated by multiple pathways, and klotho may modulate several of these pathways. Klotho was shown to inhibit the IGF-1 pathway and the Wnt pathway. These two pathways may converge at the GSK3β signaling molecule. Thus, the IGF-1 pathway inhibits GSK3β. Similarly, activated Wnt pathway also inhibits GSK3β. Therefore, inhibition of both pathways by klotho promotes GSK3β activity and reduces cell viability. Furthermore, klotho modulates the activity of bFGF, which activates ERK1/2. Similarly, both IGF-1 and bFGF may elicit ERK1/2 phosphorylation. Thus, although klotho seems to operate through many, unrelated pathways, perhaps some of these activities are connected. TGF-β was also shown to activate the ERK1/2 pathway. Klotho may access cells through the circulation, as it is a hormone. However, although it is silenced in many cancer types, it is expressed, at least to some extent, in some cancer cells. It may operate in the cytoplasm, or as a membrane-bound protein. However, it may be cleaved through ADAM10 and ADAM17 sheddasses and act from outside the cell on the different receptors. Klotho is cleaved to the entire extracellular domain, and also into the subdomains KL1 and KL2. KL1 can effectively inhibit cancer cell growth, and inhibit the IGF-1, bFGF, and Wnt pathways.

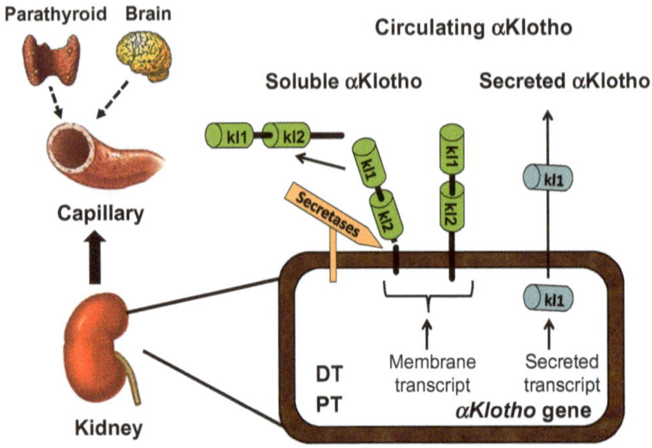

J.A. Neyra and M.C. Hu, Fig. 1 Source of circulatory αKlotho. αKlotho protein is expressed in a few organs, but the kidney is a main resource of circulating αKlotho under physiological conditions. The contribution of parathyroid gland and brain is not clear. Both renal proximal (PT) and distal tubules (DT) express membrane αKlotho protein and may also produce a secreted αKlotho protein which only contains kl1 domain and is directly secreted into the blood circulation. Extracellular domain of membrane αKlotho-containing kl1 and kl2 repeats is shed and cleaved by α and β-secretases, and released into the blood circulation.

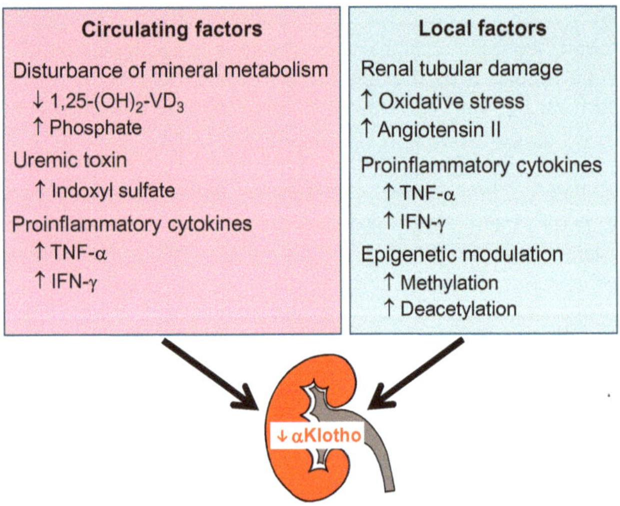

J.A. Neyra and M.C. Hu, Fig. 2 Circulating and local renal factors involved in the reduction of αKlotho expression in the kidney. In acute and chronic kidney disease, a variety of circulating factors including disturbed mineral metabolism, and accumulation of indoxyl sulfate and proinflammatory cytokines (*left panel*), can downregulate renal αKlotho expression. On the other hand, the elevation of reactive oxygen species, Ang II, and inflammatory cytokines in the diseased kidney can also downregulate renal αKlotho expression (*right panel*). Epigenetic modulation of αKlotho promoter via hypermethylation and deacetylation can reduce αKlotho expression and contribute to αKlotho deficiency in chronic kidney disease.

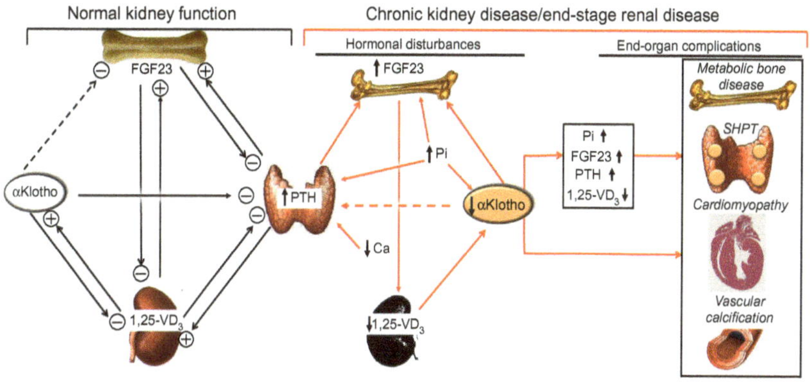

J.A. Neyra and M.C. Hu, Fig. 3 Proposed physiological role of αKlotho in mineral metabolism and pathophysiological consequences of αKlotho deficiency in CKD. In the setting of normal kidney function with normal αKlotho levels (*left panel*), αKlotho may suppress FGF23 production and release from the bone. But there is no direct evidence to prove it. αKlotho functions as coreceptor of FGFR to allow FGF23 to suppress PTH production and release from parathyroid gland. PTH stimulates and increases plasma levels of FGF23 and 1,25-$(OH)_2$-vitamin D_3. Increased 1,25-$(OH)_2$-vitamin D_3 further stimulates FGF23, and directly and indirectly suppresses PTH levels. Increased 1,25-$(OH)_2$-vitamin D_3 also stimulates αKlotho production in the kidney. Taken together, through several negative- or positive-feedback loops, αKlotho functions as both a phosphate and calcium regulatory hormone to directly or indirectly suppress PTH, 1,25-$(OH)_2$-vitamin D_3, and FGF23 production and release. αKlotho's action on the kidney is to prevent renal Pi retention and to prevent renal Ca loss. In CKD and ESRD (*right panel*), the network is deranged (*red arrows*). Renal αKlotho is decreased followed by decrease in plasma αKlotho. The downregulation of αKlotho increases FGF23 production via unknown mechanism, which in turn suppresses 1,25-$(OH)_2$-vitamin D_3 production in the kidney. Whether low plasma αKlotho renders parathyroid gland resistant to the suppressive effect of FGF23 on PTH production is not proven. However, decreased FGFR1/3 and αKlotho expression in the uremic parathyroid gland could make the gland resistant to FGF23, and triggers and/or promotes secondary hyperparathyroidism (SHPT). Low plasma Ca also participates in SHPT development. Hyperphosphatemia amplifies the high FGF23 and PTH levels, and low αKlotho levels in the blood. The high plasma PTH, Pi, and FGF23, and low plasma 1,25-$(OH)_2$-vitamin D_3 and αKlotho in concert contribute to the development of complications such as metabolic bone disease, SHPT, cardiomyopathy, and vascular calcification. *Dash line*: unproven putative roles of αKlotho. *Ca*, ion calcium; *CKD*, chronic kidney disease; *ESRD*, end-stage renal disease; *FGFR*, FGF receptor; *Pi*, phosphate; *PTH*, parathyroid hormone; *SHPT*, secondary hyperparathyroidism; *1,25-VD3*, 1,25-$(OH)_2$-vitamin D_3.

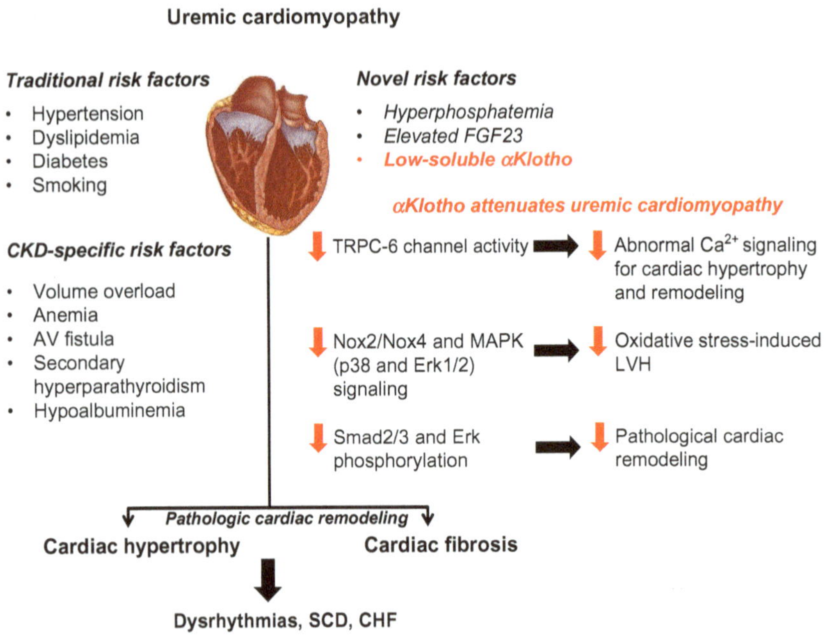

J.A. Neyra and M.C. Hu, Fig. 4 Risk factors for uremic cardiomyopathy and proposed mechanisms of attenuation of pathological cardiac remodeling by αKlotho. αKlotho deficiency is a novel risk factor for uremic cardiomyopathy. Soluble αKlotho deficiency is an intermediate mediator of the pathological cardiac remodeling observed in CKD. Experimental αKlotho overexpression (*Tg-Kl*) suppressed phosphorylation of Smad2/3 and Erk, which are known to be involved in uremic cardiac fibrosis. Furthermore, αKlotho may protect the heart against stress-induced cardiac hypertrophy by inhibiting TRPC-6 channel-mediated abnormal Ca^{2+} signaling in the heart or against uremic solute indoxyl sulfate-induced myocardial hypertrophy probably by suppressing NADPH oxidase Nox2/Nox4-derived reactive oxygen species production and its downstream signaling. *CHF*, congestive heart failure; *Erk*, extracellular signal-regulated kinase; *LVH*, left ventricular hypertrophy; *MAPK*, mitogen-activated protein kinases; *Nox*, NADPH oxidase; *SCD*, sudden cardiac death; *TRPC-6*, transient receptor potential canonical-6.

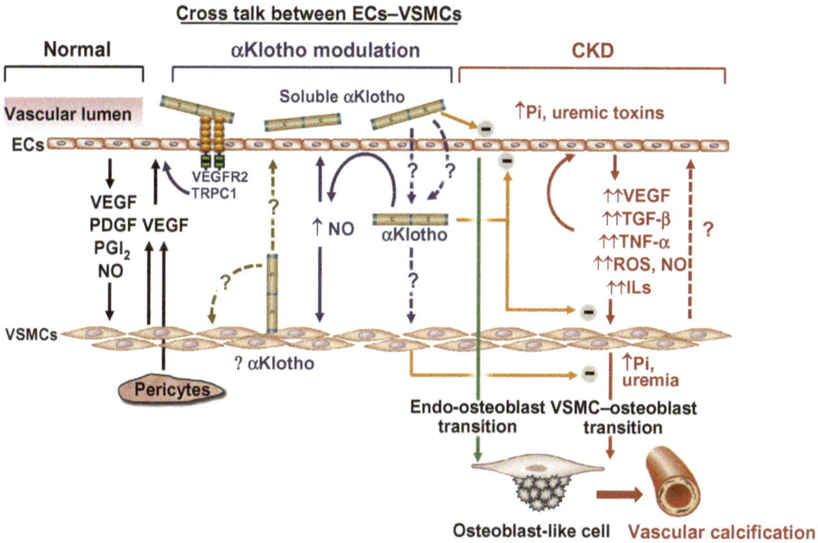

J.A. Neyra and M.C. Hu, Fig. 5 Proposed model of αKlotho as intermediate of endothelial cells (ECs)–vascular smooth muscle cells (VSMCs) cross talk. *Left panel*: Normal ECs–VSMCs cross talk. Normal ECs modulate VSMCs growth via release of growth factors (*black line* in *left panel*). *NO*, nitric oxide; *PDGF*, platelet-derived growth factor; *PGI_2*, prostaglandin I_2; *VEGF*, vascular endothelial growth factor. VEGF released from pericytes and/or VSMCs could also regulate endothelial cell. *Right panel*: In CKD, uremic toxins including high plasma Pi, damage ECs and induce release of growth factors, proinflammatory cytokines, and profibrotic factors, which exacerbate ECs injury and also induce VSMCs transition to osteoblast and promote vascular calcification in medial layer (*red solid lines* in *right panel*). Impaired ECs also directly contributes to vascular calcification through endo-osteoblast transition (*green line* in *right panel*). Whether dedifferentiated or damaged VSMCs could further modulate function of endothelial cells is speculative (*red dash line* in *right panel*). *Central panel*: αKlotho is a vascular protective protein. Whether resident aortic αKlotho protein in VSMCs functions in autocrine and/or paracrine mode to modulate VSMCs and/or ECs remains to be clarified (*brown dash line* in *central panel*). The mechanisms of how αKlotho is able to reach the VSMCs from the circulation and function as an endocrine factor remain to be defined (*blue dash line* in *central panel*). αKlotho, regardless of source, could increase NO production from ECs and NO consequently modulates VSMCs and ECs function in an autocrine mode (*blue solid line* in *middle panel*). αKlotho could protect EC from high phosphate and other uremic toxins and also attenuate oxidative stress and proinflammatory cytokines-induced cell senescence and apoptosis in VSMC (*orange line* in *central panel*). αKlotho also directly inhibits osteoblast transition induced by hyperphosphatemia and uremic milieu (*orange line* in *central panel*). Current experimental and clinical observations suggest that both ECs and VSMCs endothelium may be potential targets of soluble αKlotho to protect the vasculature from vascular calcification in CKD. *ECs*, endothelial cells; *Pi*, phosphorus; *VSMCs*, vascular smooth muscle cells.

FSC
www.fsc.org
MIX
Papier | Fördert
gute Waldnutzung
FSC® C083411

Kolibri 360 GmbH
Ferdinand-Jühlke-Straße 7
99095 Erfurt, Deutschland
produktsicherheit@kolibri360.de